Medicinal Roots and Tubers for Pharmaceutical and Commercial Applications

The root and tuber are vital parts of medicinal plants providing mechanical support, producing critical growth regulators, and storing food. Bioactive compounds obtained from plant roots and tubers demonstrate health benefits presenting antioxidative, antimicrobial, hypoglycaemic, hypocholesterolaemic, and immunomodulatory properties. Roots of many medicinal plants have been used for the treatment of disease and formulation of drugs, and they are also known for their commercial value, being used as an ingredient in the pharmaceutical and cosmetic industries. *Medicinal Roots and Tubers for Pharmaceutical and Commercial Applications* provides information on the medicinal properties of roots and tubers and various phytochemicals derived from them.

Features

- Presents exhaustive information on plant roots and tubers including *Glycyrrhiza glabra*, *Curcuma longa*, *Beta vulgaris*, *Zingiber officinale*, *Boesenbergia pandurata*, *Houttuynia cordata*, *Eutrema japonicum*, and *Withania somnifera*.
- Explains the roles of secondary metabolites isolated from roots and tubers and features information on their pharmaceutical and commercial applications.
- Discusses opportunities for and future prospects of different roots and tubers for their industrial applications.

A volume in the *Exploring Medicinal Plants* series, this book provides information on phytochemicals derived from medicinal plant roots and tubers. This is valuable information for scientists, researchers, and students working on medicinal plants, economic botany, chemistry, biotechnology, pharmaceuticals, and many other interdisciplinary subjects.

EXPLORING MEDICINAL PLANTS

Series Editor

Azamal Husen

Wolaita Sodo University, Ethiopia

Medicinal plants render a rich source of bioactive compounds used in drug formulation and development; they play a key role in traditional or indigenous health systems. As the demand for herbal medicines increases worldwide, supply is declining as most of the harvest is derived from naturally growing vegetation. Considering global interests and covering several important aspects associated with medicinal plants, the Exploring Medicinal Plants series comprises volumes valuable to academia, practitioners, and researchers interested in medicinal plants. Topics provide information on a range of subjects including diversity, conservation, propagation, cultivation, physiology, molecular biology, growth response under extreme environment, handling, storage, bioactive compounds, secondary metabolites, extraction, therapeutics, mode of action, and healthcare practices.

Led by Azamal Husen, PhD, this series is directed to a broad range of researchers and professionals consisting of topical books exploring information related to medicinal plants. It includes edited volumes, references, and textbooks available for individual print and electronic purchases.

Traditional Herbal Therapy for the Human Immune System, *Azamal Husen*

Environmental Pollution and Medicinal Plants, *Azamal Husen*

Herbs, Shrubs and Trees of Potential Medicinal Benefits, *Azamal Husen*

Phytopharmaceuticals and Biotechnology of Herbal Plants, *Sachidanand Singh, Rahul Datta, Parul Johri, and Mala Trivedi*

Omics Studies of Medicinal Plants, *Ahmad Altaf*

Exploring Poisonous Plants: Medicinal Values, Toxicity Responses, and Therapeutic Uses, *Azamal Husen*

Plants as Medicine and Aromatics: Conservation, Ecology, and Pharmacognosy *Mohd Kafeel Ahmad Ansari, Bengu Turkyilmaz Unal, Munir Ozturk and Gary Owens*

Sustainable Uses of Medicinal Plants, *Learnmore Kambizi and Callistus Bvenura*

Medicinal Plant Responses to Stressful Conditions *Arafat Abdel Hamed Abdel Latef*

Aromatic and Medicinal Plants of Drylands and Deserts: Ecology, Ethnobiology and Potential Uses *David Ramiro Aguillón Gutiérrez, Cristian Torres León, and Jorge Alejandro Aguirre Joya*

Secondary Metabolites from Medicinal Plants: Nanoparticles Synthesis and their Applications *Rakesh Kumar Bachheti, Archana Bachheti*

Aquatic Medicinal Plants *Archana Bachheti, Rakesh Kumar Bachheti, and Azamal Husen*

Antidiabetic Medicinal Plants and Herbal Treatments *Azamal Husen*

Ethnobotany and Ethnopharmacology of Medicinal and Aromatic Plants: Steps Towards Drugs Discovery *Adnan Mohd, Mitesh Patel and Mejdi Snoussi*

Wild Mushrooms and Health Diversity, Phytochemistry, Medicinal Benefits, and Cultivation *Kamal Ch. Semwal, Steve L. Stephenson, and Azamal Husen*

Medicinal Roots and Tubers for Pharmaceutical and Commercial Applications *Rakesh Kumar Bachheti and Archana Bachheti*

Medicinal Roots and Tubers for Pharmaceutical and Commercial Applications

Edited by
Rakesh Kumar Bachheti
Archana Bachheti

CRC Press
Taylor & Francis Group
Boca Raton London New York

CRC Press is an imprint of the
Taylor & Francis Group, an informa business

First edition published 2024
by CRC Press
6000 Broken Sound Parkway NW, Suite 300, Boca Raton, FL 33487-2742

and by CRC Press
4 Park Square, Milton Park, Abingdon, Oxon, OX14 4RN

CRC Press is an imprint of Taylor & Francis Group, LLC

ISBN: 9781032280387 (hbk)
ISBN: 9781032280394 (pbk)
ISBN: 9781003295037 (ebk)

DOI: 10.1201/b22924

Typeset in Times New Roman
by Deanta Global Publishing Service, Chennai, India

Contents

Preface vii
Editors viii
List of Contributors x

Chapter 1 *Curcuma longa* (Turmeric): A Magical Rhizome 1

Fekade Beshah Tessema, Yilma Hunde Gonfa, Mesfin Getachew Tadesse, Archana Bachheti, Azamal Husen, and Rakesh Kumar Bachheti

Chapter 2 Medicinal and Commercial Application of *Zingiber officinale* 16

Meseret Zebeaman, Mesfin Getachew Tadesse, Rakesh Kumar Bachheti, Archana Bachheti, Rahel Gebeyhu, and Kundan Kumar Chaubey

Chapter 3 Bioactive Compounds and Pharmaceutical Importance of *Houttuynia cordata* Rhizome 27

Atreyi Pramanik, Aashna Sinha, Kundan Kumar Chaubey, and Deen Dayal

Chapter 4 Chemical Constituents and Medicinal Uses of *Eutrema japonicum* 39

Versha Parcha, Pankaj Bhandari, Sukanya Chhetri, Uday Kumar, and Raju Chandra

Chapter 5 *Boesenbergia pandurata:* An Overview of Medicinal Use, Ethnopharmacology, and Phytochemistry 48

Addisu Tamir Wassie, Archana Bachheti, Azamal Husen, and Rakesh Kumar Bachheti

Chapter 6 Phylogeny, Phytochemistry, Traditional Uses and Pharmaceutical Properties of *Thapsia* spp. Roots 64

Khaled Taïbi, Leila Aït Abderrahim, Mohamed Boussaid, Fadhila Taïbi, Kada Souana, Mohamed Achir, and Abdelkader Tadj

Chapter 7 Secondary Metabolites, Ethnopharmacology, and Commercial Application of *Glycyrrhiza glabra* 74

Noureddine Chaachouay, Abdelhamid Azeroual, Bouchaib Bencharki, and Lahcen Zidane

Chapter 8 Medicinal and Commercial Uses of *Angelica* Roots 89

Anuj Kandwal, Rakesh Kumar Bachheti, Sonali Purohit, Archana Bachheti, and Arun Kumar Khajuria

Chapter 9 Traditional and Modern Medicinal Uses of *Cichorium intybus* Roots.................... 101

Munir Ozturk, Volkan Altay, and Mehdi Younessi-Hamzekhanlu

Chapter 10 Therapeutic Importance of *Withania somnifera* (Ashwagandha) in Medicine........ 117

Sonali Purohit, MC Purohit, Arun Kumar Khajuria, Rakesh Kumar Bachheti, and Anuj Kandwal

Chapter 11 Phytoconstituents and Medicinal Importance of *Chlorophytum borivilianum* Tuber.. 130

Sisay Awoke, Melaku Assefa, and Mesfin Getachew

Chapter 12 A Comprehensive Review on the Medicinal Use of *Piper methysticum* (Kava) 140

Versha Parcha, Pankaj Bhandari, Sukanya Chhetri, Uday Kumar, and Deepak Kumar

Chapter 13 Medical Benefits and Side-effects of *Lepidium meyenii* Root................................ 160

Deepak Kumar Verma, Vinay Pathak, Sapna Yadav, Sadiya Sameer, Navneet Kumar, Aashna Sinha, Kundan Kumar Chaubey, Krishan Raj Singh, Gaurav Bhardwaj, Alazar Essayas, and Rakesh Kumar Bachheti

Chapter 14 Phytopharmacological Aspects of *Valeriana officinalis* Root................................ 169

Pratibha Kumari, Anupam Bhatt, and Dipika Rana

Chapter 15 Photochemistry, Medicinal and Commercial Applications of *Beta vulgaris* L....... 179

M.C. Purohit, Anuj Kandwal, Sonali Purohit, Archana Bachheti, and Arun Kumar Khajuria

Chapter 16 Radish: Health Benefits and Medicinal Uses 188

Sugam Gupta, Bhavya Mudgal, Devvret Verma, Debasis Mitra, Archana Bachheti, and Rakesh Kumar Bachheti

Index 205

Preface

Humans rely on plants for a variety of requirements, including food, shelter, and health care. It has been observed that, worldwide, most people still rely on and use plants for their health care needs. People still utilize traditional medications despite the accessibility of contemporary drugs today because the former are well tolerated, compatible, and have fewer side-effects than the latter. Even today, antibiotics, contraceptives, anticancer medications, and treatments for ulcers are made mostly from bioactive chemicals derived from medicinal plants. These bioactive chemicals can be obtained from different parts of plants, including roots and tubers. Some examples of medicinal plants where the roots or tubers contain numerous pharmaceutical and other commercial applications are *Glycyrrhiza glabra, Houttuynia cordata, Curcuma longa, Berberis aristata, Beta vulgaris, Zingiber officinale, Alpinia galangal, Boesenbergia pandurata, Asparagus racemosus, Coleus forskohlii, Withania somnifera, Cichorium intybus, Chlorophytum borivilianum, Piper methysticum, Valeriana officinalis* and so on.

Chapter 1 gives an overview of the use of turmeric to treat different diseases and their commercial applications. *Zingiber officinale*, also known as ginger, is widely used as a food flavouring, spice, and food additive, and Chapter 2 overviews the medicinal and commercial uses of *Z. officinalis*. Chapters 3 and 4 discuss the chemical constituents and medicinal uses of *Houttuynia cordata* rhizomes and *Eutrema japonicum*, respectively. The rhizome of *Boesenbergia pandurata* contains various flavonoid compounds with a wide range of pharmacological properties, including antiviral, anti-ulcer, antioxidant, antiparasitic, and antibacterial activities, and Chapter 5 aims to provide an overview of the medicinal and pharmaceutical properties of *B. pandurata*. Phytochemistry, traditional uses, and pharmaceutical properties of *Thapsia* spp. roots are discussed in Chapter 6, and the secondary metabolites, ethnopharmacology, and commercial application of *Glycyrrhiza glabra* are discussed in Chapter 7. The use of the root of *Angelica* in traditional and commercial contexts is covered in Chapter 8. The root of *Cichorium intybus*, known for various bio-activities due to the presence of phytochemicals, and traditional and modern medicinal uses of *Cichorium intybus* roots are discussed in detail in Chapter 9. The therapeutic value of *Withania somnifera*, also known as Indian ginseng or Ashwagandha, in conventional and contemporary systems of treatment is covered in Chapter 10. *Chlorophytum borivilianum* is a most eminent medicinal plant, commonly known as Safed Musli and recognized for its various biological activities, and the phytoconstituents and medicinal importance of the *C. borivilianum* root are covered in Chapter 11. A comprehensive review of the medicinal use of *Piper methysticum* is presented in Chapter 12, and the pharmacological, therapeutic, and phytochemical characteristics of *Valeriana officinalis* roots are examined in Chapter 13. The medicinal impacts, side-effects, doses, and mechanism of action of *Lepidium meyenii* root are discussed in Chapter 14. The phytochemistry, therapeutic, and commercial uses of *Beta vulgaris* are covered in Chapter 15, whereas Chapter 16 focuses on the diverse varieties of radish that may be grown over the different seasons and the medicinal potential of roots of this plant, including its effects as antioxidants, in protection against cancer, and as an antidiabetic.

In this book, the reader will find a wealth of knowledge on medicinal roots and tubers for pharmaceutical and commercial applications, along with significant references. The book has 16 chapters, and the Table of Contents effectively reflects its extensive examination of the topic's various aspects.

We are extremely grateful to all the distinguished authors who wrote chapters and donated their time and expertise to this edited book. Additionally, we are happy to thank the anonymous reviewer who made a specific contribution to improving the chapters; revision comments and criticism from subject experts and readers are welcome.

Rakesh Kumar Bachheti
Archana Bachheti

Editors

Rakesh Kumar Bachheti graduated from Hemwati Nandan Bahuguna University, Garhwal, India, in 1996. He completed his MSc in Organic Chemistry from the same university in 1998. Later, in 2001, he underwent a one-year Post Graduate Diploma in Pulp and Paper Technology from Forest Research Institute, Dehradun, India. Subsequently, he pursued his Ph.D. in Organic Chemistry from Kumaun University, Nainital, India, which he obtained in 2007. Dr. Bachheti has a total research and teaching experience of 20 Years. Before joining Addis Ababa Science and Technology University (AASTU) in Ethiopia, he worked as Dean of Project (Assistant) at Graphic Era University in Dehradun, India. It's worth noting that Graphic Era University is accredited with an 'A+' grade by the National Assessment and Accreditation Council (NAAC). Dr. Bachheti is also a Research Fellow at INTI International University Persiaran Perdana BBN, Putra Nilai, Nilai, Negeri Sembilan, Malaysia, and Adjunct Faculty at the Department of Allied Sciences, Graphic Era Hill University, Dehradun, India.

Dr. Bachheti has an impressive research background and has presented papers at various international and national conferences. He has also been an active member of important committees like the Internal Quality Assurance Cell (IQAC) and the Anti-ragging Committee. His research interests primarily focus on natural products for industrial applications, biofuels and bioenergy, green synthesis of nanoparticles and their applications, and pulp and paper technology. His passion for natural products is evident in all of his research endeavors. As an academic mentor, Dr. Bachheti has successfully advised 40 MSc and 5 Ph.D. students to completion. Additionally, numerous undergraduates have conducted research in his laboratory under his guidance. He actively contributes to curriculum development for BSc/MSc/Ph.D. programs. Dr. Bachheti is also recognized for his editorial work, having edited 6 books and authored 110 publications on various aspects of natural product chemistry and nanotechnology. He has contributed twenty book chapters published by prestigious publishers like Springer, Elsevier, Tylor and Francis and Nova Publisher.

Archana (Joshi) Bachheti completed her BSc in 1997 and MSc in 1999 from HNB Garhwal University. She earned her Ph.D. from Forest Research Institute, Dehradun, India, in 2006. Throughout her career, she has been actively involved in research projects and consultancy work in various areas, including ecorestoration and development of wasteland, physico-chemical properties of Jatropha curcus seed oil in relation to altitudinal variation, and serving as a consultant ecologist for a project funded by a government agency. Currently, Dr. Archana is a Professor at Graphic Era University in Dehradun, India. With an extensive experience of more than 17 years, she has served in different capacities within academia in India and has also provided expertise internationally. Her teaching portfolio includes subjects like Ecology and Environment, Environmental Science, Freshwater Ecology, Disaster Management, and Bryophytes and Pteridophytes. Dr. Archana's research interests cover a broad and interdisciplinary field of plant ecology. Her focus areas include ecorestoration, green chemistry, especially the synthesis of nanomaterials, and exploring the medicinal properties of plants. Her research has encompassed ecological amelioration of degraded lands, studying physical and chemical properties of plant oils, and delving into plant-based nanomaterials. Throughout her career, she has been actively involved in mentoring students and researchers. She guided one Ph.D. student and supervised three scholars. Additionally, she has provided guidance to graduate and undergraduate students in their research projects. Driven by her fascination with forest biodiversity, Dr. Archana has maintained a passion for exploring the values of biodiversity and how it can contribute to social upliftment. As an editor, she has contributed to six books and has published over 80 research articles in both international and national journals. She has also authored sixteen book chapters. Furthermore, she has organized several National seminars/conferences at Graphic Era University, India.

List of Contributors

Leila Aït Abderrahim
Faculty of Life and Natural Sciences
University of Tiaret
Algeria

Mohamed Achir
Faculty of Life and Natural Sciences
University of Tiaret
Algeria

Volkan Altay
Department of Biology
Faculty of Arts & Sciences
Hatay Mustafa Kemal University
Turkiye

Melaku Assefa
Department of Chemistry, Dessie, Wollo University
Ethiopia

Sisay Awoke
Department of Chemistry, Dessie, Wollo University
Ethiopia

Abdelhamid Azeroual
Agri-Food and Health Laboratory (AFHL)
Hassan First University
Morocco

Bouchaib Bencharki
Plant, Animal Production and Agro-industry Laboratory
Ibn Tofail University
Morocco

Pankaj Bhandari
Department of Pharmaceutical Chemistry & Chemistry
Dolphin (PG) Institute of Biomedical & Natural Sciences
India

Gaurav Bhardwaj
Department of Biotechnology
Institute of Applied Sciences and Humanities
GLA University
India

Anupam Bhatt
CSIR–Institute of Himalayan Bioresource Technology
India

Mohamed Boussaid
Faculty of Life and Natural Sciences
University of Tiaret
Algeria

Noureddine Chaachouay
Interdisciplinary Research Laboratory in the Sciences
Hassan First University
Morocco

Raju Chandra
Department of Pharmaceutical Chemistry & Chemistry
Dolphin (PG) Institute of Biomedical & Natural Sciences
India

Kundan Kumar Chaubey
Division of Research and Innovation
Uttaranchal University
India

Sukanya Chhetri
Department of Pharmaceutical Chemistry & Chemistry
Dolphin (PG) Institute of Biomedical & Natural Sciences
India

Deen Dayal
Department of Biotechnology
Institute of Applied Sciences and Humanities
GLA University
India

Alazar Essayas
Department of Biotechnology
Wollo University
Ethiopia

Rahel Gebeyhu
Armauer Hansen Research Institute
Ethiopia

Yilma Hunde Gonfa
Nanotechnology Center of Excellence
Addis Ababa Science and Technology University
Ethiopia

Sugam Gupta
Department of Applied Science & Engineering
Tula's Institute
Dhoolkot, Dehradun
Uttarakhand

Azamal Husen
Wolaita Sodo University
Ethiopia

Anuj Kandwal
Department of Chemistry
Harsh Vidya Mandir (P.G.) College
India

Arun Kumar Khajuria
Department of Botany
Cluster University of Jammu
India

Deepak Kumar
Department of Pharmaceutical Chemistry & Chemistry
Dolphin (PG) Institute of Biomedical & Natural Sciences
India

Navneet Kumar
College of Paramedical Sciences
Teerthanker Mahaveer University
India

Uday Kumar
Department of Pharmaceutical Chemistry & Chemistry
Dolphin (PG) Institute of Biomedical & Natural Sciences
India

Pratibha Kumari
School of Biological & Environmental Sciences
Shoolini University
Solan, H.P.
India

Debasis Mitra
Department of Microbiology
Raiganj University
India

Bhavya Mudgal
Department of Biotechnology
Graphic Era University
India

Munir Ozturk
Centre for Environmental Studies and Botany Department
Ege University
Turkiye

Versha Parcha
Department of Pharmaceutical Chemistry & Chemistry
Dolphin (PG) Institute of Biomedical & Natural Sciences
India

Vinay Pathak
College of Paramedical Sciences
Teerthanker Mahaveer University
India

Atreyi Pramanik
Department of Biochemistry
All-India Institute of Medical Sciences
India

M.C. Purohit
Department of Chemistry
Hemvati Nandan Bahuguna Garhwal University
India

Sonali Purohit
Department of Shalakya Tantra
Shivalik Institute of Ayurved and Research
Dehradun
India

Dipika Rana
School of Biological & Environmental Sciences
Shoolini University
Solan, H.P.
India

Sadiya Sameer
Department of Biotechnology
Institute of Applied Sciences and Humanities
GLA University
India

Krishan Raj Singh
Department of Biotechnology
Institute of Applied Sciences and Humanities
GLA University
India

Aashna Sinha
Department of Biotechnology
Institute of Applied Sciences and Humanities
GLA University
India

Kada Souana
Faculty of Life and Natural Sciences
University of Tiaret
Algeria

Mesfin Getachew Tadesse
Bio-process and Biotechnology Center of Excellence
Ethiopia

Fadhila Taïbi
Faculty of Life and Natural Sciences
University of Tiaret
Algeria

Khaled Taïbi
Faculty of Life and Natural Sciences
University of Tiaret
Algeria

Fekade Beshah Tessema
Department of Chemistry
Faculty of Natural and Computational Science
Woldia University
Ethiopia

Devvret Verma
Department of Biotechnology
Graphic Era University
India

Addisu Tamir Wassie
Department of Industrial Chemistry
Addis Ababa Science and Technology University
Ethiopia

Sapna Yadav
Department of Biotechnology
Institute of Applied Sciences and Humanities
GLA University
India

Mehdi Younessi-Hamzekhanlu
Department of Forestry and Medicinal Plants
University of Tabriz
Iran

Meseret Zebeaman
Department of Industrial Chemistry
Addis Ababa Science and Technology University
Ethiopia

Lahcen Zidane
Plant, Animal Productions and Agro-industry Laboratory
Ibn Tofail University
Morocco

1 *Curcuma longa* (Turmeric)
A Magical Rhizome

Fekade Beshah Tessema, Yilma Hunde Gonfa, Mesfin Getachew Tadesse, Archana Bachheti, Azamal Husen, and Rakesh Kumar Bachheti

1.1 INTRODUCTION

Curcuma longa L. (syn. *Curcuma domestica*; *Curcuma aromatica*) (Zingiberaceae). Ird; Erd (Amh). Turmeric (Eng) (Dagne, 2011). It is a leafy, stemless, ginger-like plant with broad, hairless leaves sprouting from ground level and oblong spikes covered in beautiful yellow and white flowers. The smooth, fleshy rhizomes are bright orange on the inside. The plant is a long-used cult symbol believed to have come originally from India. It is cultivated in most tropical regions of the world, including China, India, Indonesia, Madagascar, and Malaysia (Ben-Erik & Michael, 2017). Turmeric has several medicinal purposes, including the prevention of stroke and cataracts, the treatment of athlete's foot, bunions, diabetes, gallstones, and gout, as well as the relief of pain, liver issues, headaches, oedema, and ulcers (Duke, 1997). In Ethiopia, *C. longa* is used to treat weeping eyes (Fullas, 2006). Commission E, a scientific advisory board of the German Federal Institute for Drugs and Medical Devices, authorized the use of turmeric root to treat dyspeptic disorders. The German Standard License recommends turmeric tea infusion for digestive issues (Blumenthal et al., 2000; Dagne, 2011).

The dried powder/paste (Chew et al., 2022), essential oils (EOs) (Guerrini et al., 2023), and extracts (Amin et al., 2022; Mariano et al., 2022) of the rhizome of *C. longa* are used in traditional medicine (Figure 1.1). The rhizome of *C. longa* has strong anti-inflammatory, antibacterial, antifungal, and antioxidant effects. Amin et al. (2022) examined the anti-inflammatory and antioxidant activity of the hexane and ethanolic extracts of *C. longa* against lipopolysaccharide (LPS)-induced inflammation in buffalo mammary epithelial cells (BuMECs) (Amin et al., 2022). The following sections in this chapter will cover each activity in more depth.

The key ingredient in *C. longa* is curcumin. It has a wide range of medicinal uses to treat a variety of illnesses. Due to its antioxidant activity, curcumin possesses potent antibacterial and antifungal properties, as well as remarkable wound-healing abilities. Since ancient times, curcumin has been used extensively to treat a wide range of illnesses and ailments, most notably to treat wounds (Farhat et al., 2023). Curcumin is a Class IV medication according to the Biopharmaceutics Classification System, but it has minimal gastrointestinal epithelial permeability and low water solubility (Alolga et al., 2022).

Its application in medicine is restricted by its limited oral bioavailability, low water solubility, and rapid metabolism. Curcumin's medicinal and pharmacological benefits are enhanced by designing appropriate drug delivery methods to deliver it. The most recently reported curcumin-loaded delivery systems for wound-healing purposes are chiefly hydrogels (Nikolić et al., 2023), films (Gopinath et al., 2004), wafers (Adel et al., 2021), and sponges (Nguyen et al., 2013). Using *N*-isopropyl-methacrylamide and *N*-isopropyl-acrylamide as the building blocks and ethylene

DOI: 10.1201/b22924-1

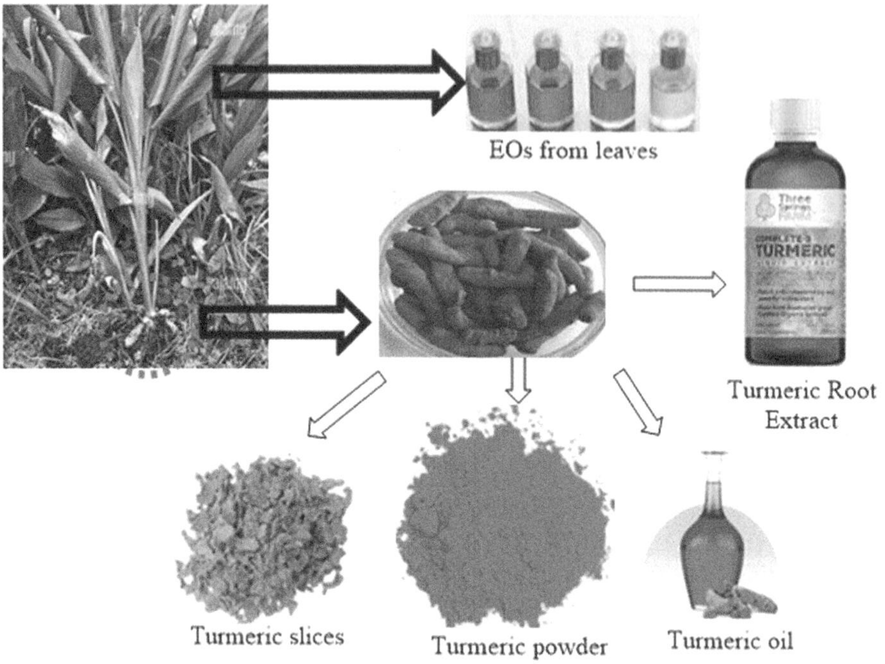

FIGURE 1.1 *Curcuma longa* products used in traditional medicine.

glycol dimethacrylate as the crosslinker, curcumin was combined with 2-hydroxypropyl-cyclodextrin to create a thermosensitive hydrogel (Nikolić et al., 2023). In addition, curcumin nanoformulations, like nanohydrogels, nanoparticles, and nanofibers, that offer greater solubility, bioavailability, and sustained release, are also recommended. These formulations enhance the effects of curcumin on wound healing by promoting the various stages of healing with the help of a low-molecular-weight carrier (Sideek et al., 2023). To increase solubility, for instance, a modified nanogel delivery system, with poly(N-isopropylacrylamide) (PNIPAM) and β-cyclodextrin grafted onto hyaluronic acid (PNCDHA), was used (Kaewruethai et al., 2023). The combination of curcumin with black pepper also increases the absorption and bioavailability of curcumin up to 20-fold (Górnicka et al., 2023). The therapy of cancer is especially well-suited to curcumin nanoformulations based on poly(lactic-co-glycolic acid) (PLGA), cyclodextrin assembly, and magnetic nanoparticles (Hafez Ghoran et al., 2022).

By fostering synergistic effects in the control of several malignant pathways, curcumin increased the activity of other chemotherapy medications and radiotherapy (Younes et al., 2022). The co-administration of rosemary essential oil and curcumin increased both their hepatoprotective effects and the amount of curcumin in plasma, demonstrating a synergistic interaction between the two natural products (Mohamed et al., 2022). The interaction between losartan and *C. longa* herb interaction altered the drug's pharmacokinetics and pharmacodynamics (Ahad et al., 2023).

With the aims of increasing its effectiveness in quenching reactive oxygen species (ROS) and lessening its negative effects, curcumin can be thought of as a prospective treatment alternative (Ghareghomi et al., 2021). Inhibiting cell proliferation, preventing invasion and migration, inducing cell apoptosis, activating ferroptosis, lowering inflammation, stimulating ROS formation, and controlling gut microbiota and adjuvant therapy are some of its mechanisms of action (Yang et al., 2022).

Here, we shall attempt to discuss the ethnobotanical uses, biological activities, phytochemical investigations, and commercial applications of the *C. longa* rhizome.

1.2 ETHNOBOTANICAL USE

Ethnobotanical uses of turmeric plant parts and identification of the major components of the rhizomes are mainly concerned with food spice preparations and their medicinal uses. Some of these uses in different countries are summarized in the table below (Table 1.1). Mostly, the rhizome is found to be used traditionally for different purposes.

1.3 PHYTOCHEMISTRY/CHEMICAL COMPOSITION

The primary components of the orange-yellow volatile oil found in the rhizome of turmeric are tumerone and zingiberene. A yellow extract, containing curcumin and other closely related diaryl heptanoids, known as curcuminoids, is produced by extracting the rhizome, using solvents like acetone or alcohol (Dagne, 2011).

TABLE 1.1
Ethnomedicinal Uses of Turmeric Plant Parts

Plant Part Used (Country)	Preparation	Use	Reference
Rhizome (Ethiopia)	Dried and ground	In the preparation of *alicha wot* and poultry feed	Demissew (1993)
Not mentioned	Not specified	Antihyperglycaemic effect in STZ-induced diabetic rats	Regassa (2013)
Rhizome (Brazil)	Powder	Used against intestinal and stomach ailments	Agra et al. (2008)
Rhizome (Canada)	Pounded rhizome	Treatment of arthritis, retained placenta	Lans et al. (2006)
Rhizome (Iran)	Not specified	Treatment of gall stones, contusions, digestive, emmenagogue	Amiri and Joharchi (2013)
Leaves, rhizomes (Iraq)	Decoction for both internal and external uses	Facial massage, fat-burning, spice, arthritis pain, antiviral and anticancer agent	Ahmed (2016)
Rhizome (Korea)	Prepared as soup for oral use	Blood circulation, gastroenteric disorder	Kim and Song (2011)
Rhizome (Morocco)	The powder is taken orally	Calefacient, condiment, digestive, and tonic	Merzouki et al. (2000)
Whole plant (India)	Not specified	Blood purification	Kala (2005)
Not mentioned	Wound dressing	Wound healing	Peter (2012)
Turmeric juice or rhizome	Different preparations in combination with other plant extracts and other components of traditional medicine	Wide range of ailments	Velayudhan et al. (2012)
Not mentioned	Not specified	Against Alzheimer's disease	Peter (2012)
Rhizome	Not specified	Antioxidants, to lower blood sugar and cholesterol, are used to increase the NADPH/NADP ratio, enhance glucose peroxidase activity, and decrease glucose inflow through the polyol pathway.	Ponnusamy et al. (2011)

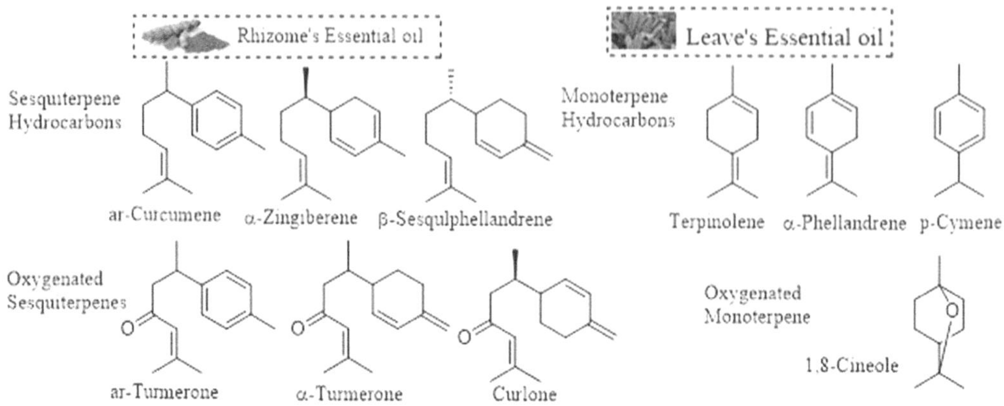

FIGURE 1.2 Major compounds found in the essential oils from rhizomes and leaves of turmeric. Adapted from Ibáñez and Blázquez (2021).

The results of the analysis of Indian turmeric were as follows: carbohydrates (69.4%), moisture (13.1%), protein (6.3%), fat (5.1%), mineral matter (3.5%), fibre (2.6%), and carotene (50 IU vitamin A-equivalents/100 g). The dry rhizome used to make the EO (5.8%) contained the following compounds: turmerones (sesquiterpenes) (53.0%), zingiberene (25%), α-phellandrene (1%), cineol (1%), sabinene (0.6%), and borneol (0.5%). From the volatile distillate, a ketone, $C_{13}H_{20}O$, and an alcohol, known as tolylmethyl carbinol, were extracted. Curcumin is a crystalline, coloured substance with a yield of 0.6% and a melting point of 180 to 183°C. Its chemical formula is $C_{21}H_{20}O_6$. The yellow-red colour is visible following its dissolution in strong sulphuric acid. According to Kapoor (2017), the coloured substance is curcumin, a member of the dicinnamoyl methane group. Additionally, the rhizome contains turmeric oil, a fragrant oil. Curcuminoids were discovered in rhizomes by chromatographic analysis (Kapoor, 2017). In another phytochemical study, Ikpeama et al. (2014) showed that the plant contains 1.08% tannins, 0.45% saponins, 0.40% flavonoids, 0.08% phenolics, and 0.03% sterols.

Two hundred and thirty-five compounds have been isolated or identified from the leaves, flowers, roots, and rhizomes of *C. longa*, including 109 sesquiterpenes, 68 monoterpenes, 22 diarylheptanoids and diarylpentanoids, eight phenylpropenes and other phenolic compounds, five diterpenes, four sterols, three triterpenoids, two alkaloids, and 14 other compounds. With a minimum limit to the concentrations, curcumin and its derivatives, demethoxycurcumins, can be used as marker compounds for the quality control of ('curcumin') products like rhizomes, powders, and extracts. The quality of turmeric oil and oleoresin products can be managed by using the three main ketonic sesquiterpenes (ar-, α-, and β-turmerones) (Li et al., 2011). *C. longa* oleoresin is a viscous, dark yellow liquid that combines 37%–55% curcuminoids and 25% EO. It primarily functions as a food colouring, medication, and dietary supplement (Fuloria et al., 2022). Figure 1.2 shows the main phytochemicals present in the EOs of rhizomes and leaves of turmeric.

1.4 PHARMACOLOGICAL EFFECTS

As mentioned in the Introduction, curcumin is a major phytoconstituent of turmeric and a nutraceutical with multiple pharmacological effects that have been demonstrated in both experimental and clinical settings. Some of the major pharmacological activities of curcumin and other products of turmeric extracts are discussed in the following subsections.

1.4.1 Antioxidant Activity

2,2-Diphenyl-1-picrylhydrazyl (DPPH) radical scavenging, metal chelating, and other lipid peroxidation assays have all been used to determine antioxidant activities. Fresh rhizomes have greater antioxidant qualities than dry ones due to their higher EO and ethanol oleoresin contents (Singh et al., 2010). By promoting the activity of antioxidant enzymes, dietary turmeric reduces lipid peroxidation (Reddy & Lokesh, 1994). Pure curcumin and isolated turmeric extracts have very high antioxidant activity but showed very little efficacy against the mycobacteria under study by Çıkrıkçı et al. (2008). Turmerin, a water-soluble peptide, is a powerful antioxidant, a DNA protective against oxidative damage, an antimutagen against lipid peroxide-mediated mutagenicity, and an arachidonic acid release inhibitor (Srinivas & Place, 1992). Guerrini et al. (2023) reported the mutagen-protective capacity of the EOs from *C. longa*.

The natural antioxidant curcumin, which has a wide range of biological functions, is nontoxic and extremely promising in pharmaceutical terms (Verma et al., 2018). Investigations have demonstrated that *C. longa* and its active components lowered lung, renal, hepatic, cardiovascular, and neuro-toxicity, mostly by reducing inflammatory cytokines and increasing antioxidant activities and anti-apoptotic effects (Hosseini & Hosseinzadeh, 2018). Due to their high contents of polyphenols, flavonoids, tannins, and ascorbic acid, as well as their considerable DPPH free radical-scavenging activities and ferric-reducing antioxidant power (FRAP) values, Bangladeshi turmeric cultivars are promising sources of natural antioxidants (Tanvir et al., 2017). Even at high doses, standardized turmeric powder and curcumin extract are safe for human consumption; nevertheless, further research is required on the effects of various curcumin formulations, particularly in humans (Soleimani et al., 2018).

1.4.2 Antimicrobial Activities

1.4.2.1 Antibacterial Activity

Turmeric is an effective antibiotic for treating fevers, sore throats, and septicaemia, and is also excellent for external use as a topical pain reliever for bruises, infections, sprains, and soreness (Pole, 2006). Most of the microorganisms linked with food were actively inhibited by turmeric extracts (Ram Kumar & Jain, 2010). The discovery of antibacterial action highlighted turmeric's potential to lower bacterial contamination in both food and humans (Irshad et al., 2018). An effective producer of silver nanoparticles, the fungus *Penicillium* sp., was effectively extracted from healthy turmeric leaves (Singh et al., 2013). Antibacterial effectiveness against pathogenic Gram-negative bacteria was also tested on the extracellular production of silver nanoparticles (AgNPs). Endophytic fungus-generated AgNPs showed strong antibacterial activity. Good results were seen in the antibacterial activity against multidrug-resistant (MDR) *Staphylococcus aureus* and *Escherichia coli*, at 80 μL of AgNPs, with maximum zones of inhibition of 17 mm and 16 mm, respectively (Singh et al., 2014).

Turmeric has antibacterial components that have the potential to be very useful for the development by pharmaceutical companies as products for the treatment of a variety of ailments (Ikpeama et al., 2014). Cholagogues (agents which promote bile flow) from *C. longa* include curcumin, as a component, and the pigment choleretic acid, which also has antimicrobial properties (Kapoor, 2017). An alcoholic extract of the *C. longa* rhizome showed strong antiprotozoal action against *Entamoeba histolytica*.

1.4.2.2 Antifungal Activity

The only two fungi against which the ethanolic turmeric extract was effective were the food contaminants *Rhizopus stolonifer* and *Mucor* sp. (Ram Kumar & Jain, 2010). There is preliminary evidence that oral and topical turmeric/curcumin supplements and products may have therapeutic advantages for skin health (Vaughn et al., 2016). Curcumin was discovered to have no antifungal activity by Apisariyakul et al. (1995). In *in-vitro* settings, turmeric oil can suppress dermatophytes and pathogenic moulds. In guinea pigs, *Trichophyton rubrum*-induced erythaema and scale were significantly reduced by turmeric oil (Apisariyakul et al., 1995). Turmeric oil's main component, (aromatic) ar-turmerone, has potent antidermatophytic efficacy. It can be used as a quality indicator for turmeric oil, the active component in turmeric creams and other antifungal treatments. Turmeric cream, which contained 6% (w/w) turmeric oil, was found to be suitable for further development as a specific antidermatophytic agent (Jankasem et al., 2013).

1.4.2.3 Antiviral Activity

According to an *in-silico* study, curcumin from turmeric has strong anti-COVID-19 activity by blocking the primary protease enzyme of the SARS-CoV-2 virus (Rajagopal et al., 2020). By preventing HIV-1 long terminal repeat (LTR) promoter-driven gene expression, curcumin exhibits antiviral potential against HIV, with no impact on host cell viability (Ashraf & Sultan, 2017).

1.4.3 Antidiabetic Activity

The traditional Ayurvedic drug 'Rajanyamalakadi' contains extracts from the plants *Salacia oblonga*, *Curcuma longa*, and *Emblica officinalis*, that have been shown to have potent antioxidant, antidiabetic, and hypolipidaemic properties (Faizal et al., 2009). There was a 50% decrease in the urine glucose concentration, which is generally high in those with chronic diabetes, in diabetic patients treated with *C. longa* powder and milk. Scientific research and in-depth analysis demonstrated the benefits of freeze-dried *C. longa* rhizome powder diluted in milk as an efficient and safe antidiabetic dietary supplement (Rai et al., 2010). An extract of *C. longa* in isopropanol contains podocarpic acid ($C_{17}H_{22}O_3$), curlone ($C_{15}H_{22}O$), and cinnamic acid ($C_9H_8O_2$), all of which are predicted to inhibit the human pancreatic amylase (HPA) enzyme, hence slowing the rate of starch hydrolysis and lowering blood glucose levels (Ponnusamy et al., 2011).

Curcuminoids controlled the expression of genes related to glycolysis when turmeric oleoresin was ingested. Additionally, it was claimed that turmeric EO and curcuminoids synergistically controlled the expression of genes related to peroxisomal α-oxidation in response to turmeric oleoresin ingestion (Honda et al., 2006). Following the findings of other research showing *C. longa* to lower plasma glucose, Wickenberg et al. (2010) set out to investigate its potential as a functional food component for treating type 2 diabetes. According to their research, consumption of *C. longa* increased postprandial serum insulin levels, but, in healthy participants, it had no impact on GI or plasma glucose levels. The findings imply that *C. longa* might impact insulin secretion (Wickenberg et al., 2010).

1.4.4 Anti-cancer Activity

Dietary materials like curcumin and docosahexaenoic acid have been shown to inhibit the growth of breast cancer cells (Altenburg et al., 2011). Curcumin is widely known for its synergistic effects. Because curcumin impacts various molecular targets, it may be helpful in the treatment and prevention of several illnesses, including cancer (Kocaadam & Şanlier, 2015).

Curcumin inhibits various pathways involved in the development of cancer and tumours, while also being non-toxic and diverse in its actions (Wilken et al., 2011). The elimination of cancer stem cells and the functionally induced autophagy of cancer cells by curcumin are of particular interest. These results reveal important strategies that can destroy cancer cells and stop cancer from returning, suggesting a novel method for lung cancer treatment (Ye et al., 2012). Curcumin triggers the DNA damage response, setting up these nutraceuticals for therapeutic use in prostate cancer chemoprevention (Horie, 2012). *C. longa* extract and curcumin demonstrated the ability to reduce the impacts of many recognized carcinogens and mutagens in diverse physiological tissues, according to a study using *in-vitro* and *in-vivo* models. As a result of cell cycle arrest, curcumin (50 mM) increased the volume of colorectal HT-29 cancer cells and caused apoptosis in human kidney tumour cells (Kössler et al., 2012)

1.4.5 Anti-inflammatory Activity

Six human clinical trials have shown that curcumin is safe and that it has anti-inflammatory effects (Chainani-wu et al., 2003). In a rat model of human rheumatoid arthritis, the turmeric rhizome was superior to ginger rhizome and indomethacin in lowering the inflammatory immune response and oxidative stress (Ramadan et al., 2011). Bones and joints in rats undergoing collagen-induced inflammation had inflammatory and degenerative changes that were seen using macroscopic and radiological techniques, changes which were shown to be beneficially arrested by orally administered extract of *C. longa* extract (Anna et al., 2011). *C. longa* is among the most studied anti-arthritic plant extracts and is recognized as a valuable remedy against joint diseases (Mariano et al., 2022).

According to Arora et al. (2010), rats treated with petroleum ether extracts of turmeric rhizomes and two of its column fractions – crystalline solid and viscous oil – had strong anti-inflammatory efficacies superior to those of hydrocortisone acetate and phenylbutazone. Additionally, the volatile oil significantly reduced inflammation in oedema and arthritis that was created experimentally (Kapoor, 2017). *C. longa* can prevent the synthesis of neutrophils and prostaglandin derivatives of arachidonic acid during inflammation. This may be a good indicator of its anti-inflammatory capabilities. In order to enhance curcumin's absorption and anti-inflammatory properties, bromelain, a protease enzyme which can be obtained from pineapple, is widely employed (Jacob et al., 2007).

1.4.6 Cardiovascular Effects

Numerous preclinical studies have shown that curcumin has various cardioprotective effects (Singletary, 2010). By its raktavardhaka and blood-building properties, turmeric nourishes the heart. Turmeric improves heart function by boosting blood flow and lowering total cholesterol (Pole, 2006). One of the processes through which turmeric's curcuminoids and sesquiterpenoids have hypoglycaemic effects is via activating peroxisome proliferator-activated receptor-gamma, PPAR-γ (Nishiyama et al., 2005). By altering the calcium influx mediated by L-type calcium channels in smooth vascular muscle, it was shown that cyclocurcumin, a hitherto unrecognized curcuminoid in *C. longa*, could have anticontractile effects (Kim et al., 2017).

1.4.7 Hepatoprotective Effects

A scientific and systemic investigation revealed that the freeze-dried rhizome powder of *C. longa* diluted in milk beneficially affected lipidaemia and hepatoprotection (Rai et al., 2010). In rats with thioacetamide-induced liver cirrhosis, the alcoholic extract of *C. longa* rhizomes had a hepatoprotective effect by halting the negative chain of events (Salama et al., 2013). Douichene et al. (2020) proposed the preventive role of a low concentration of an aqueous extract of *C. longa* to avoid paracetamol-induced liver toxicity in mice. Hepatotoxicity is caused by reactive metabolites, which are accompanied by a decline in chemical parameters in hepatic tissues (Douichene et al., 2020). Ibrahim et al. (2020) showed that curcuminoids might be viewed as a cutting-edge contender for creating a novel approach to treating liver illnesses. Curcuminoids improved the histological alterations in carbon tetrachloride-exposed rats and increased total protein and albumin contents (Ibrahim et al., 2020).

1.4.8 Neuroprotective Effects

Turmerone, α-phellandrene, 1,8-cineole, arturmerone, and curlone were identified as the major components of the EO extracted from fresh *C. longa* rhizomes. Neuropharmacological experiments provided evidence that may support the folklore use of this extract by demonstrating that the plant's EO demonstrated considerable CNS-depressant action as well as hypothermic, sedative, anxiolytic, and anticonvulsant effects in mice (Oyemitan et al., 2017). The development and progress of Alzheimer's disease may be slowed by curcumin. Curcumin appeared to show anti-inflammatory qualities in *in-vitro* experiments that may prevent neurodegeneration (Singletary, 2010). *In-vivo* and *in-vitro* studies have demonstrated the preventive effect of curcumin towards several neurodegenerative disorders (Benameur et al., 2022).

Drug resistance can be effectively inhibited by curcumin. Curcumin exhibits an astounding ability to block the rise in P-glycoprotein and its mRNA brought on by the chemotherapy medication Adriamycin (ADM) (Xu et al., 2011).

1.5 OTHER USES

1.5.1 Feed Supplement in Poultry

Powdered turmeric rhizome was found to contain considerable amounts of vitamins, minerals, and nutrients that could enhance a domestic animal's physiological performance. It also included moderate amounts of phytochemicals that could have pharmacological effects on humans and animals. Turmeric had the following quantities of various phytochemicals (in mg/g): cardiac glycosides (14.61), phytate (10.30), alkaloids (10.04), tannins (1.87), saponins (1.36), flavonoids (0.68), steroids (0.99), and terpenoids (0.54). According to the results of the vitamin studies, turmeric rhizome included the following vitamins (mg/g of sample weight): vitamin A (3.44), vitamin B_2 (1.20), vitamin C (0.84), vitamin D (0.64), vitamin E (0.39), vitamin B_3 (0.32), and vitamin B_1 (0.09) (Imoru et al., 2018; Durrani et al., 2008). Due to its high safety and pharmacological qualities, turmeric can be utilized as a natural growth booster in chicken diets (Khan et al., 2012). The performance of broiler chickens is improved overall when turmeric is added to the diet at a concentration of 0.5% (Durrani et al., 2008).

1.5.2 Dye

In 600 BC, in an Assyrian herbal, turmeric was identified as a source of a colourant. Before the invention of litmus paper, laboratories used turmeric to check the alkalinity of various scents. In the Middle Ages, turmeric was referred to as 'Indian saffron' in Europe. As a spice and a yellow food colourant, turmeric has been used in cooking for a long time. It is one of the main components of curry powder. The orange-yellow, waxy, short turmeric rhizomes are ground into a fragrant, fine powder used to make curry powder. It is a colourful, multipurpose product that combines the qualities of spice and a brilliant yellow dyestuff. The flavour is warm and harsh. It is used as a condiment in Ethiopia to make *alicha wot*. Many different foods are prepared with it, such as pastry (*sambousa*), mild beef stew (*ye siga alicha*), mild fried beef stew (*t'ibs alicha*), spiced ground beef stew (*minchet abish alicha*), curried fish stew (*yasa alicha*), pumpkin mild sauce (*ye duba alicha*), split pea stew (*ye kik alicha*), and lamb stew (*ye beg-alicha*) (Fullas, 2003). The powdered rhizomes of the turmeric plant are extracted using a solvent to produce the food colour curcumin, which is then purified by crystallization (WHO, 2004). The dried finger rhizomes of turmeric are ground into a yellow or red-yellow powder, which can be used as a colour for cosmetics, textiles, and food (Li et al., 2011). Turmeric oil is what gives this spice its distinctive flavour and aroma (Orellana-Paucar & Machado-Orellana, 2022).

Turmeric oleoresin is an organic turmeric extract used as a food spice and colourant. Oleoresin yield is between 7.0 to 14.0%. The main colouring agent, curcumin, makes up one-third of high-quality oleoresin. The curcuminoids (2.5–6%) found in the rhizomes give the yellow colour. These natural antioxidants include diferuloylmethane (curcumin I), demethoxycurcumin (curcumin II), and bisdemethoxycurcumin (curcumin III) (Peter, 2012). The uses of turmeric as a spice and as a vegetable dye are the significant functions that make it so well recognized (Ben-Erik & Michael, 2017).

1.5.3 As a Spice

Due to its widespread use in Indian households as a spice, colouring source, food preservative, and remedy for various illnesses, *C. longa* is known as 'The Golden Spice of India' (Fuloria et al., 2022). The presence of physiologically significant phytochemicals demonstrates the value of turmeric as a spice with numerous culinary and health benefits, and curcumin's role as a DNA protectant and a potent antioxidant (Irshad et al., 2018). The primary uses of turmeric are as a culinary colourant and spice (Dagne, 2011). *Alicha wot* is prepared from dried, ground rhizome (Demissew, 1993).

Oils of turmeric spices are made by steam distilling or supercritical CO_2 extracting dried rhizomes (ground turmeric) or leaves on their own or with other pastes or curry powder (Li et al., 2011). The amount of curcumin, total phenolic content, and antioxidant activity of bread were all significantly boosted by the addition of turmeric powder (Lim et al., 2011).

1.5.4 As a Food Additive

Ingesting curcumin alone does not produce the associated health advantages due to its low water solubility and poor bioavailability, the latter appearing to be principally driven by poor absorption, quick metabolism, and rapid elimination from the body. Numerous substances can improve bioavailability. For instance, black pepper's main active component, piperine, increases curcumin's bioavailability by 2000% when combined with curcumin in a mixture (Hewlings & Kalman, 2017). The food colourant curcumin (E 100) has been approved for use as a food additive in the EU after being tested by the Joint FAO/WHO Expert Committee on Food Additives (JECFA) in 2004. Three main colouring agents, in varying proportions, make up the 1,7-diphenylhepta-1,6-diene-3,5-dione food dye curcumin. The main components of the product are curcumin ((1E,6E)-1,7-bis-(4-hydroxy-3-methoxyphenyl)-1,6-heptadiene-3,5-dione) with the chemical formula $C_{21}H_{20}O_6$ and a molecular weight of 368.39 g/mol, as well as its desmethoxy and bis-desmethoxy derivatives with the chemical formulas $C_{20}H_{18}O_5$ and $C_{19}H_{16}O_4$ and molecular weights of 338.39 and 308.39, respectively (ANS, 2010). These applications, together with the other uses mentioned and very few side-effects, make turmeric an impressive pharmacological/health product.

1.5.5 Perfumery

A small amount of the turmeric EO is produced from dry turmeric tubers, providing spicy, peppery notes, and is used sparingly in perfumery at a rate of 2–6%.

1.6 COMMERCIAL APPLICATIONS

The only *Curcuma* species that is widely farmed and traded worldwide is turmeric, *C. longa* (Burapan et al., 2020). The multi-purpose major compound curcumin and the rhizome in powder, extracts, and EO forms are used both traditionally and as industrial products. Mostly, the use of turmeric root products as a spice, food additive, and food supplement are commercialized. After the processing of turmeric rhizome, many products are made from it which are available for commercial use. Curcumin is commercially available in multiple forms, including capsules, energy drinks, soaps, tablets, ointments, and cosmetics. It is also reported that *C. longa* is the best source of ω-3 fatty acid and α-linolenic acid (2.5%) (Fuloria et al., 2022). Some of the health benefits, medicinal preparations, and cosmetics from turmeric commercial applications are shown in Figure 1.3.

Therefore, this plant can be considered as a commercially important plant and is propagated for cultivation in agricultural cropping systems for its industrial exploitation (Khan et al., 2016). Compost addition and alley cropping were demonstrated to be efficient and long-lasting agro-ecological systems for enhancing turmeric yield and quality for its various applications and uses (Soliman et al., 2023).

1.7 SIDE-EFFECTS

The effectiveness and safety of curcumin have been supported by numerous studies on healthy human volunteers (Kocaadam & Şanlier, 2015). Although its safety is well known, a few side-effects

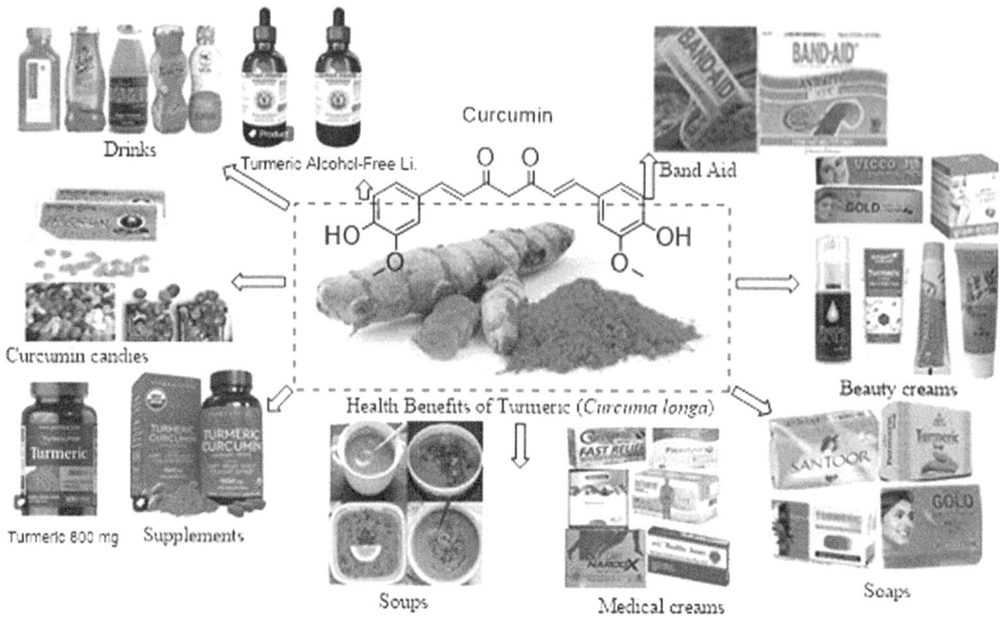

FIGURE 1.3 Health benefits of commercial products based on *Curcuma longa* powder and extract. Adapted from Prasad et al. (2014).

have been documented. Seven of twenty-four subjects who received 500–12,000 mg in a dose-response trial and were subsequently observed for 72 hours reported experiencing diarrhoea, headaches, rashes, and yellow stools (Lao et al., 2006). In another study, some patients who received 0.45 to 3.6 g of curcumin per day for one to four months suffered from diarrhoea and nausea, in addition to higher serum levels of lactic dehydrogenase and alkaline phosphatase. (Sharma et al., 2004). In albino rats, petroleum ether, alcoholic, and aqueous extracts of the rhizome had 80%, 60%, and 100% antifertility effects, respectively. It was discovered that curcumin caused a sudden, temporary drop in blood pressure. The mucin content in the stomach juice of rabbits was significantly increased in response to treatment with turmeric powder (Kapoor, 2017).

1.8 CONCLUSION

Since ancient times, Siddha, Unani, and Ayurvedic medicinal systems have all used turmeric. A review of the literature reveals that turmeric, whether consumed as a powder, an extract, or one of its purified constituents, has various pharmacological effects and few adverse effects. Curcumin has three functional reactive groups, namely two phenolic groups (including a methoxy group on a phenyl ring) and the 1,3-diketone moiety, which are essential for many of its pharmacological effects. Fortified with curcumin or turmeric, many products have been introduced in domestic and foreign markets for treatment of various ailments. Even though this plant has been the subject of extensive research, there is still room for improvement in product development. Compared to other phyto-antioxidants, curcumin is a potent, natural antioxidant that is safe and non-toxic.

REFERENCES

Adel, I.M., Elmeligy, M.F., Abdelrahim, M.E.A., Maged, A., Abdelkhalek, A.A., Abdelmoteleb, A.M.M. et al. (2021) Design and characterization of spray-dried proliposomes for the pulmonary delivery of curcumin. *International Journal of Nanomedicine*, 16, 2667–2687. https://doi.org/10.2147/ijn.s306831

Agra, M.F. de, Silva, K.N., Basílio, I.J.L.D., Freitas, P.F. de & Barbosa-Filho, J.M. (2008) Survey of medicinal plants used in the region Northeast of Brazil. *Brazilian Journal of Pharmacognosy*, 18, 472–508.

Ahad, A., Raish, M., Abdelrahman, I.A., Bin Jardan, Y.A., Alam, M.A., Al-Mohizea, A.M. et al. (2023) Changes in pharmacokinetics and pharmacodynamics of losartan in experimental diseased rats treated with Curcuma longa and Lepidium sativum. *Pharmaceuticals*, 16. https://doi.org/10.3390/ph16010033

Ahmed, H.M. (2016) Ethnopharmacobotanical study on the medicinal plants used by herbalists in Sulaymaniyah Province, Kurdistan, Iraq. *Journal of Ethnobiology and Ethnomedicine*, 12. https://doi.org/10.1186/s13002-016-0081-3

Alolga, R.N., Wang, F., Zhang, X., Li, J., Tran, L.S.P. & Yin, X. (2022) Bioactive compounds from the zingiberaceae family with known antioxidant Activities for possible therapeutic uses. *Antioxidants*, 11. https://doi.org/10.3390/antiox11071281

Altenburg, J.D., Bieberich, A.A., Terry, C., Harvey, K.A., Vanhorn, J.F., Xu, Z. et al. (2011) A synergistic antiproliferation effect of curcumin and docosahexaenoic acid in SK-BR-3 breast cancer cells: unique signaling not explained by the effects of either compound alone. *BMC Cancer*, 11, 1–16.

Amin, I., Rashid, S.M., Shubeena, S., Hussain, I., Ahmad, S.B., Mir, M.U.R. et al. (2022) TLR4/NFκB-mediated anti-inflammatory and antioxidative effect of hexanic and ethanolic extracts of *Curcuma longa* L. in buffalo mammary epithelial cells. *Separations*, 9. https://doi.org/10.3390/separations9120414

Amiri, M.S. & Joharchi, M.R. (2013) Ethnobotanical investigation of traditional medicinal plants commercialized in the markets of Mashhad, Iran. *Avicenna Journal of Phytomedicine*, 3, 254–271.

Anna, T.K., Suhana, E., Ramli, M., Das, S. & Othman, F. (2011) Original article anti-inflammatory effect of *Curcuma longa* (turmeric) on collagen-induced arthritis: An anatomico-radiological. *Clinical Therapeutics*, 162, 201–207.

ANS. (2010) Scientific opinion on the re-evaluation of curcumin (E 100) as a food additive: EFSA panel on food additives and nutrient sources added to food. *European Food Safety Authority*, 8, 1679. https://doi.org/10.2903/j.efsa.2010.1679

Apisariyakul, A., Vanittanakom, N. & Buddhasukh, D. (1995) Antifungal activity of turmeric oil extracted from *Curcuma longa* (Zingiberaceae). *Journal of Ethnopharmacology*, 49, 163–169.

Arora, N., Banerjee, A.K. & Murty, U.S.N. (2010) In silico characterization of Shikimate Kinase of Shigella flexneri: A potential drug target. *Interdiscip Sci Comput Life Sci 2*, 280–290. https://doi.org/10.1007/s12539-010-0012-2

Ashraf, K. & Sultan, S. (2017) A comprehensive review on *Curcuma longa* Linn: Phytochemical, pharmacological, and molecular study. *International Journal of Green Pharmacy*, 2017, 671–685.

Ben-Erik, V.W. & Michael, W. (2017) *Medicinal plants of the World: An illustrated scientific guide to important medicinal plants and their uses*, 2nd edn. Briza Publications, Pretoria, South Africa.

Benameur, T., Giacomucci, G., Panaro, M.A., Ruggiero, M., Trotta, T., Monda, V. et al. (2022) New promising therapeutic avenues of curcumin in brain diseases. *Molecules* 27. https://doi.org/10.3390/molecules27010236

Blumenthal, M., Arzneimittel, B für, Medizinprodukte (eds) (2000) *Herbal medicine, expanded commission E monographs*. German: Integrative Medicine Communications.

Burapan, S., Kim, M., Paisooksantivatana, Y., Eser, B.E. & Han, J. (2020) Thai curcuma species: Antioxidant and bioactive compounds. *Foods*, 9, 1–11. https://doi.org/10.3390/foods9091219

Chainani-wu, N. (2003) Safety and anti-inflammatory activity of curcumin: A component of turmeric (*Curcuma longa*). *Journal of Alternative and Complementary Medicine*, 9, 161–168.

Chew, Y., Khor, M., Xu, Z., Lee, S., Keng, J., Sang, S. et al. (2022) *Cassia alata*, *Coriandrum sativum*, *Curcuma longa* and *Azadirachta indica*: Food ingredients as complementary and alternative therapies for atopic dermatitis-a comprehensive review Yik-Ling. *Molecules*, 27. https://doi.org/10.3390/molecules27175475

Çıkrıkçı, S., Mozio, E. & Yılmaz, H. (2008) Biological activity of curcuminoids isolated from *Curcuma longa*. *Records of Natural Products*, 2, 19–24.

Dagne, E. (2011) Natural Database for Africa (NDA) Version 2.0, August 2011, Addis Ababa, Ethiopia.

Demissew, S. (1993) A description of some essential oil bearing plants in ethiopia and their indigenous uses. *Journal of Essential Oil Research*, 5, 465–479. https://doi.org/10.1080/10412905.1993.9698266

Douichene, S., Rached, W. & Djebli, N. (2020) Hepato-protective effect of *Curcuma longa* against paracetamol-induced chronic hepatotoxicity in swiss mice. *Jordan Journal of Biological Sciences*, 13, 275–279.

Duke, J.A. (1997) *The Green pharmacy*. Rodale Press Inc., New York.

Durrani, F.R., Ismail, M., Sultan, A., Suhail, S.M., Chand, N. & Durrani, Z. (2008) Effect of different levels of feed added turmeric (*Curcuma longa*) on the Performance of broiler chicks. *American Journal of Agricultural and Biological Sciencesm*, 1, 9–12.

Faizal, P., Suresh, S., Kumar, R.S. & Augusti, K.T. (2009) A study on the hypoglycemic and hypolipidemic effects of an ayurvedic drug Rajanyamalakadi in diabetic patients. *Indian J Clin Biochem* 24, 82–87.

Farhat, F., Sohail, S.S., Siddiqui, F., Irshad, R.R. & Madsen, D.Ø. (2023) Curcumin in wound healing — a bibliometric analysis. *Life* 13, 1–15. https://doi.org/10.3390/life13010143
Fullas, F. (2006) *Interactions of Ethiopian Herbal Medicines and Spices with Conventional Drugs: A Conventional Guide*. F. Fullas.
Fullas, F. (2003) *Spice plants in ethiopia: Their culinary and medicinal applications*. RPH, PHD, Iowa.
Fuloria, S., Mehta, J., Chandel, A., Sekar, M., Rani, N.N.I.M., Begum, M.Y. et al. (2022) A comprehensive review on the therapeutic potential of *Curcuma longa* Linn. in relation to its major active constituent curcumin. *Frontiers in Pharmacology*, 13, 1–27. https://doi.org/10.3389/fphar.2022.820806
Ghareghomi, S., Rahban, M., Moosavi-Movahedi, Z., Habibi-Rezaei, M., Saso, L. & Moosavi-Movahedi, A.A. (2021) The potential role of curcumin in modulating the master antioxidant pathway in diabetic hypoxia-induced complications. *Molecules*, 26, 1–26. https://doi.org/10.3390/molecules26247658
Gopinath, D., Ahmed, M.R., Gomathi, K., Chitra, K., Sehgal, P.K. & Jayakumar, R. (2004) Dermal wound healing processes with curcumin incorporated collagen films. *Biomaterials* 25, 1911–1917. https://doi.org/10.1016/S0142-9612(03)00625-2
Górnicka, J., Mika, M., Wróblewska, O., Siudem, P. & Paradowska, K. (2023) Methods to improve the solubility of curcumin from turmeric. *Life*, 13, 1–13. https://doi.org/10.3390/life13010207
Guerrini, A., Tacchini, M., Chiocchio, I., Grandini, A., Radice, M., Maresca, I. et al. (2023) A comparative study on chemical compositions and biological activities of four amazonian ecuador essential oils: *Curcuma longa* L. (Zingiberaceae*), Cymbopogon citratus* (DC.) Stapf, (Poaceae), *Ocimum campechianum* Mill. (Lamiaceae), and . antibiotics 12. https://doi.org/https://doi.org/10.3390/antibiotics12010177
Hafez Ghoran, S., Calcaterra, A., Abbasi, M., Taktaz, F., Nieselt, K. & Babaei, E. (2022) Curcumin-based nanoformulations: a promising adjuvant towards cancer treatment. *Molecule*, 27. https://doi.org/10.3390/molecules27165236
Hewlings, S.J. & Kalman, D.S. (2017) Curcumin: A review of its effects on human health. *Foods* 6, 1–11. https://doi.org/10.3390/foods6100092
Honda, S., Aoki, F., Tanaka, H., Kishida, H., Nishiyama, T., Okada, S. et al. (2006) Effects of ingested turmeric oleoresin on glucose and lipid metabolisms in obese diabetic mice : a dna microarray study. *Journal of Agricultural and Food Chemistry*, 54, 9055–9062.
Horie, S. (2012) Chemoprevention of prostate cancer : Soy Isoflavones and curcumin. *Korean Journal of Urology*, 53, 665–672. https://doi.org/10.4111/kju.2012.53.10.665
Hosseini, A. & Hosseinzadeh, H. (2018) Biomedicine & pharmacotherapy antidotal or protective effects of curcuma longa (turmeric) and its active ingredient, curcumin, against natural and chemical toxicities hemireview. *Biomedicine & Pharmacotherapy*, 99, 411–421. https://doi.org/10.1016/j.biopha.2018.01.072
Ibáñez, M.D. & Blázquez, M.A. (2021) *Curcuma longa* l. rhizome essential oil from extraction to its agri-food applications: A review. *Plants*, 10, 1–31. https://doi.org/10.3390/plants10010044
Ibrahim, J., Kabiru, A.Y., Abdulrasheed-Adeleke, T., Lawal, B. & Adewuyi, A.H. (2020) Antioxidant and hepatoprotective potentials of curcuminoid isolates from turmeric (*Curcuma longa*) Rhizome on CCl_4-induced hepatic damage in Wistar Rats. *Journal of Taibah University for Science*, 14, 908–915. https://doi.org/10.1080/16583655.2020.1790928
Ikpeama, A., Onwuka, G.I. & Nwankwo, C. (2014) Nutritional composition of turmeric (*Curcuma longa*) and its antimicrobial properties. *International Journal of Scientific Engineering and Research*, 5, 1085–1089.
Imoru, A., Onibi, G.E. & Osho, I.B. (2018) Nutritional and biochemical compositions of turmeric (*Curcuma longa* Linn) Rhizome powder – A promising animal feed additive. *International Journal of Scientific Engineering and Research*, 9, 424–429.
Irshad, S., Muazzam, A., Shahid, Z. & Dalrymple, M.B. (2018) *Curcuma longa* (Turmeric): An auspicious spice for antibacterial, phytochemical and antioxidant activities. *Pakistan Journal of Pharmaceutical Sciences*, 31 (Supplement), 2689–2696.
Jacob, A., Wu, R., Zhou, M. & Wang, P. (2007) Mechanism of the anti-inflammatory effect of curcumin : PPAR– γ activation. *PPAR Research*. https://doi.org/10.1155/2007/89369
Jankasem, M., Wuthi-udomlert, M. & Gritsanapan, W. (2013) Antidermatophytic properties of ar-turmerone, turmeric oil, and *Curcuma longa* preparations. *ISRN Dermatology*. https://doi.org/10.1155/2013/250597
Kaewruethai, T., Lin, Y., Wang, Q. & Luckanagul, J.A. (2023) The dual modification of pnipam and β-cyclodextrin grafted on hyaluronic acid as self-assembled Nanogel for curcumin delivery. *Polymers (Basel)*, 15. https://doi.org/10.3390/polym15010116
Kala, C.P. (2005) Ethnomedicinal botany of the Apatani in the Eastern Himalayan Region of India. *Journal of Ethnobiology and Ethnomedicine*, 1, 1–8. https://doi.org/10.1186/1746-4269-1-11
Kapoor, L.D. (2017) *CRC handbook of ayurvedic medicinal plants*. CRC Press.

Khan, M., Al-Saleem, M.S.M. & Alkhathlan, H.Z. (2016) A detailed study on chemical characterization of essential oil components of Two *Plectranthus* species grown in Saudi Arabia. *Journal of Saudi Chemical Society*, 20, 711–721. https://doi.org/10.1016/j.jscs.2016.03.006

Khan, R.U., Naz, S., Javdani, M., Nikousefat, Z., Selvaggi, M, Tufarelli, V. et al. (2012) The use of turmeric (*Curcuma longa*) in poultry feed. *World's Poultry Science Journal*, 68, 97–103. https://doi.org/10.1017/S0043933912000104

Kim, H. & Song, M. (2011) Analysis and recordings of orally transmitted knowledge about medicinal plants in the southern mountainous region of Korea. *Journal of Ethnopharmacology*, 134, 676–696. https://doi.org/10.1016/j.jep.2011.01.024

Kim, K., Kim, J.-J., Jung, Y., Noh, J.-Y., Syed, A.S., Kim, C.Y. et al. (2017) Cyclocurcumin, an Antivasoconstrictive Constituent of *Curcuma longa* (Turmeric). *Journal of Natural Products*, 80, 196–200. https://doi.org/10.1021/acs.jnatprod.6b00331

Kocaadam, B. & Şanlier, N. (2015) Curcumin, an active component of turmeric (*Curcuma longa*), and its effects on health. *Critical Reviews in Food Science and Nutrition*, 8398. https://doi.org/10.1080/10408398.2015.1077195

Kössler, S., Nofziger, C., Jakab, M., Dossena, S. & Paulmichl, M. (2012) Curcumin affects cell survival and cell volume regulation in human renal and intestinal cells. *Toxicology*, 292, 123–135. https://doi.org/10.1016/j.tox.2011.12.002

Lans, C., Turner, N., Brauer, G., Lourenco, G. & Georges, K. (2006) Ethnoveterinary medicines used for horses in Trinidad and in British Columbia, Canada. *Journal of Ethnobiology and Ethnomedicine*, 2. https://doi.org/10.1186/1746-4269-2-31

Lao, C.D., Iv, M.T.R., Normolle, D., Heath, D.D., Murray, S.I., Bailey, J.M. et al. (2006) Dose escalation of a curcuminoid formulation. *BMC Complementary Medicine and Therapies*, 6, 4–7. https://doi.org/10.1186/1472-6882-6-10

Li, S., Yuan, W., Deng, G., Wang, P., Yang, P. & Aggarwal, B. (2011) Chemical composition and product quality control of turmeric (*Curcuma longa* L.). *Pharmaceutical Crops*, 2, 28–54.

Lim, H.S., Park, S.H., Ghafoor, K., Hwang, S.Y. & Park, J. (2011) Quality and antioxidant properties of bread containing turmeric (*Curcuma longa* L.) cultivated in South Korea. *Food Chemistry*, 124, 1577–1582. https://doi.org/10.1016/j.foodchem.2010.08.016

Mariano, A., Bigioni, I., Misiti, F., Fattorini, L., d'Abusco, A.S. & Rodio, A. (2022) The nutraceuticals as modern key to achieve erythrocyte oxidative stress fighting in osteoarthritis. *Current Issues in Molecular Biology*, 44, 3481–3495. https://doi.org/10.3390/cimb44080240

Merzouki, A., Ed-derfoufi, F. & Mesa, J.M. (2000) Contribution to the Knowledge of Rifian traditional medicine II : Folk medicine in Ksar Lakbir district (NW morocco). *Fitoterapia*, 71, 278–307.

Mohamed, M.E., Younis, N.S., El-Beltagi, H.S. & Mohafez, O.M. (2022) The synergistic hepatoprotective activity of rosemary essential oil and curcumin: The role of the MEK/ERK pathway. *Molecules*, 27, 1–18. https://doi.org/10.3390/molecules27248910

Nguyen, V.C., Nguyen, V.B. & Hsieh, M.F. (2013) Curcumin-Loaded Chitosan/Gelatin Composite Sponge for Wound Healing Application. *International Journal of Polymer Science*. https://doi.org/10.1155/2013/106570

Nikolić, L., Urošević, M., Nikolić, V., Gajić, I., Dinić, A., Miljković, V. et al. (2023) The formulation of curcumin: 2-hydroxypropyl-β-cyclodextrin complex with smart hydrogel for prolonged release of curcumin. *Pharmaceutics*, 15, 1–28. https://doi.org/10.3390/pharmaceutics15020382

Nishiyama, T., Mae, T., Kishida, H., Tsukagawa, M., Mimaki, Y. & Uroda, M.I.K. et al. (2005) Curcuminoids and sesquiterpenoids in turmeric (*Curcuma longa* L.) suppress an increase in blood glucose level in type 2 diabetic KK-A y Mice. *Journal of Agricultural and Food Chemistry*, 53, 959–963.

Orellana-Paucar, A.M. & Machado-Orellana, M.G. (2022) Pharmacological profile, bioactivities, and safety of turmeric oil. *Molecules*, 27, 1–16. https://doi.org/10.3390/molecules27165055

Oyemitan, I.A., Elusiyan, C.A., Onifade, A.O., Akanmu, M.A., Oyedeji, A.O. & Mcdonald, A.G. (2017) Neuropharmacological profile and chemical analysis of fresh rhizome essential oil of *Curcuma longa* (Turmeric) cultivated in Southwest Nigeria. *Toxicology Reports*, 4, 391–398. https://doi.org/10.1016/j.toxrep.2017.07.001

Peter, K.V. (ed) (2012) *Handbook of herbs and spices* Volume 2, Second. Woodhead Publishing Limited, New Delhi.

Pole, S. (2006) *Ayurvedic medicine: The principles of traditional practice*. Churchill Livingstone., London and Philadelphia

Ponnusamy, S., Ravindran, R., Zinjarde, S., Bhargava, S. & Kumar, A.R. (2011) Evaluation of Traditional Indian Antidiabetic Medicinal Plants for Human Pancreatic Amylase Inhibitory Effect In Vitro. *Evidence-Based Complementary and Alternative Medicine*. https://doi.org/10.1155/2011/515647

Prasad, S., Tyagi, A.K. & Aggarwal, B.B. (2014) Recent developments in delivery, bioavailability, absorption and metabolism of curcumin: The golden pigment from golden spice. *Cancer Research and Treatment*, 46, 2–18. https://doi.org/10.4143/crt.2014.46.1.2

Rai, P.K., Jaiswal, D., Mehta, S., Rai, D.K., Sharma, B. & Watal, G. (2010) Effect of *Curcuma longa* freeze dried rhizome powder with Milk in STZ induced diabetic Rats. *Indian Journal of Clinical Biochemistry*, 25, 175–181.

Rajagopal, K., Varakumar, P., Baliwada, A. & Byran, G. (2020) Activity of phytochemical constituents of *Curcuma longa* (Turmeric) and *Andrographis paniculata* against coronavirus (COVID-19): An in Silico approach. *Future Journal of Pharmaceutical Sciences,* 6. https://doi.org/10.1186/s43094-020-00126-x

Ram Kumar, P. & Jain, P. (2010) Comparative studies on the antimicrobial activity of balck pepper (*Piper nigrrum*) and turmeric (*Curcuma longa*) extracts. *International Journal of Applied Biology and Pharmaceutical Technology*, 1, 492–501.

Ramadan, G., Al-kahtani, M.A. & El-sayed, W.M. (2011) Anti-inflammatory and Anti-oxidant Properties of *Curcuma longa* (Turmeric) Versus *Zingiber officinale* (Ginger) Rhizomes in Rat Adjuvant-Induced Arthritis. *Inflammation*, 34. https://doi.org/10.1007/s10753-010-9278-0

Reddy, A.C.P. & Lokesh, B.R. (1994) Effect of dietary turmeric (*Curcuma longa*) on iron-induced lipid peroxidation in the rat liver. *Food and Chemical Toxicology,* 32, 279–283. https://doi.org/10.1016/0278-6915(94)90201-1

Regassa, R. (2013) Assessment of indigenous knowledge of medicinal plant practice and mode of service delivery in Hawassa City, Southern Ethiopia. *Journal of Medicinal Plants Research*, 7, 517–535. https://doi.org/10.5897/JMPR012.1126

Salama, S.M., Abdulla, M.A., Alrashdi, A.S., Ismail, S. & Alkiyumi, S.S. (2013) Hepatoprotective effect of ethanolic extract of *Curcuma longa* on thioacetamide induced liver cirrhosis in rats. *BMC Complementary and Alternative Medicine,* 13(56), 1–17 .

Sharma, R.A., Euden, S.A., Platton, S.L., Cooke, D.N., Shafayat, A., Hewitt, H.R. et al. (2004) Phase I clinical trial of oral curcumin: Biomarkers of systemic activity and compliance. *Clinical Cancer Research*, 10, 6847–6854.

Sideek, S.A., El-Nassan, H.B., Fares, A.R. & Elmeshad, A.N. (2023) Different curcumin-loaded delivery systems for wound healing applications : A comprehensive review. *Pharmaceutics*, 15, 1–21. https://doi.org/10.3390/pharmaceutics15010038

Singh, D., Rathod, V., Ninganagouda, S., Herimath, J. & Kulkarni, P. (2013) Biosynthesis of silver nanoparticle by endophytic fungi pencillium sp. isolated from *Curcuma longa* (turmeric) and its antibacterial activity against pathogenic gram negative bacteria. *Journal of Pharmacy Research*, 7, 448–453. https://doi.org/10.1016/j.jopr.2013.06.003

Singh, D., Rathod, V., Ninganagouda, S., Hiremath, J., Singh, A.K. & Mathew, J. (2014) Optimization and characterization of silver nanoparticle by endophytic fungi penicillium sp . Isolated from *Curcuma longa* (Turmeric) and application studies against MDR *E . coli* and *S. aureus. Bioinorganic Chemistry and Applications.* https://doi.org/10.1155/2014/408021

Singh, G., Kapoor, I.P.S., Singh, P., Heluani, C.S. De, Lampasona, M.P. De & Catalan, C.A.N. (2010) Comparative study of chemical composition and antioxidant activity of fresh and dry rhizomes of turmeric (*Curcuma longa* Linn.). *Food and Chemical Toxicology,* 48, 1026–1031. https://doi.org/10.1016/j.fct.2010.01.015

Singletary, K. (2010) Turmeric: An overview of potential health benefits. *Nutrition Today,* 45, 216–225.

Soleimani, V., Sahebkar, A. & Hosseinzadeh, H. (2018) Turmeric (*Curcuma longa*) and its major constituent (curcumin) as Nontoxic and Safe Substances: Review. *Phyther Res*, 1–11. https://doi.org/10.1002/ptr.6054

Soliman, Y.M., Soliman, W.S. & Abbas, A.M. (2023) Alley cropping and organic compost: An efficient and sustainable agro-ecological strategy for improving turmeric (Curcuma longa L.) growth and attributes. *Agriculture*, 13, 1–14. https://doi.org/10.3390/agriculture13010149

Srinivas, L. & Place, J. (1992) Turmerin: A water soluble antioxidant from turmeric (*Curcuma longa*) peptide. *Archives of Biochemistry and Biophysics*, 292, 617–623.

Tanvir, E.M., Hossen, S., Hossain, F., Afroz, R., Gan, S.H., Khalil, I. et al. (2017) Antioxidant properties of popular turmeric (*Curcuma longa*) varieties from Bangladesh. *Journal of Food Quality.* https://doi.org/10.1155/2017/8471785

Vaughn, A.R., Branum, A. & Sivamani, R.K. (2016) Effects of turmeric (*Curcuma longa*) on Skin Health : A systematic review of the clinical evidence. *Phyther Res*, 30(8), 1243–1264.

Velayudhan, K.C., Dikshit, N. & Nizar, M.A. (2012) Ethnobotany of turmeric (*Curcuma longa* L.). *Indian Journal of Traditional Knowledge*, 11, 607–614.

Verma, R.B., Kumari, P., Maurya, R.K., Kumar, V., Verma, R. & Singh, R.K. (2018) Medicinal properties of turmeric (*Curcuma longa* L.): A review. *International Journal of Chemical Studies*, 6, 1354–1357.

WHO. (2004) Evaluation of certain food additives and contaminants: Sixty-first report of the joint FAO/WHO expert committee on food additives. Geneva.

Wickenberg, J., Ingemansson, S.L. & Hlebowicz, J. (2010) Effects of *Curcuma longa* (turmeric) on postprandial plasma glucose and insulin in healthy subjects. *Nutrition Journal*, 9, 1–5. https://doi.org/10.1186/1475-2891-9-43

Wilken, R., Veena, M.S., Wang, M.B. & Srivatsan, E.S. (2011) Curcumin: A review of anti-cancer properties and therapeutic activity in head and neck squamous Cell carcinoma. *Mol Cancer*, 10. https://doi.org/10.1186/1476-4598-10-12

Xu, D., Tian, W. & Shen, H. (2011) Curcumin prevents induced drug resistance: A novel function ? *Chinese Journal of Cancer Research,* 23, 218–223. https://doi.org/10.1007/s11670-011-0218-9

Yang, Z.J., Huang, S.Y., Zhou, D.D., Xiong, R.G., Zhao, C.N., Fang, A.P. et al. (2022) Effects and mechanisms of curcumin for the prevention and management of Cancers: An updated review. *Antioxidants*, 11, 1–24. https://doi.org/10.3390/antiox11081481

Ye, M., Li, Y., Yin, H. & Zhang, J. (2012) Curcumin: updated molecular mechanisms and intervention targets in human lung cancer. *International Journal of Molecular Sciences*, 13, 3959–3978. https://doi.org/10.3390/ijms13033959

Younes, M., Mardirossian, R., Rizk, L., Fazlian, T., Khairallah, J.P., Sleiman, C. et al. (2022) The synergistic effects of curcumin and chemotherapeutic drugs in inhibiting metastatic, invasive and proliferative pathways. *Plants*, 11. https://doi.org/10.3390/plants11162137

2 Medicinal and Commercial Application of *Zingiber officinale*

Meseret Zebeaman, Mesfin Getachew Tadesse, Rakesh Kumar Bachheti, Archana Bachheti, Rahel Gebeyhu, and Kundan Kumar Chaubey

2.1 INTRODUCTION

Ginger, the rhizome or root of *Zingiber officinale*, is mainly distributed in Asia. The ginger plant family, the Zingiberaceae, has around 1300 species. *Z. officinale* is a known food additive, spice, and flavouring agent, and, more importantly, it is a common medicinal plant in many parts of the world (Zhang et al., 2021). Since antiquity, ginger has been used to treat various unrelated ailments such as arthritis, rheumatism, sprains, muscular aches, pains, sore throats, cramps, constipation, indigestion, vomiting, hypertension, dementia, fever, infectious diseases, and helminthiasis. The main biological activities of ginger are immunomodulatory, antitumorigenic, anti-inflammatory, antiapoptotic, antihyperglycaemic, antilipidaemic, and anti-emetic (Zhang et al., 2022). The history of ginger as a medicinal plant is described in the old traditional medicinal books of the world, which included Egypt (Ebers Papyrus) (Li & Weng, 2017), India (Ayurvedic) (Dissanayake et al., 2020), and China (Shennong Bencaojing, or the divine farmers Materia Medica) (Zhang et al., 2021).

The leading producers of this plant are India, China, and Nigeria (Dhanik et al., 2017). The rhizome, the thick underground stem, is the main edible part of the plant. Even though there are many species of ginger in the genus *Zingiber*, due to their physical appearance they are categorized as either red or common ginger (Figure 2.1) (Zhang et al., 2022). So far, the chemical constituents in the ginger rhizome consist of around 307 chemical compounds, including 194 volatile oils, 85 gingerols, and 28 diarylheptanoid compounds. The major class of chemical found in the plant and which gives the pungent odour to the rhizome is the vanilloids, which includes gingerol, zengerone, shogaol, and their derivatives. Fresh ginger is rich in 6-gingerol, but dried ginger is rich in dehydrated 6-gingerol, which is 6-shogaol (Liu et al., 2019). Due to the presence of these various chemicals in ginger, it is believed that these chemicals are responsible for the plant's medicinal effects, through either synergistic or additive effects. This chapter is important to researchers who work on natural products, functional foods, and ethnomedicine. Basically, the objective of this chapter is to underline the most research work done on ginger, which are highly cited on Google Scholar. Specifically, the chemical constituents, their biosynthesis, and their medicinal uses are discussed in detail. Some of the herbal ginger-derived drugs which are on the market are also discussed.

2.2 BOTANICAL DESCRIPTION

The scientific name of ginger is *Zingiber officinale*, and it belongs to the ginger family, the Zingiberaceae. This plant has a thickened 'root' (actually an underground stem) known as the rhizome, and it is a perennial plant. During cultivation of the plant, it reaches up to 90 cm high. The

 DOI: 10.1201/b22924-2

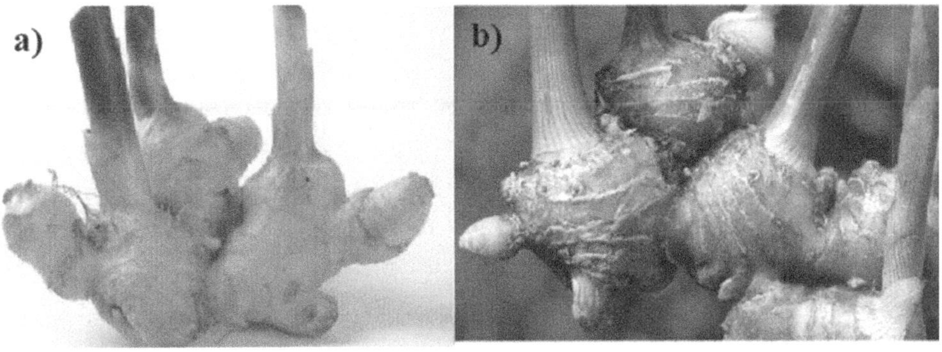

FIGURE 2.1 Images of a) common ginger and b) red ginger.

pungent odour of ginger originates from the rhizomes, which are thick-lobed and have a yellowish flesh after peeling (cf. Figure 2.1). The herb develops several lateral shoots in clumps, which begin to dry when the plant matures. Flowers are rare, relatively small, calyx superior, with a corolla of three sub-equal oblong to lanceolate connate greenish segments (Dhanik et al., 2017; Zhang et al., 2021).

Warm and humid climates are suitable conditions for ginger to grow. Similarly, rain-fed or irrigated conditions are good conditions for cultivating ginger at sea level up to an altitude of above 1500 MSL (m above sealevel) (Dhanik et al., 2017).

2.3 CHEMICAL CONSTITUENTS

Ginger chemical constituents are discussed by comparing the two common types of ginger namely (white) common and red ginger. Red ginger has 164 chemicals, whereas common ginger has around 300 chemicals, and the chemistry behind why red ginger becomes red is still not known. Similarly, red ginger turns to black upon drying whereas common ginger does not. The other main difference between common and red ginger is that red ginger is not edible, whereas common ginger is (Liu et al., 2019; Zhang et al., 2022). However, red ginger has greater medicinal uses than common ginger in different parts of the world.

A review by Zhang et al. (2022) showed red ginger extracted with methanol at 76.9°C for 3.4 h yielded 2.8 g bioactive compounds. The review also indicated that the major bioactive compounds were vanilloids. The concentrations of vanilloids are higher in red ginger than in common white ginger. Various spectroscopic methods are used to determine the chemical constituents, including UV-Vis, NMR, HPLC, and MALDI-TOFMS. The different types of vanilloids are shown in Figure 2.2. The vanilloids are biosynthesized from the amino acid phenylalanine, as shown in Figure 2.3.

The other bioactive chemical constituents of gingers are polyphenols (primarily flavonoids) and essential oils (monoterpenes and sesquiterpenes). A study by Jan et al. (2022) revealed that ethanol or water extracts of dried ginger have much higher concentrations of flavonoids than did fresh ginger. To determine the flavonoid content, they use HPLC and IR. Similarly, they use SEM to study the morphology of the crude extract. Figure 2.4 shows some of the different phenolic and flavonoid compounds in ginger.

The other main constituents of ginger are essential oils. Specifically, monoterpenes and sesquiterpenes are the major constituents (Figure 2.5). A study conducted by Kalhoro et al. (2022) determined that essential oil of citrol is the dominant oil in ginger. They use the GC-MS spectroscopy method for characterization and the microwave-assisted hydro-distillation method for extraction of the oils.

2.4 MEDICINAL USES

From antiquity up to the present day, the use of ginger in health care is well known. Scientists are looking into ginger for anti-cancer drugs because the current commercial drugs have side-effects.

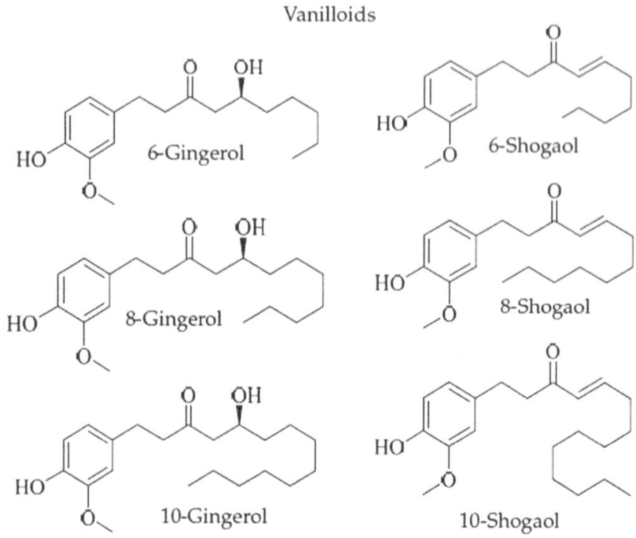

FIGURE 2.2 Vanilloids: bioactive components of ginger.

For instance, a study by Wang et al. (2021) indicates that silver nanoparticles (AgNP) generated by green synthesis using *Z. officinale* showed low cell viability and high anti-human pancreatic cancer activity. The nanoparticle activity was dose-dependent and the particles are active at 295, 312, and 220 mg/mL against the cancer cell lines AsPC-1, PANC-1, and MIA PaCa-2, respectively. Based on such findings, the green silver nanoparticles obtained could be used as an anticancer drug.

2.4.1 Antimicrobial Activity

The most effective bioactive molecules in red ginger responsible for its antimicrobial activity are monoterpenes, specifically β-caryophyllenes. The other ginger essential oils showing antimicrobial activity were ar-curcumene, zingiberen, β-bisabolene, β-sesquiphellandrene, and camphene (Rialita et al., 2018; Zhang et al., 2022). Similarly, a study by Ashour et al. (2022) showed that this plant could be used in dentistry. After synthesizing silver nanoparticles dipped in *Z. officinale* extract (ZOE-AgNPs), they were combined with conventional glass ionomer cement (GIC), lyophilized miswak, or chlorhexidine diacetate (CHX) to treat against oral microbes. The authors found high antimicrobial efficacy in GIC combined with ZOE-AgNPs and chlorhexidine. Similarly, they observed a comparable antimicrobial activity of GIC with ZOE-AgNPs and chlorhexidine, compared with chlorhexidine alone. For characterizing the ZOE-AgNPs, they used FESEM, IR, and XRD; to assess antimicrobial activity, they used the disc diffusion method. A review study by Beristain-Bauza et al. (2019) concludes that the main bioactive compounds responsible for broad antimicrobial activity in ginger against different microorganisms are monoterpenoids, sesquiterpenoids, phenolic compounds, and their derivatives, aldehydes, ketones, alcohols, and esters. This makes the ginger plant an interesting antimicrobial substitute for commercial drugs. At the same time, the authors indicated that ginger is 'generally recognized as safe' by the FDA and WHO. The other application of the essential oil of ginger is its use in food packaging. They synthesized a polymer using electron spinning and encapsulated the essential oil within the prepared polymer. When they evaluated its antibacterial activity against five bacteria, *Listeria monocytogenes, Staphylococcus aureus, Escherichia coli, Salmonella typhimurium*, and *Pseudomonas aeruginosa*. They found that the fibres containing 12% ginger essential oil showed high potential microbial growth inhibition and concluded that it could be applied as part of food packaging material (da Silva et al., 2018). Table 2.1 shows a summary of antimicrobial and other medicinal uses of ginger.

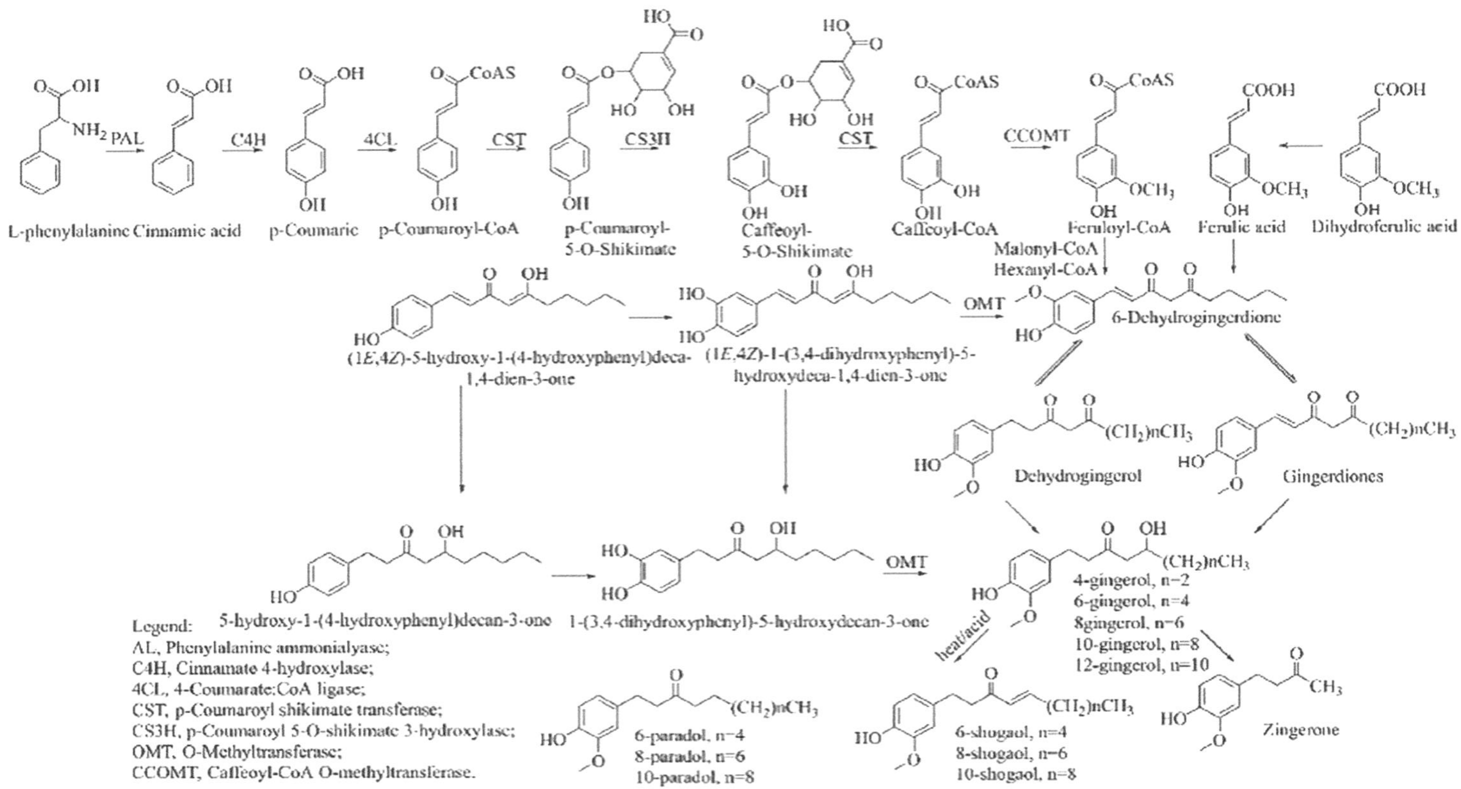

FIGURE 2.3 Biosynthesis of the vanilloids from the amino acid L-phenylalanine.

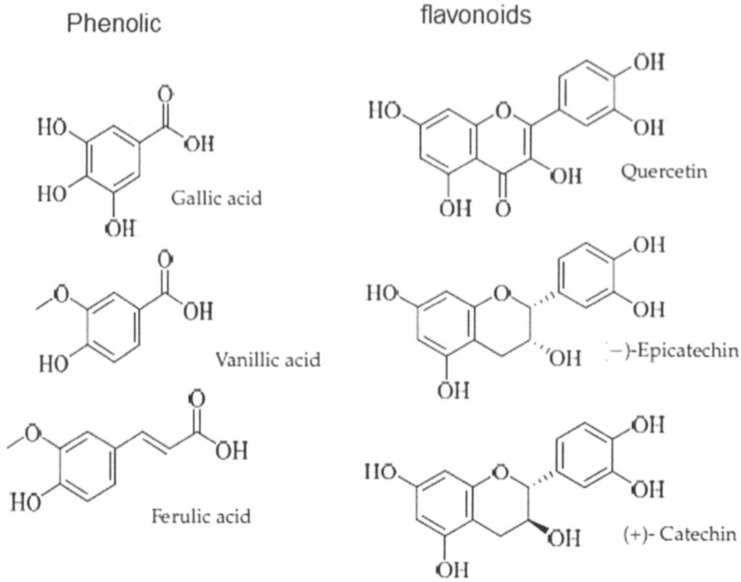

FIGURE 2.4 Phenolic and flavonoid constituents of ginger.

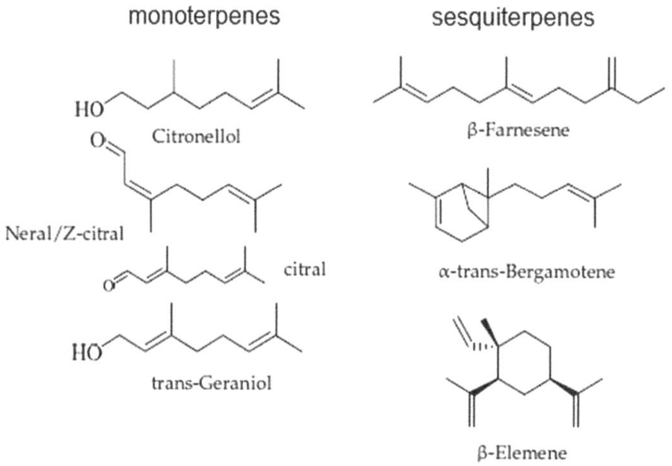

FIGURE 2.5 Essential oils (monoterpenes and sesquiterpenes) found in ginger.

2.4.2 Antioxidant

Research indicates that ginger has strong antioxidant potential by either scavenging or stopping the generation of free radicals. Due to this, ginger rhizome is used as a safe herbal medicine. A recent *in vivo* study by Zammel et al. (2021) indicates that aqueous ginger extracts improved the inflammation and oxidative stress status induced by a carrageenan-based acute inflammation model as outlined by anti-oedematous, antioxidant, and anti-inflammatory activities. To evaluate the antioxidant activity, they measured markers of inflammation, such as haematological parameters, fibrinogen, and C-reactive protein. A parallel *in-silico* computational study showed that a ginger bioactive molecule had high affinities to bind to Toll-like receptor 6 (TLR6).

A comparative study was also carried out by Ghafoor et al. (2020). The study clearly showed that freeze-dried ginger had a higher total phenolic content and antioxidant activity than either oven- or microwave-dried ginger. Freeze-dried ginger had significantly ($p<0.005$) higher total phenolic concentration (931.94 mg gallic acid equivalents (GAE)/100 g) and antioxidant activity (82.00%) than the other drying techniques. The main phenolic compounds in dried ginger are catechin, gallic acid, and 3,4-dihydroxybenzoic. To measure the antioxidant activity of the aqueous methanolic extract of fresh ginger, they used the DPPH (1,1-diphenyl-2-picrylhydrazyl) assay. Another study, conducted by Ali et al. (2018), showed that a chloroform: methanol (CM) extract of ginger showed a higher phenolic concentration than in a petroleum ether (PE) extract. Similarly, they indicated that ginger rhizome had a higher phenolic concentration than callus of ginger. The phenolic concentration found in CM extract of rhizomes was 60.34±0.43 mg GAE/g, compared with 33.6 ±0.07 GAE/g in callus. The other chemical difference was that flavonoids (40.25 ± 0.21 mg quercetin/ g) were detected in CM extracts of rhizomes but not callus. From CM and PE extract of both rhizomes and callus of ginger, PE extract showed the higher antioxidant activity, with a half-maximal inhibitory concentration (IC_{50}) value of 8.29 ± 1.73 l g/mL. Table 2.1 shows a summary of antioxidant activity and other medicinal uses in ginger.

2.4.3 Anticancer

AgNPs generated by green synthesis using ginger showed good anti-colon-cancer activity. The study, by Venkatadri et al. (2020), indicated that 25 to 500 mg/mL AgNPs dipped in ginger extract expressed cytotoxicity against HT-29 cells. They also exhibited an IC_{50} value of 150.8 mg/mL. The *in vitro* anticancer activity measurement method they used was the MTT assay. They concluded that AgNPs dipped in the ginger extract have anti-colon cancer activity and characterized the AgNPs dipped in ginger extract using UV–visible spectroscopy, FTIR, XRD, TEM, and SEM analyses.

The existence of carboxyl (-C=O) and amine (N-H) functional groups in the nanoparticles were confirmed by FTIR. The sharp peak observed in UV-Vis spectroscopy at 350–430 nm is called the surface plasmon response sharp peak, which indicates the formation of AgNPs. While an XRD study confirmed the presence of four spectral peaks at 111, 200, 220, and 311, a scanning electron microscopy (SEM) study revealed that the nanoparticles were spherical in shape with a diameter of 42–61 nm. Similarly, a transmission electron microscopy (TEM) study showed a size range, 20–51 nm, similar to that revealed by SEM.

The other ginger phytochemicals responsible for anticancer activity are the vanilloids. The red ginger constituent dehydrozingerone inhibited cell proliferation of KB mouth cancer and A549 lung cancer cell lines. Similarly, an *in-vivo* study on mice revealed that 6-gingerol inhibits cell proliferation in an ovarian cancer cell line. The cancer cell lines they used for this study included C1 and C2 (Zhang et al., 2022).

2.4.4 Hypoglycaemic or Antidiabetic

Diabetes mellitus is one of the serious metabolic disorder diseases and causes high levels of morbidity and mortality annually. Natural products are nowadays used to combat the development of diabetes. Among the natural products, ginger extract is available on the market. The review by Alam et al. (2019) listed a total of 12 mechanisms of action by which plant extracts generally cause their antidiabetic activities. These are inhibition of intestinal alphaglucosidase and alpha-amylase, lens aldose reductase, oxidative stress protection, inhibition of formation of advanced glycation end-products, inhibition of aldose reductase, lowering of plasma glucose levels, altering enzyme activity of hexokinases and glucose-6-phosphate, synthesis and release of insulin, postprandial hyperglycemia inhibition, stimulation of GLUT-4, decreasing activity of G6P, and lowering the level of skeletal hexokinases. Similarly, an *in-vitro* study by Noipha & Ninla-Aesong (2018) revealed that an aqueous extract of ginger enhanced glucose uptake of L6 myotube cells. This means that the glucose level in the blood comes to a standard level. In their *in-vitro* experiment, they determined the levels of glucose transporter proteins (GLUT1 and GLUT4) and the corresponding mRNA levels to evaluate antidiabetic activity of the ginger extract. They found that, after co-culture for 24 h, 400μg/mL

TABLE 2.1
Summary of Medicinal Use and Activity Studies on Ginger from 2012 to 2022

Medicinal Use	Evaluation Method	Use	Mechanism of Action	References
Obesity	Body composition and anthropometric parameters, and clinical trials	Weight loss	Increasing thermogenesis, increasing lipolysis	Ebrahimzadeh Attari et al. (2018)
Ulcer	Clinical trials and blood sampling	Ulcer prevention	Reduced malondialdehyde	Nikkhah-Bodaghi et al. (2019)
Immunomodulatory	*In-vivo* measurement of immunoglobulin level,	Improvement of immunity	Increase immunoglobulin level	Ahmadifar et al. (2019)
Anti-viral (anti-chikungunya virus)	MTT assay; cytopathic effects and cell viability amount	Prevention of disease caused by chikungunya virus	Increased viability of Vero cell line	Kaushik et al. (2020)
Antibacterial	Disc diffusion	Prevention of tooth decay	Inhibition of bacterial growth	Ashour et al. (2022)
Neuropathic pain	*In-vivo* study: peripheral mono-neuropathy	Decreased neuropathic pain	Inhibit neuroinflammation	Borgonetti et al. (2020)
Antioxidant	DPPH, ABTS, and NORI assays	Prevention of cell damage that causes rheumatoid arthritis	Scavenging radicals	Murugesan et al. (2020)
Anti-arthritic	Membrane stabilization, anti-proteinase, and protein denaturation reserve assays	Prevention of rheumatoid arthritis	Inhibition of enzymes	Murugesan et al. (2020)

of ginger extract significantly ($p<0.05$) enhanced glucose uptake with a value of 208.03±10.65% in L6 myotubes. The controlling groups which inhibit ginger-based increases in glucose uptake were the protein synthesis inhibitor, 3.5 μM cycloheximide, 1 μM wortmannin (phosphatidylinositol 3-kinase (PI3 kinase) inhibitor) and 15 nM rapamycin (mammalian target of rapamycin (mTOR) inhibitor). A concentration of 400 μg/mL of ginger extract significantly ($p<0.05$) increased GLUT1 levels by 1.6-, 2.03-, and 2.25-fold at 4, 8, and 24 h culture times, respectively; similarly, at $p < 0.05$, the mRNA level increased by 1.22-, 1.45-, 1.91-, 2.32- and 2.2-fold at 1, 2, 4, 8, and 24 h culture times, respectively. Therefore *Z. officinale* rhizome extract increases the glucose transport activity of L6 myotubes by enhancing *GLUT1* expression.

2.4.5 Miscellaneous Medicinal Uses

One medicinal use of ginger, which is not much studied, is its use in combating Alzheimer's disease, a neurological disorder mostly noted in elderly people. The cause of the incidence is related to several types of cellular damage, such as apoptosis, oxidative stress, some neurotransmitter imbalances, amyloid-beta aggregation, tau protein hyperphosphorylation, and inflammatory responses. A pre-clinical study review on ginger (Talebi et al., 2021) revealed that ginger has biological activities in which the anti-apoptosis property is the one responsible for anti-cancer activity, rather than the antioxidant and anti-inflammatory activities. A clinical trial on the use of ginger by a patient affected by breast cancer revealed that ginger lessened nausea and vomiting during chemotherapy treatment (Saneei Totmaj et al., 2019). Ginger extract can also be used as an adjuvant medicine to combat migraine. A clinical study by Martins et al. (2019) carried out on migraine patients revealed (Saneei Totmaj et al., 2019) that the addition of ginger to non-steroidal anti-inflammatory drugs may contribute to the reduced intensity of the migraine attack. They conducted the clinical experiment by following a double-blind placebo-controlled randomized clinical trial. The results indicated that patients treated with ginger showed better clinical response after 1, 1.5, and 2 h with significant p-values of less than 0.04, 0.01, and 0.04, respectively. Furthermore, ginger treatment resulted in reduced pain and improved functional status at all times assessed. The other medicinal uses of ginger that need to be studied in detail are its immunomodulatory, anti-vomiting, and improved post-prandial digestion activities; other effects are listed in Table 2.1.

2.5 MEDICINAL PRODUCTS

The rhizomes of ginger are the preparation mainly used in medicinal applications. According to the Chinese Pharmacopoeia (ChP), three medicinal products of ginger rhizomes are found on the market, namely fresh ginger, dried ginger, and ginger processed by stir-frying. Whereas fresh ginger is used for treating wind, the common cold, and gastrofrigid vomiting, dried ginger treats deficiency-cold syndromes of the spleen and stomach and Yang-collapse. In particular, fresh ginger is being used as a common 'broad-spectrum antiemetic' for treating nausea and vomiting in travel sickness and pregnancy. More than 30 Chinese patented drugs (Zhang et al., 2021) containing ginger are included in the Chinese Pharmacopoeia, Chinese Materia Medica preparations, and National Chinese Patent Medicine Standards. Interestingly, fresh ginger juice is often employed as an adjuvant in the preparation of some herbal medicines to enhance their therapeutic effects or to change drug properties. In addition, ginger juice used as a body cleanser in Greece for treating physical imbalances and as an indispensable component of curry powder in India.

Similarly, red ginger has gained traction as a medicinal plant, especially in Asia, where it is known for its medicinal effects: dispelling wind from the body, relieving indigestion, improving blood circulation, and relieving inflammation. In Southeast Asia, red ginger is locally known as halia bara, halia merah, or jahe merah. Currently, the red ginger products marketed in Southeast Asia include foodstuffs such as dried ginger pieces, pickled ginger, and red ginger extracts in instant beverages such as tea and coffee. Furthermore, red ginger extracts are also added to body lotions,

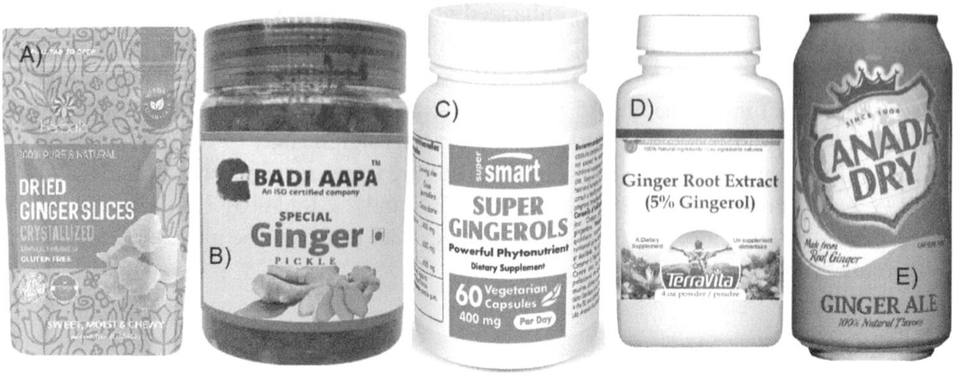

FIGURE 2.6 Ginger food and medicinal products available on the market. Food and beverages: dried (A) and pickled ginger (B), ginger juice (E). Medicinal supplements: gingerols (C) and ginger root extracts (D).

creams, ointments, and capsules (Zhang et al., 2022). Figure 2.6 shows the ginger-based medicinal and food products available on the market.

2.6 CONCLUSION

Ginger has been a widely used herbal medicine and food additive since antiquity, being described in old traditional medicinal books of Egypt (Ebers Papyrus), India (Ayurvedic) and China (Shennong Bencaojing, or the divine farmer's Materia Medica). Traditionally, common ginger is used as a 'broad-spectrum anti-emetic'. India, China, and Nigeria are the leading countries that produce and prepare ginger for the world market. The major ginger chemical constituents are vanilloids, essential oils, and polyphenols. Vanilloids cause the pungent odour and anticancer activity of ginger, whereas monoterpenes are responsible for the antibacterial activity of ginger, and phenolics present in ginger are potent antioxidants.

Similarly, red ginger is used to drive out wind from the body, treat indigestion, improve blood circulation, and heal inflammation. For the above reasons, ginger medicinal and food products are available on the market as dried and pickled ginger, ginger juice, and gingerol medicinal supplements. In China, a total of 30 herbal drugs that are made from ginger are listed in the Chinese Pharmacopoeia and hence are available on the market.

REFERENCES

Ahmadifar, E., Sheikhzadeh, N., Roshanaei, K., Dargahi, N. & Faggio, C. (2019) Can dietary ginger (*zingiber officinale*) alter biochemical and immunological parameters and gene expression related to growth, immunity and antioxidant system in zebrafish (*danio rerio*)? *Aquaculture*, 507, 341–348.

Alam, F., Shafique, Z., Amjad, S.T. & Bin Asad, M.H.H. (2019) Enzymes inhibitors from natural sources with antidiabetic activity: A review. *Phytotherapy Research*, 33(1), 41–54.

Ali, A.M.A., El-Nour, M.E.M. & Yagi, S.M. (2018) Total phenolic and flavonoid contents and antioxidant activity of ginger (zingiber officinale rosc.) rhizome, callus and callus treated with some elicitors. *Journal of Genetic Engineering and Biotechnology*, 16(2), 677–682.

Ashour, A.A., Basha, S., Felemban, N.H., Enan, E.T., Alyamani, A.A. & Gad El-Rab, S.M. (2022) Antimicrobial efficacy of glass ionomer cement in incorporation with biogenic zingiber officinale capped silver-nanobiotic, chlorhexidine diacetate and lyophilized miswak. *Molecules*, 27(2), 528.

Beristain-Bauza, S.D.C., Hernández-Carranza, P., Cid-Pérez, T.S., Ávila-Sosa, R., Ruiz-López, I.I. & Ochoa-Velasco, C.E. (2019) Antimicrobial activity of ginger (zingiber officinale) and its application in food products. *Food Reviews International*, 35(5), 407–426.

Borgonetti, V., Governa, P., Biagi, M., Pellati, F. & Galeotti, N. (2020) Zingiber officinale roscoe rhizome extract alleviates neuropathic pain by inhibiting neuroinflammation in mice. *Phytomedicine*, 78, 153307.

da Silva, F.T., da Cunha, K.F., Fonseca, L.M., Antunes, M.D., El Halal, S.L.M., Fiorentini, Â.M. et al. (2018) Action of ginger essential oil (*zingiber officinale*) encapsulated in proteins ultrafine fibers on the antimicrobial control in situ. *International Journal of Biological Macromolecules*, 118, 107–115.

Dhanik, J., Arya, N. & Nand, V. (2017) A review on *zingiber officinale*. *Journal of Pharmacognosy and Phytochemistry*, 6(3), 174–184.

Dissanayake, K.G.C., Waliwita, W. & Liyanage, R. (2020) A review on medicinal uses of *zingiber officinale* (ginger). *International Journal of Health Sciences and Research*, 10(6), 142–148.

Ebrahimzadeh Attari, V., Malek Mahdavi, A., Javadivala, Z., Mahluji, S., Zununi Vahed, S. & Ostadrahimi, A. (2018) A systematic review of the anti-obesity and weight lowering effect of ginger (*zingiber officinale* roscoe) and its mechanisms of action. *Phytotherapy Research*, 32(4), 577–585.

Ghafoor, K., Al Juhaimi, F., Özcan, M.M., Uslu, N., Babiker, E.E. & Ahmed, I.A.M. (2020) Total phenolics, total carotenoids, individual phenolics and antioxidant activity of ginger (*zingiber officinale*) rhizome as affected by drying methods. *LWT*, 126, 109354.

Jan, R., Gani, A., Dar, M.M. & Bhat, N.A. (2022) Bioactive characterization of ultrasonicated ginger (*zingiber officinale*) and licorice (glycyrrhiza glabra) freeze dried extracts. *Ultrasonics Sonochemistry*, 88, 106048.

Kalhoro, M.T., Zhang, H., Kalhoro, G.M., Wang, F., Chen, T., Faqir, Y. & Nabi, F. (2022) Fungicidal properties of ginger (*zingiber officinale*) essential oils against phytophthora colocasiae. *Scientific Reports*, 12(1), 1–10.

Kaushik, S., Jangra, G., Kundu, V., Yadav, J.P. & Kaushik, S. (2020) Anti-viral activity of *zingiber officinale* (ginger) ingredients against the chikungunya virus. *Virusdisease*, 31(3), 270–276.

Li, F.-S. & Weng, J.-K. (2017) Demystifying traditional herbal medicine with modern approach. *Nature Plants*, 3(8), 1–7.

Liu, Y., Liu, J. & Zhang, Y. (2019) Research progress on chemical constituents of *zingiber officinale* roscoe. *BioMed Research International*, 2019.

Martins, L.B., Rodrigues, A.M.D.S., Rodrigues, D.F., Dos Santos, L.C., Teixeira, A.L. & Ferreira, A.V.M. (2019) Double-blind placebo-controlled randomized clinical trial of ginger (*zingiber officinale* rosc.) addition in migraine acute treatment. *Cephalalgia*, 39(1), 68–76.

Murugesan, S., Venkateswaran, M.R., Jayabal, S. & Periyasamy, S. (2020) Evaluation of the antioxidant and anti-arthritic potential of *zingiber officinale* rosc. By in vitro and in silico analysis. *South African Journal of Botany*, 130, 45–53.

Nikkhah-Bodaghi, M., Maleki, I., Agah, S. & Hekmatdoost, A. (2019) *Zingiber officinale* and oxidative stress in patients with ulcerative colitis: A randomized, placebo-controlled, clinical trial. *Complementary Therapies in Medicine*, 43, 1–6.

Noipha, K. & Ninla-Aesong, P. (2018) Antidiabetic activity of *zingiber officinale* roscoe rhizome extract: An in vitro study. *HAYATI Journal of Biosciences*, 25(4), 160–160.

Rialita, T., Nurhadi, B. & Puteri, R. (2018) Characteristics of microcapsule of red ginger (*zingiber officinale* var. Rubrum) essential oil produced from different arabic gum ratios on antimicrobial activity toward escherichia coli and staphylococcus aureus. *International Journal of Food Properties*, 21(1), 2500–2508.

Saneei Totmaj, A., Emamat, H., Jarrahi, F. & Zarrati, M. (2019) The effect of ginger (*zingiber officinale*) on chemotherapy-induced nausea and vomiting in breast cancer patients: A systematic literature review of randomized controlled trials. *Phytotherapy Research*, 33(8), 1957–1965.

Talebi, M., Ilgün, S., Ebrahimi, V., Talebi, M., Farkhondeh, T., Ebrahimi, H. et al. (2021) *Zingiber officinale* ameliorates Alzheimer's disease and cognitive impairments: Lessons from preclinical studies. *Biomedicine & Pharmacotherapy*, 133, 111088.

Venkatadri, B., Shanparvish, E., Rameshkumar, M., Arasu, M.V., Al-Dhabi, N.A., Ponnusamy, V.K. et al. (2020) Green synthesis of silver nanoparticles using aqueous rhizome extract of *zingiber officinale* and curcuma longa: In-vitro anticancer potential on human colon carcinoma ht-29 cells. *Saudi Journal of Biological Sciences*, 27(11), 2980–2986.

Wang, Y., Chinnathambi, A., Nasif, O. & Alharbi, S.A. (2021) Green synthesis and chemical characterization of a novel anti-human pancreatic cancer supplement by silver nanoparticles containing *zingiber officinale* leaf aqueous extract. *Arabian Journal of Chemistry*, 14(4), 103081.

Zammel, N., Saeed, M., Bouali, N., Elkahoui, S., Alam, J.M., Rebai, T. et al. (2021) Antioxidant and anti-inflammatory effects of *zingiber officinale* roscoe and allium subhirsutum: In silico, biochemical and histological study. *Foods*, 10(6), 1383.

Zhang, M., Zhao, R., Wang, D., Wang, L., Zhang, Q., Wei, S. et al. (2021) Ginger (*Zingiber officinale* Rosc.) and its bioactive components are potential resources for health beneficial agents. *Phytotherapy Research*, 35(2), 711–742.

Zhang, S., Kou, X., Zhao, H., Mak, K.-K., Balijepalli, M.K. & Pichika, M.R. (2022) *Zingiber officinale var. rubrum*: Red ginger's medicinal uses. *Molecules*, 27(3), 775.

3 Bioactive Compounds and Pharmaceutical Importance of *Houttuynia cordata* Rhizome

Atreyi Pramanik, Aashna Sinha, Kundan Kumar Chaubey, and Deen Dayal

LIST OF ABBREVIATIONS

Art	Artemisinin
AFB_1	Aflatoxin B_1
CHM	Compound *Houttuynia* mixture
HCE	*H. cordata* extracts
HCWE	*H. cordata* water extracts
HCT	*Houttuynia cordata* Thunb.
HSV-1	Herpes simplex virus type 1
Htty	*Houttuynia*
Httn	Houttuynin
LPS	Lipopolysaccharides
Polysacc.	Polysaccharides
RNS	Reactive nitrogen species
ROS	Reactive oxygen species
SARS-CoV	SARS-coronavirus
SARS	Severe Acute Respiratory Syndrome

3.1 INTRODUCTION

In Asia, including China, Japan, and India, the perennial herbaceous plant *Houttuynia cordata* Thunb. (HCT) thrives in the wild in damp shaded areas. HCT possesses anti-inflammatory, antiviral, antibacterial, immunostimulant, and other properties (Li & Zhao, 2015). The *H. cordata* plant has a spreading rhizome and is a fragrant therapeutic herb. It has broad ovate-cordate leaves that are 4–8 cm long and 3–6 cm wide and grows to a height of 20–50 cm. Petiole is 1–3.5 cm long, glabrous, and the stipular sheath is 1–2.5 cm long, ¼ to ½ as long as the petiole, typically ciliate. Leaf blades are broadly ovate or ovate-cordate, 4–10 cm long, 2.5–6 cm wide, thinly papery, thickly glandular, and typically glabrous (Kumar et al., 2014). The use of herbal plants for medicinal purposes has a long history, from down the ages until today. This is a practice followed for over 8,000 years in Eastern and far Eastern countries like India and China. For several reasons, viz. costly drug discovery using chemical synthesis, the efficiency of herbal medicines in treating complex disorders, and the durability of host-pathogen target interactions, herbal medicine remains a choice

DOI: 10.1201/b22924-3

of interest to the present day (Jiangang et al., 2013). In India and other Asian regions, approximately 95% of basic health care treatments are achieved with medicinal plants as they are rich in a range of chemical components, with diverse biological activities, which efficiently prevent or treat common illnesses. In addition to whole medicinal plants, plant extracts and derivatives like flavonoids, saponins, polysaccharides, and alkaloids have also been shown to exhibit promising potential for anti-ageing product development. Like the discovery of artemisinin (Art), an anti-malarial semisynthetic derivative drug isolated from the wormwood plant *Artemisia annua*, HCT is a perennial medicinal herb, belonging to the Saururaceae family, that is widely used for both medicinal and edible purposes, being grown in moist soil and humid conditions and harvested in the Yarlung Zangbo Valley in Assam, India (Wu et al., 2021).

3.2 BIOACTIVE COMPOUNDS OF *H. CORDATA*

The main abundant bioactives of *H. cordata* are volatile oils, flavonoids and water-soluble polysaccharides. An overview of the bioactive compounds and their pharmaceutical uses is shown in Figure 3.1. *H. cordata* contains high levels of volatile oils, decanoyl aldehyde, myrcene, lauryl aldehyde, α-pinene, D-limonene, methyl nonyl ketone (2-undecanone), organic acids, such as palmitic acid, linoleic acid, and aspartic acid, flavonoids, such as quercetin, isoquercetin, afzelin, hyperin, hyperoside, rutin, and cordarine, as well as water-soluble polysaccharides. Most of the flavonoids are in the form of glycosides, combined with the sugar rhamnose. Apart from this, a new blend of *Houttuynia*-specific flavonoids, namely houttuynin, hyperin, hyperoside, houttuynoids A-E, houttuynoside A, and houttuynamide A have been extracted (Wu et al., 2021). Recent research on *H. cordata* has reported the presence of amino acids, vitamins, and mineral elements, namely K, Zn, Fe, Cu, and Mn, that play a vital role in the body's physiology. Therefore, this herb, loaded with nutrients and bioactives, is considered to be highly conducive to health (Yang & Jiang, 2009a) The presence of compounds like 2-undecanone, myrcene, ethyledecanoate, ethyledodecanoate,

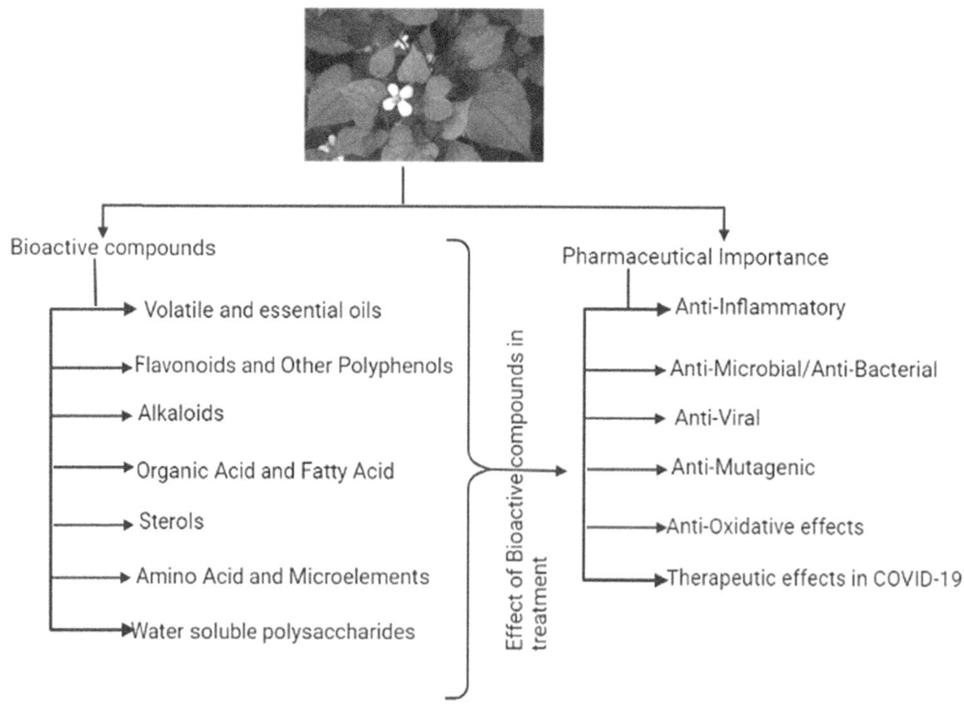

FIGURE 3.1 Overview of bioactive compounds of H. cordata and their pharmaceutical importance.

2-tridecanol, and decanal were reported to be at higher concentrations in the aboveground stems than in the underground rhizomes of the plant. The ketone 2-undecanone is an indispensable bioactive compound that displays various types of biological activity and is used as a standard marker for quality control of *H. cordata.* Fatty acids, such as linolenic acid, linoleic acid, oleic acid, palmitic acid, and stearic acid, and phenolic acids, such as quinic acid derivatives, caffeic acid derivatives, neo-chlorogenic acid, cryptochlorogenic acid are abundant in *H. cordata.* Additionally, alkaloids, particularly phenanthrolactam-derived compounds, like aristololactam and piperolactam, are main bioactives controlling the physical action of houttuynia-containing herbs (Wu et al., 2021).

3.3 BIOACTIVE COMPOUNDS

3.3.1 Essential Oils

The volatile oils of *H. cordata* are extracted using several procedures like supercritical carbon-dioxide extraction, steam distillation, petroleum ether extraction, solid-phase micro-extraction, flash evaporation, and simultaneous distillation-extraction. The main components of volatile oils are depicted in Figure 3.2 (Jiangang et al., 2013) Another principal component of *H. cordata* is decanoyl aldehyde; although it is unstable and easily oxidizable, it is also known to have pharmacological activities. Volatile oils from *H. cordata* are obtained from both wild and cultivated plants, as well as distinct places of the plants, like rhizomes, leaf stalks, and leaves. The volatile oils represent a low fraction of the composition of medicinal plants but have extensive presentation in the plant kingdom (Yang & Jiang, 2009b). Research reveals that essential oils are chiefly subdivided into three categories: aliphatic, aromatic, and terpenes. These oils are extracted from both dried and fresh plant tissue, yielding 2-undecanone, β-myrcene, and β-pinene as major components of the volatile oil (Pan et al., 2021). Li et al. (2013), in their study, reported that the oils have prominent inflammation-inhibitory effects, acting by suppressing lipopolysaccharides (LPS) to a major extent. Furthermore *H. cordata* essential oils also decrease the production of nitric oxide and tumour necrosis factor-alpha (TNF-α), thereby lowering inflammation (Li et al., 2013).

3.3.2 Flavonoids and Other Polyphenols

One of the most widely distributed natural products in the plant kingdom are the flavonoids. Found in the aerial parts of *H. cordata*, flavonoids are used as anti-hepatotoxic, anti-allergic, anti-inflammatory, anti-osteoporotic, and even anti-tumour activity (Table 3.1). Leaves contain a higher concentration of flavonoids than the rhizomes. According to a Hung et al. (2015) study, ethanol extracts from the herb's aerial portions caused 70% suppression of the generation of inflammatory biomarkers interleukin-6 (IL-6) and nitric oxide (NO) in lung cancer epithelial cells (A549) and alveolar

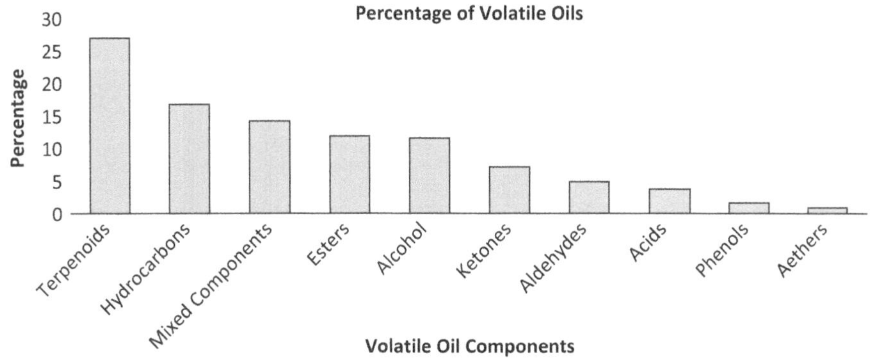

FIGURE 3.2 Main components of essential oils of *H. cordata.*

TABLE 3.1
List of Flavonoids Obtained from *H. cordata* and Their Method of Extraction

Sl. No	Name of Flavonoids	Class	Part of Plant (*H. cordata*) Used	Method Used for Extraction	Reference
1	Quercetin-3-*O*-β-D-galactoside-7-O-β-D-glucoside	Flavonoids	Isolated for the first time	Pressurized liquid extraction, hot water soaking, macroporous exchange resin, ultrasound-assisted extraction	Chaves et al. (2020)
2	Kaempferol-3-*O*-α-l-rhamnopyranosyl-(1 →6)-β-D-glucopyranoside	Flavonoids	Entire plant		
3	Quercitrin	Flavonoids	Leaves and stems		
4	Hyperin	Flavonoids	Entire plant		
5	Quercetin-3-*O*-α-l-rhamnopyranosyl-7-*O*-β-D-glucopyranoside	Flavonoids	Isolated for the first time		
6	Quercetin-3-*O*-β-D-galactoside-7-*O*-β-D-glucoside	Flavonoid glycosides	Entire plant	NA	Fu et al. (2013)
7	Kaempferol 3-*O*-[α-L-rhamnopyranosyl-(1→ 6)-β-D-glucopyranoside]	Flavonoid glycosides	Entire plant	NA	Jiangang et al. (2013)
8	Quercetin 3-*O*-α-L-rhamnopyranosyl-7-*O*-β-D-glucopyranoside (N.B: Quercitrins were first isolated from this herb)	Flavonoid glycosides	Entire plant, leaves, and stems	Column chromatography	Jiangang et al. (2013)
9	Chlorogenic acid methyl ester	Polyphenols	Entire plant	Hot soaking, ethyl acetate, Soxhlet, ultrasonic, microwave-assisted and pressurized liquid column chromatography	Bahadur Gurung et al. (2021); Jiangang et al. (2013)
10	4-[(2E)-3-(ζ-D-glucopyranosyloxy)-2-buten-1-y l]-4-hydroxy-3,5,5-trimethyl-2-cyclohexen-1-one, 2-(4-hydroxyphenyl)ethyl-β-D-glucopyranoside	Polyphenols	Entire plant	HPLC-mass spectrometry was used to determine structure	Jiangang et al. (2013)
11	2-(3,4-dihydroxyphenyl) ethyl-ζ-D-glucopyranoside	Polyphenols	Entire plant	HPLC-mass spectrometry was used to determine structure	Jiangang et al. (2013)
12	4-(ζ-D-glucopyrano-syloxy)-3-hydroxybenzoic	Polyphenols	Entire plant	HPLC-mass spectrometry was used to determine structure	
13	Catechin	Polyphenols	Leaves	HPLC-mass spectrometry was used to determine structure	
14	Procyanidin B	Polyphenols	Leaves	HPLC-mass spectrometry was used to determine structure	
15	Houttuynamide A and houttuynoside A	Polyphenols	Entire plant	HPLC-mass spectrometry was used to determine structure	
16	Chlorogenic acid	Polyphenols	Aerial tissue, entire plant, leaves, roots, and stem	HPLC-mass spectrometry was used to determine structure	

macrophages, respectively. A study in mice revealed that the major *Houttuynia* flavonoids afzelin, hyperoside, and quercitrin reduced LPS-induced lung inflammation, with quercitrin being the most potent flavonoid extracted from the leaves. Thus, flavonoids from *H. cordata* are used to treat lung inflammation (Fu et al., 2013; Hung et al., 2015; Yang & Jiang, 2009b).

3.3.3 Alkaloids

Alkaloids like aporphine, pyridine, aristolactam A, aristolactam B, piperolactam A, and norcepharacdione B have been isolated from *H. cordata*, along with houttuynamide B, C, and houttuynin, which were extracted from the aerial portions of the herb. All these isolates proved to have inhibitory effects on nitric oxide production in RAW 264.7 macrophage-like cells triggered by LPS (Ahn et al., 2017; Fu et al., 2013; Jong & Jean, 1993).

3.3.4 Organic Acids and Fatty Acids

Among the many bioactive compounds which have been extracted from this plant are organic acids and fatty acids (Palmitic acid, Stearic acid, heptanoic acid, nonanoic acid, undecanoic acid, octanoic acid, hexanoic acid, lauric acid, capric acid, heptadecanoic acid, tetradecanoic acid, tridecanoic acid, pentadecanoic acid, octadecenoic acid, hexadecenoic acid, octadecadienoic acid, aspartic acid, glutamic acid, Chlorogenic acid, crypto-chlorogenic acid, quinic acid, caffeic acid, Oleic acid, Succinic acid etc.) A list of organic compounds that have been extracted and characterized from *H. cordata*, using different types of chromatography, mass spectrometry, and nuclear magnetic resonance, is given in Table 3.2.

3.3.5 Sterols

Sterols are other bioactive compounds that are obtained from this herb. Although different parts of the plant have been ascribed various medicinal values, sterols are extracted from the whole plant, including the rhizome. A list of sterols (Stigmast-4-en-3-one, 3β-hydroxystigmast-5-en-7-one, 5α-stigmastane-3,6-dione, stigmast-4-ene-3,6-dione, Stigmast-3,6-dione, sitoindoside I, daucosterol, β-Sitosterol, β-sitosterolglucoside, brassicasterol, stigmasterol, spinasterol, stigmast-4-ene-3,6-dione) obtained from *H. cordata* is shown in Table 3.3 (Jiangang et al., 2013).

3.3.6 Amino Acids and Microelements

H.cordata has been reported to contain more than 20 amino acids in addition to microelements like Zn, Fe, Cu, and Mn. Of all the amino acids studied, the highest concentration observed was for glutamic acid, followed by leucine and aspartic acid. The concentration of the microelements obtained depends upon different factors, like the season of harvest, cultivars, growing conditions, and test facilities. The membranous envelope of *H. cordata* holds high concentrations of Zn and Cu, and their proportions vary during fruiting (Fu et al., 2013; Yang & Jiang, 2009a). A comprehensive list of amino acids and microelements obtained from the medicinal herb is given in Table 3.4.

3.3.7 Water-Soluble Polysaccharides

Numerous therapeutic properties have been attributed to botanical polysaccharides and it is believed that the mechanisms involved help in innate immunity modulation, particularly macrophage function. Literature reviews identify the enhancement or activation of macrophage immune responses leading to immunomodulatory effects or anti-tumour activity as the primary effects of botanical polysaccharides (Schepetkin & Quinn, 2006). Polysaccharides obtained from plants, epiphytes, and extracts of animals are valuable additives for the food and drug industries. The water-soluble

TABLE 3.2
List of Organic Acids Obtained from *H. cordata*

Name of Organic Acids	Source	Technique used	Reference
Palmitic acid	Aerial parts of *H.cordata*	Benzene fraction and phytochemical separation by gas chromatography	Bauer et al. (1996); Fu et al. (2013)
Stearic acid, heptanoic acid, nonanoic acid, undecanoic acid, octanoic acid, hexanoic acid, lauric acid, capric acid, heptadecanoic acid, tetradecanoic acid, tridecanoic acid, pentadecanoic acid, octadecenoic acid, hexadecenoic acid, octadecadienoic acid, aspartic acid, glutamic acid,	*H.cordata*	gas chromatography	Fu et al. (2013)
Chlorogenic acid, crypto-chlorogenic acid, quinic acid, caffeic acid.	*H.cordata*	Mass spectra and fragmentation patterns	Fu et al. (2013); Nuengchamnong et al. (2009)
Oleic acid		Benzene fraction and phytochemical examination	Bauer et al. (1996); Fu et al. (2013)
Succinic acid	Dried rhizome of *H.cordata*	Solvent, silica gel, and Sephadex LH-20 column chromatography	Fu et al. (2013)

components of *H. cordata* have heteroglycan components, as determined by polyacrylamide gel electrophoresis and thin-layer chromatography. The heteroglycan components are derivatives of glucose, fructose, arabinose, galactose, xylose, rhamnose, and an unknown pentaglucose. (Yang & Jiang, 2009b). A polysaccharide was obtained from a water extract of the plant, with a molecular weight of 43 KDa and comprising principally galactouronic acid, galactose, glucose, and xylose in different ratios (Cheng et al., 2019). The polysaccharides of *H. cordata* could reduce the accumulation of hydroxyl and superoxide radicals by 50% (Yang & Jiang, 2009b).

3.4 PHARMACOLOGICAL IMPORTANCE OF BIOACTIVE COMPOUNDS

3.4.1 Anti-Inflammatory Activity

H. cordata preparations exhibit a diverse range of pharmaceutical effects such as anaphylaxis, enzyme-inhibitory, anti-mutagenic, antimicrobial, antiviral, and insecticidal activities, and a depressant effect on the central nervous system (Yang & Jiang, 2009b).The water extract has anti-allergic activities achieved by inhibiting IgE-mediated systemic passive cutaneous anaphylaxis, which has been studied in mice. Antigen-induced phosphorylation of Syk, Lyn, LAT, Gab2, and

TABLE 3.3
List of Sterols Obtained from *H. cordata*

List of Sterols	Part of Plant	Methods	Reference
Stigmast-4-en-3-one, 3β-hydroxystigmast-5-en-7-one, 5α-stigmastane-3,6-dione, stigmast-4-ene-3,6-dione	Entire plant	NA	Fu et al. (2013)
Stigmast-3,6-dione,	Dried rhizome of *H. cordata*	Solvent, silica gel, and Sephadex LH-20 column chromatography	
sitoindoside I, daucosterol		NA	
β-Sitosterol,	Entire plant	NA	
β-sitosterolglucoside,		NA	
brassicasterol,		NA	
stigmasterol,		NA	
spinasterol,		NA	
stigmast-4-ene-3,6-dione		NA	

TABLE 3.4
List of Amino Acids and Elements Obtained from *H. cordata*

List of Amino Acids	List of Elements	References
Valine	Iron	Jiangang et al. (2013); Yang and Jiang (2009)
Glutamic acid	Magnesium	
Aspartic acid	Manganese	
Isoleucine	Potassium	
Proline	Copper	
Leucine	Zinc	
Glycine	Calcium	
Serine	Phosphonium ion	
Lysine		
Cystine		
Tyrosine		
Methionine		
Phenylalanine		
Histidine		
Threonine		
Tyrosine		
Arginine		
Hydroxyproline		
Citrulline		

PLC2 was reduced by the water extract, as was antigen-induced phosphorylation of Akt and MAP kinases (ERK1/2 and JNK1/2 but not p38 MAP kinase) (Han et al., 2009; Jiangang et al., 2013). Experiments have proven that expression of the IgE receptor FcεRI, which plays a pivotal role in IgE-mediated allergic reactions and usually expresses a high affinity for human basophilic KU812F cells, and the binding activity of IgE were repressed when the cells were cultured with an *H. cordata* extract. (Jiangang et al., 2013) Therefore, studies suggest that the water extract could be a possible therapy for inflammatory allergic disease through cytokine cascade inhibition and multiple events of FceRI-dependent signalling cascades in mast cells.

3.4.2 Anti-microbial activity

Despite the fact that *H. cordata* preparations have numerous pharmacological effects, the underlying mechanism of these activities is unknown. *H. cordata* has anti-microbial properties against different strains of bacteria. Previous studies have reported that essential oils from *H. cordata*, containing aldehydes, such as decanal, dodecanal, 3-oxo-dodecanal, and houttuynin exhibit anti-bacterial action against Gram-positive bacteria, including *Staphylococcus aureus* (Sekita et al., 2016). Bacteriocidal effects have been observed against Gram-negative bacteria with volatile essential oil components, namely methyl n-nonyl ketone, β-myrene, β-ocimene, 1-decanol and houttuynin, and fraction 5, containing decanal, bornyl acetate, fenchene, and decanoic acid (Jiangang et al., 2013). Li et al. (2013) reported that the essential oils had a Minimum Inhibitory Concentration (MIC) value against different bacterial species ranging between 0.0625×10^{-3} to 4×10^{-3} mg/mL. In addition, although the whole herb is considered to be medicinal, the water and ethanol extracts of the rhizomes have a particularly strong bacteriostatic action on *S. aureus, Escherichia coli*, and yeast, with appreciable MIC values, whereas the leaf extract had a stronger antibacterial action on *Bacillus dysenteriae* (Fu et al., 2013; Yang & Jiang, 2009b). Yeast and mould growth is suppressed by the application of houttuynin isolated from the rhizome of *H. cordata*. The azole heterocycle 3-nonyl-pyrazole hindered the growth of the bacteria *S. aureus* and *Bacillus subtilis*, and the fungi *Trichophyton* spp., *Zygosaccharomyces salsus*, and *Aspergillus niger* (Jiangang et al., 2013; Kim et al., 2008). In addition, root and leaf extracts showed inhibitory effects against microbes, oil extracts have been shown to possess inhibitory properties towards β-haemolysis, *Streptococcus, S. aureus, P. aeruginosa*, and *E.coli*, whereas the volatile flavour compounds like aldehydes, phenols, terpenes, alcohols, ketones, hydrocarbons, furans, and esters exhibit robust antibacterial activities against *Bacillus cereus, B. subtilis, Vibrio cholera*, and *Vibrio parahaemolyticus* (Jiangang et al., 2013; Kwon et al., 1996). As well as the *H. cordata* ethanol extract, which has antibacterial, antibiofilm, and anti-inflammatory effects against *S. aureus*, as well as other bacterial species, a water extract of a *H. cordata* poultice (prepared from submerging fresh leaves) exhibits antimicrobial and antibiofilm properties against some oral microorganisms, such as *Fusobacterium nucleatum, Streptococcus mutans, and Candida albicans*. Despite this extract having no cytotoxic effects, it had inhibitory effects on IL-8 and CCL20 levels in *Porphyromonas gingivalis* lipopolysaccharide-stimulated human oral keratinocytes (Sekita et al., 2016).

3.4.3 Anti-Viral Activity

H. cordata has preventive properties against human immunodeficiency virus type 1, HSV-1, influenza virus, herpes simplex virus (HSV) type 2, SARS-coronavirus, enterovirus-71, porcine epidemic diarrhoea virus, dengue virus type 2, and mouse hepatitis virus (Sit et al., 2017). Steam distillates of *H.cordata*, containing volatiles like methyl n-nonyl ketone, lauryl aldehyde, and capryl aldehyde, exhibit direct antiviral effects on HSV-1, influenza virus (IFV), mumps viruses (IFVA, IFVB, MPV), and human immunodeficiency virus type 1 but not poliovirus and coxsackie-virus, without displaying any cell cytotoxicity (Jiangang et al., 2013). These essential oils exhibit antiviral activities against enveloped viruses by interfering with the functions of the envelope proteins (Hayashi et al., 1995). HSV infection is a latent infection in humans and therefore repeated use of acyclovir leads to the development of drug-resistant strains of the virus. Hence, in a crucial requirement of a drug for use against HSV, water extracts of *H. cordata* have been promising in treatments against HSV-1, herpes simplex virus type 2, and acyclovir-resistant HSV-1, acting principally by obstructing viral binding and penetration at the beginning of infection. As soon as there is viral

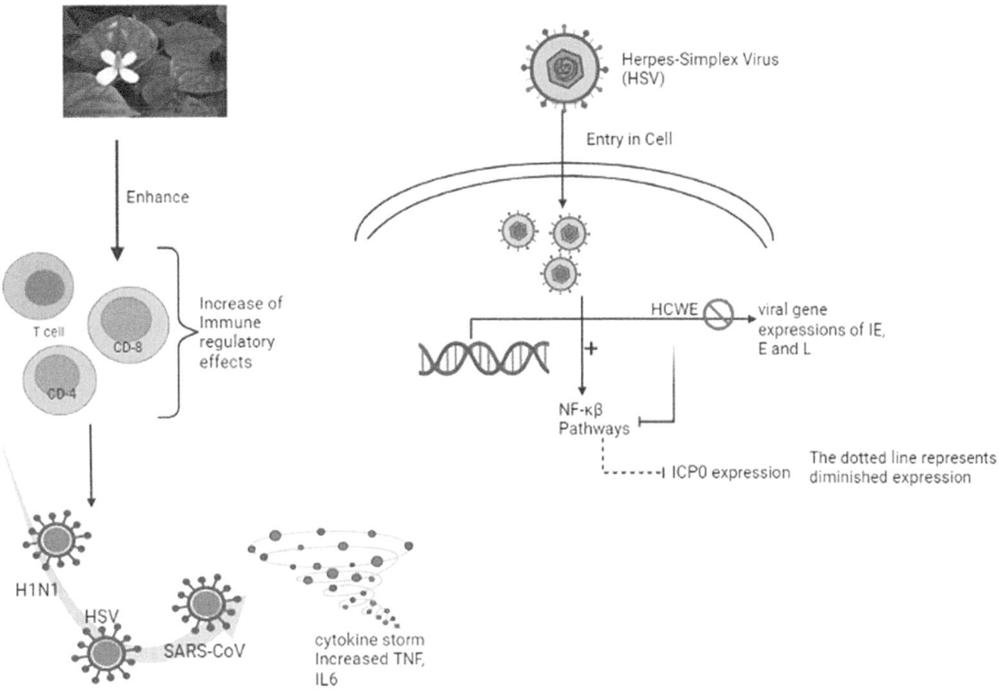

FIGURE 3.3 The antiviral effects of *H. cordata*. The left-hand-side figure shows an overall pictorial representation of *H. cordata* antiviral effects, whereas the right-hand-side figure depicts the mechanism of inhibition by *H. cordata* extracts of a particular virus (HSV).

entry, viral gene expressions of IE, E and L are stalled by the water extracts. Additionally, HSV-induced NF-κ β activation is also suppressed by *H. cordata* water extracts, leading to diminished NF-κ β-mediated *ICP0* transcription, which is one of the important HSV IE genes involved in binding several cellular proteins as well as viral promoters, ensuring amplified expression of viral E and L genes (Figure 3.3). HSV glycoprotein D (gD) is another critical protein necessary for binding of the virus on host cells, cell-to-cell spreading, and viral entry into host cells; therefore, HSV gD is an ideal drug target for the development of anti-HSV drugs. Water extracts, containing quercetin and rutin, exhibit anti-HSV activities targeted at inhibiting viral binding, penetration, and host NF-κB activation. On the other hand, NF-κB activity is also suppressed by 2-Amino-3-methylimidazole [4,5-f]quinoline (IQ) (Hung et al., 2015; Jiangang et al., 2013; Wu et al., 2021).

3.4.4 Antimutagenic activity

The methanol extracts of freeze-dried *H. cordata* exhibited a much greater antimutagenic effect, than did extracts of sun-dried *H. cordata*. But the inhibitory activity towards *N*-methyl *N'*-nitro-*N*-nitrosoguanidine by methanol extracts *H. cordata* showed a relatively lower inhibition than was achieved with aflatoxin B1 (AFB_1). On the contrary, for *N*-methyl *N'*-nitro-*N*-nitrosoguanidine, the aqueous fraction of the methanol extract of *H. cordata* displayed anti-mutagenic effects whereas the chloroform fraction did not (Kil et al., 2005; Yang & Jiang, 2009b).

3.4.5 Antioxidative Effects

Oxidative stress is a result of oxidized fats which contain harmful toxic lipid peroxidation products, leading to the production of free radicals like reactive oxygen species (ROS) and reactive nitrogen species (RNS). The free radicals are end-products of continuous human metabolism. The free radicals have destructive effects which, under unwanted pathological conditions, lead to cardiovascular diseases, mutagenesis, and also life-threatening cancers. The mutagenicity of AFB1, benzo(a)pyrene (B(a)P), and OFO were inhibited by both water and methanol extracts of *H. cordata*. Water extracts of *H. cordata* have a greater antioxidant power, as well as polyphenol content, than methanol extracts. The ability of *H. cordata* extracts to fight off free radicals may also have an antimutagenic impact. An additional factor which plays a crucial role in mutagenesis is the liver microsome-based cytochrome P450 (CYP450). Polyphenols have the ability to lessen a number of CYP450-mediated malignancies or chronic diseases because they were discovered to limit the activity of CYP450. The ability of *H. cordata* extracts to prevent mutations may also be a result of the capacity of polyphenols to inhibit CYP450 activity. A strong association between the antimutagenic effect and the polyphenol content was found, leading to the idea that the high polyphenol content of *H. cordata* extract contributes to its anti-mutagenic effects. The factor protecting *H. cordata* from oxidative stress damage and mutation is probably its high polyphenol content (Chen et al., 2003). In their study, Fischedick et al. (2012) have confirmed that nuclear factor E2-related factor 2 (NRF2), a key transcription factor, up-regulates a series of antioxidants like NAD (P) H:quinone oxidoreductase 1 (NQO1). Treatment with houttuynin, apart from resisting various diseases and ailments, also results in protection against skin diseases. *H. cordata* extracts stimulate aryl hydrocarbon receptor (AHR): NRF2 with increased expression of CYP1A1 (from the cytochrome P450 family of enzymes) and NQO1, resulting in suppression of ROS generation in human keratinocytes. Moreover, AHR activation also helps to boost expression of an epidermal barrier protein filaggrin. This antioxidant, in addition to AHR-NRF2 signalling, aids in treatment of barrier-disrupted skin diseases (Doi et al., 2014).

3.4.6 *H. cordata* in COVID-19

The lethal form of SARS pneumonia, caused by the deadly virus SARS-CoV, infected vast numbers of people worldwide. Unfortunately, due to a lack of appropriate medicines, *H. cordata* was a medicine of choice by Chinese scientists to face SARS challenges as it is typically used for pneumonia treatment. Studies in mice revealed anti-viral properties by exhibiting inhibitory effects on SARS-CoV3C like protease (3CL^{pro}) and the RNA-dependent RNA polymerase enzyme that helps in transcription and replication of the viral genome (Lau et al., 2008). The phytocompounds show inhibitory effects on the principal replication proteins of COVID, namely the SARS-CoV-2 main protease, papain-like proteases, and ADP-ribose phosphatase that also controls replication (Das et al., 2022). Apart from the fact that *H. cordata* bioactive compounds have inhibitory effects against SARS-CoV-2, it is also reported that molecular docking studies also demonstrated that the phytocompound ligand 6-hydroxyondansetron (A104) has higher binding affinity towards receptors and papain-like proteases, whereas QR (A166) has a strong binding affinity towards ADRP (Bahadur Gurung et al., 2021; Das et al., 2022; Sargol Mazraedoost et al., 2020). The compound *Houttuynia* mixture (CHM) prepared from the *H. cordata* herb and other herbal plants has detoxifying effects and is also used for throat pain, pharyngitis, and tonsillitis. Molecular docking technology has paved the scientific ground for the use of CHM in COVID-19 treatment. Protein-protein interaction network analysis identified *MAPK1, MAPK3, MAPK8, TP53, TNF, IL6* as main targets of CHM for treatment of COVID-19. Mitogen-activated protein kinases are involved in cellular immune regulation, inflammation, cell proliferation, apoptosis, and signal transduction of various stress responses (Wu et al., 2021). In addition, CHM is also involved in hypoxia-inducible factor-1, vascular endothelial growth factor, that plays an important role in vascular endothelial cell division, increases permeability of blood vessels, and, most importantly, in the process of lung inflammation and TNF signal pathways. The vascular endothelial growth factor is a promising target for suppressing lung inflammation in COVID-19 infection. Studies have also proved that blocking TNF-α helps in treatment of the downstream inflammatory cascade in COVID-19. Hence,

CHM can be considered to be a promising treatment for COVID-19 by preventing inflammation of the body and regulating subsequent immunological functions (Ali et al., 2021; Wu et al., 2021).

3.5 CONCLUSION AND FUTURE PERSPECTIVES

H. cordata is a herb with considerable medicinal values and diverse biological activities. From having organ-protecting roles to relieving oxidative stresses and inflammation, the medicinal values of the herb are innumerable. The most appreciable part is that the whole plant has therapeutic standards. The Chinese herb proves to be an encouraging therapeutic source for a wide range of diseases, as well as for suppression of inflammation. Keeping in mind the current situation of the COVID-19 pandemic, phytocompounds 6-hydroxyondansetron (A104) and QR (A166) are of particular importance as they act as replication inhibitors, thereby helping to control COVID-19. A variety of extracts of *H. cordata*, with water, methanol, or steam as extractants, extract valuable chemicals, namely methyl n-nonyl ketone and lauryl aldehyde, which play important roles in combating oxidative stress, mutagenesis, and viral infections. Therefore, we can conclude in this chapter that the plant and its extracts have a wide range of pharmaceutical properties which are of potentially great help to mankind.

REFERENCES

Ahn, J., Chae, H.S., Chin, Y.W. & Kim, J. (2017) Alkaloids from aerial parts of Houttuynia cordata and their anti-inflammatory activity. *Bioorganic and Medicinal Chemistry Letters*, 27(12), 2807–2811. https://doi.org/10.1016/J.BMCL.2017.04.072

Ali, S.G., Ansari, M.A., Alzohairy, M.A., Almatroudi, A., Alomary, M.N., Alghamdi, S. et al. (2021) Natural products and nutrients against different viral diseases: Prospects in prevention and treatment of SARS-CoV-2. *Medicina*, 57(2), 169. https://doi.org/10.3390/MEDICINA57020169

Bahadur Gurung, A., Ajmal Ali, M., Lee, J., Abul Farah, M., Mashay Al-Anazi, K. & Al-Hemaid, F. (2021) Identification of SARS-CoV-2 inhibitors from extracts of Houttuynia cordata Thunb. *Saudi Journal of Biological Sciences*, 28(12), 7517–7527. https://doi.org/10.1016/J.SJBS.2021.08.100

Bauer, R., Pröbstle, A., Lotter, H., Wagner-Redecker, W. & Matthiesen, U. (1996) Cyclooxygenase inhibitory constituents from *Houttuynia cordata*. *Phytomedicine* 2(4), 305–308.

Chaves, J.O., De Souza, M.C., Da Silva, L.C., Lachos-Perez, D., Torres-Mayanga, P.C., Machado, A.P. et al. (2020) Extraction of flavonoids from natural sources using modern techniques. *Frontiers in Chemistry*, 8:507887.

Chen, Y.Y., Liu, J.F., Chen, C.M., Chao, P.Y. & Chang, T.J. (2003) A study of the antioxidative and anti-mutagenic effects of Houttuynia cordata Thunb. using an oxidized frying oil-fed model. *Journal of Nutritional Science and Vitaminology*, 49(5), 327–333. https://doi.org/10.3177/JNSV.49.327

Cheng, D., Sun, L., Zou, S., Chen, J., Mao, H., Zhang, Y. et al. (2019) Antiviral effects of Houttuynia cordata Polysaccharide extract on Murine norovirus-1 (MNV-1)-A human norovirus surrogate. *Molecules (Basel, Switzerland)*, 24(9). https://doi.org/10.3390/MOLECULES24091835

Das, S.K., Mahanta, S., Tanti, B., Tag, H. & Hui, P.K. (2022) Identification of phytocompounds from Houttuynia cordata Thunb. as potential inhibitors for SARS-CoV-2 replication proteins through GC–MS/LC–MS characterization, molecular docking and molecular dynamics simulation. *Molecular Diversity*, 26(1), 365–388. https://doi.org/10.1007/S11030-021-10226-2/FIGURES/9

Doi, K., Mitoma, C., Nakahara, T., Uchi, H., Hashimoto-Hachiya, A., Takahara, M. et al. (2014) Antioxidant Houttuynia cordata extract upregulates filaggrin expression in an aryl hydrocarbon-dependent manner. *Fukuoka Igaku Zasshi = Hukuoka Acta Medica*, 105(11), 205–213.

Fischedick, J.T., Standiford, M., Johnson, D.A., De Vos, R.C., Todorović, S., Banjanac, T. et al. (2012) Activation of antioxidant response element in mouse primary cortical cultures with sesquiterpene lactones isolated from Tanacetum parthenium. *Planta medica*, 24, 1725–1730.

Fu, J., Dai, L., Lin, Z. & Lu, H. (2013) Houttuynia cordata Thunb: A review of phytochemistry and pharmacology and quality control. *Chinese Medicine*, 4, 101–123. https://doi.org/10.4236/cm.2013.43015

Han, E.H., Park, J.H., Kim, J.Y. & Jeong, H.G. (2009) Houttuynia cordata water extract suppresses anaphylactic reaction and IgE-mediated allergic response by inhibiting multiple steps of FcεRI signaling in mast cells. *Food and Chemical Toxicology*, 47(7), 1659–1666. https://doi.org/10.1016/J.FCT.2009.04.025

Hayashi, K., Kamiya, M. & Hayashi, T. (1995) Virucidal effects of the steam distillate from Houttuynia cordata and its components on HSV-1, influenza virus, and HIV. *Planta Medica*, 61(3), 237–241. https://doi.org/10.1055/S-2006-958063

Hung, P.-Y., Ho, B.-C., Lee, S.-Y., Chang, S.-Y., Kao, C.-L., Lee, S.-S. et al. (2015) *Houttuynia cordata targets the beginning stage of herpes simplex virus infection*. https://doi.org/10.1371/journal.pone.0115475

Jiangang, F., Ling, D., Zhang, L. & Hongmei, L. (2013) *Houttuynia cordata* Thunb: A review of phytochemistry and pharmacology and quality control. *Chinese Medicine*, 2013(03), 101–123. https://doi.org/10.4236/CM.2013.43015

Jong, T.-T & Jean, M.-Y. (1993) Alkaloids from Houttuyniae Cordata. *Journal of the Chinese Chemical Society*, 40(3), 301–303. https://doi.org/10.1002/JCCS.199300045

Kil, J.-H., Kim, E.-Y., Kong, C.-S. & Kim, S.-H. (2005) Antimutagenic effects of aqueous fraction from Houttuynia cordata in Ames test system. *Journal of Cancer Prevention*, 10(1), 67–71.

Kim, G.S., Kim, D.H., Jeong Ju, L., Lee, J.J., Han, D.Y., Lee, W.M. et al. (2008) Biological and antibacterial activities of the natural herb Houttuynia cordata water extract against the intracellular bacterial pathogen Salmonella within the RAW 264.7 macrophage. *Biological and Pharmaceutical Bulletin*, 31(11), 2012–2017. https://doi.org/10.1248/BPB.31.2012

Kumar, M., Prasad, S. & Hemalatha, S. (2014) A current update on the phytopharmacological aspects of Houttuynia cordata Thunb. *Pharmacognosy Reviews*, 8(15), 22. https://doi.org/10.4103/0973-7847.125525

Kwon, H.-D., Cha, I., Lee, W.-K., Song, J.-H. & Park, I.-H. (1996) Antibacterial activity of volatile flavor components from Houttuynia cordata Thunb. *Journal of Food Science Nutrition*, 1(2), 208–213.

Lau, K.M., Lee, K.M., Koon, C.M., Cheung, C.S.F., Lau, C.P., Ho, H.M. et al. (2008) Immunomodulatory and anti-SARS activities of Houttuynia cordata. *Journal of Ethnopharmacology*, 118(1), 79. https://doi.org/10.1016/J.JEP.2008.03.018

Li, J. & Zhao, F. (2015) Anti-inflammatory functions of Houttuynia cordata Thunb. and its compounds: A perspective on its potential role in rheumatoid arthritis. *Experimental and Therapeutic Medicine*, 10(1), 3–6. https://doi.org/10.3892/ETM.2015.2467/HTML

Li, W., Fan, T., Zhang, Y., Fan, T., Zhou, P., Niu, X. et al. (2013) Houttuynia cordata Thunb. volatile oil exhibited anti-inflammatory effects in vivo and inhibited nitric oxide and tumor necrosis factor-α production in LPS-stimulated mouse peritoneal macrophages in vitro. *Phytotherapy Research*, 27(11), 1629–1639. https://doi.org/10.1002/PTR.4905

Mazraedoost, S., Behbudi, G., Mousavi, S.M. & Hashemi, S.A. (2020) Covid-19 treatment by plant compounds. *Advances in Applied NanoBio-Technologies*, 2(1), 23–33. https://doi.org/10.47277/AANBT/2(1)33

Nuengchamnong, N., Krittasilp, K. & Ingkaninan, K. (2009) Rapid screening and identification of antioxidants in aqueous extracts of Houttuynia cordata using LC–ESI–MS coupled with DPPH assay. *Food Chemistry*, 117(4):750–756.

Pan, X., Li, H., Chen, D., Zheng, J., Yin, L., Zou, J. et al. (2021) *Comparison of essential oils of Houttuynia cordata Thunb. from different processing methods and harvest seasons based on GC-MS and chemometric analysis*. https://doi.org/10.1155/2021/8324169

Schepetkin, I.A. & Quinn, M.T. (2006) Botanical Polysacc.: Macrophage immunomodulation and therapeutic potential. *International Immunopharmacology*, 6(3), 317–333. https://doi.org/10.1016/j.intimp.2005.10.005

Sekita, Y., Murakami, K., Yumoto, H., Amoh, T., Fujiwara, N., Ogata, S. et al. (2016) Preventive effects of Houttuynia cordata extract for oral infectious diseases. *BioMed Research International*, 2016. https://doi.org/10.1155/2016/2581876

Sekita, Y., Murakami, K., Yumoto, H., Mizuguchi, H., Amoh, T., Ogino, S. et al. (2016) Anti-bacterial and anti-inflammatory effects of ethanol extract from Houttuynia cordata poultice. *OUP*, 80(6), 1205–1213. https://doi.org/10.1080/09168451.2016.1151339

Sit, N.W., Chan, Y.S., Chuah, B.L., Cheng, R.J., Leong, W.M. & Khoo, K.S. (2017) Antiviral, antifungal and antibacterial activities of the chinese medicinal plants, houttuynia cordata, lobelia chinensis and selaginella uncinata. *Southeast Asian Journal of Tropical Medicine and Public Health*, 48(3), 616–627.

Wu, X.P., Wang, T.S., Yuan, Z.X., Yang, Y.F. & Wu, H.Z. (2021) Mechanism of compound Houttuynia mixture as an anti-COVID-19 drug based on network pharmacology and molecular docking. *Natural Product Communications*, 16(5). https://doi.org/10.1177/1934578X211016727

Yang, L. & Jiang, J. (2009a) *Bioactive components and functional properties of Hottuynia cordata and its applications*. 47(August 2008), 1154–1161. https://doi.org/10.3109/13880200903019200

Yang, L. & Jiang, J.G. (2009b) Bioactive components and functional properties of Hottuynia cordata and its applications. *Pharmaceutical Biology*, 47(12), 1154–1161. https://doi.org/10.3109/13880200903019200

4 Chemical Constituents and Medicinal Uses of *Eutrema japonicum*

Versha Parcha, Pankaj Bhandari, Sukanya Chhetri, Uday Kumar, and Raju Chandra

ABBREVIATIONS USED

ITCs: Isothiocyanates
AITC: allylisothiocyanate
BITC: 3-butenyl isothiocyanate
5-MPITC: 5-(methylthio) pentylisothiocyanate
6-MHITC: 6-(methylthio) hexyl isothiocyanate
6-MITC: 6-(methylsulphinyl) hexyl isothiocyanate
NMR: Nuclear Magnetic Resonance
HRESIMS: High-Resolution Electrospray Ionization–Mass Spectrometry

4.1 INTRODUCTION

Wasabi (*Eutrema japonicum* (Miq.) Koidz. [syn *Wasabia japonica*]) of the family Brassicaceae is a perennial plant native to Japan and has, since time immemorial, also been used medicinally. The paste of the rhizome is used as one of the ingredients in raw fish sashimi, sushi, and noodle (soba) (Kinae et al., 2000). Wasabi grows naturally in gravel beds in mountain streams in Japan. The *E. japonicum* plant is cultivated in two conventional ways: grown in or water in specially built flooded beds (Chadwick et al., 1993) to create enormous rhizomes for the production of leaves and petioles products. It is cultivated in Japan, China, Taiwan, and New Zealand. Regular consumption of *E. japonicum* may contribute to the longevity and health of the Japanese. By adding unique flavour, greenish colour and heat to foods, wasabi is a highly valued plant in Japanese cuisine. Now, scientific studies are being done to analyse *E. japonicum*, a superfood and effective herbal remedy.

Possessing a certain degree of firmness and freshness with a pungent smell is characteristic of *E. japonicum* but the heat element is different from what we find across chillies since it is quickly dissipated as soon as it is in the mouth and ends up with a delicious vegetable-like taste, without a fiery taste. It is considered a common condiment in the Japanese diet since the elegant appetizing appeal of a variety of foods is also achieved by *E. japonicum*. Using the aesthetic contrast of the light green colour, it may transform an everyday meal into something exceptional and attractive, making it a new flavouring ingredient for the rest of the globe, so that it has become very popular as a spice in Western *cordon bleu* cooking.

Undoubtedly, this herb has a consistent flavour but it can vary in the level of crispness achieved after consumption (Sultana et al., 2000, 2003) and can be adapted for many purposes. It is a potential consumable in wasabi-flavour processed foods, or as a savoury condiment, or alongside a dish, or in a dish. Rhizomes are used to prepare a fresh paste that can be spread on raw fish. Sometimes, soya sauce and vinegar are mixed with grated wasabi to prepare a dip for use.

DOI: 10.1201/b22924-4

Wasabi has pungent characteristics because of isothiocyanates (ITCs) such as AITC and is therefore utilized as a spice or pickling medium in Japanese meals (Yamasaki et al., 2013). Higher concentrations of ITCs, especially long-chain ITCs, are characteristic of this plant, contributing to its role as a herbal medicine. The two major components, 6-MITC and AITC, are reported to have anticancer (Manesh et al., 2003; Fuke et al., 2006, 2014; Chen et al., 2014; Lee et al., 2018; Yano et al., 2018), antioxidant (Mizuno et al., 2011), anti-inflammatory (Kim et al., 2018), neuroprotective (Subedi et al., 2017), and antimicrobial potential. Li et al. (2010) reported that other phytochemicals of *E. japonicum* also exhibit antioxidant, anticancer, anti-inflammatory, and antifungal activity. The rhizome is the most significant component of the plant since it contains the most molecules (ITCs) that are responsible for the flavours of wasabi..

4.1.1 Geographical Distribution

Wasabi is native to Japan but is cultivated widely in China and Taiwan. The wasabi rhizomes are also cultivated in rainforest regions of the United States. It is best grown in a humid and damp temperate environment. Wasabi plants take up to three years to grow to marketable size. Wasabi roots are available year-round.

4.1.2 Botanical Description

Botanical name: *Eutrema japonicum*

Family: Brassicaceae

Order: Brassicales

Wasabi is a perennial plant that grows very slowly, reaching a height of about 25–65 cm.

Leaves: The leaves are large, circular to kidney-shaped, and borne on leaf stalks

Flower: Small clusters of white flowers are formed, each with four petals arranged in a cross shape. The fruits are dry capsules.

Stem: Each leaf has a single stem, and they are edible.

Roots: The rhizomes, being fleshy, can grow up to about 1.5 inches/4 cm in diameter. Wasabi root is the long, knobby rhizome of a branchless plant. A mature wasabi rhizome is cylindrical, brown- and green-skinned, and tapered. (Figure 4.1 and Figure 4.2)

FIGURE 4.1 *Eutrema japonicum.*

FIGURE 4.2 (A) Rhizome and (B) kidney-shaped leaves of *Eutrema japonicum* with small white flowers.

4.2 CHEMICAL CONSTITUENTS

A literature survey revealed the presence of five main active phytochemical constituents which are volatile, each in conjunction with 2-butenolide derivatives (**1**) and (**17**), so the recognized compounds (**2–18**) were extracted, isolated and identified using conventional NMR and HRESIMS instrument analysis. These active compounds are AITC, BITC, 5-MPITC, 6-MHITC, and 6-MITC (Etoh et al., 1990; Kumagai et al., 1994; Ono et al., 1996; Fuke et al., 1997). The 17 known compounds (**2–18**) were identified as methyl 10-oxodecanoate (**2**), monomethylsebacate (**3**), azelaic acid (**4**), dimethyl azelate (**5**), dimethyl succinate (**6**), 4-isothiocyanatobutanoic acid (**7**), octadecanoic acid (**8**), methyl stearate (**9**), (10*E*)-12-oxo-10-octadecenoic acid methyl ester (**10**), methyl 11-oxo-9-dodecenoate (**11**), (*S*)-5-hydroxy-3,4-dimethyl-5-pentylfuran-2(5*H*)-one (**12**), 6,10,14-rimethyl-2-pentadecanone (**13**), 5-methyl-5-(4,8,12-trimethyl-tridecyl)-dihydro-furan-2-one (**14**), α-tocospiro A (**15**), stigmast-4-en-3-one (**16**), β-sitosterol (**17**), and 7-oxo-β-sitosterol (**18**), by comparing their spectroscopic data with those from standards (Park et al., 2022). Figure 4.3

4.3 HEALTH BENEFITS AND MEDICINAL VALUES

History reveals that 'The Honzowamyou', Japan's oldest botanical script, compiled in 918, is thought to be the first to mention *E. japonicum* being cultivated as a medicinal plant, the Japanese using it to prevent food poisoning from raw fish, unaware of the mechanism behind it. Soon, it became the choice of higher-class people. The rhizome is the most desired part of the plant but the whole plant is edible; in fact, all parts are highly appreciated in gastronomy. Wasabi leaf is a delicacy of this plant. Raw leaf, as in tempura or blanched, retains its wasabi flavour and that characteristic touch of spice. The flower also adds the same flavours but is more subtle and sweet. Conventionally, *E. japonicum* has been customarily used to ameliorate poor blood circulation and rheumatic arthralgia, operating as an effective pain reliever (Subedi et al., 2017).

E. japonicum has a nutritional value. A serving of 28 g fresh wasabi contains 30.5 kcals and contains 9% DV (daily value) of dietary fibre, 3% protein, and 0.3% fat, together with minerals, like zinc, calcium, magnesium, iron, manganese, phosphorus, and potassium which are essential to keep your gut and heart healthy. Additionally, it is a nutrient powerhouse, supplying vitamins A, B_1, B_2, B_3, B_6, and C, with 70% of the daily necessary amount of vitamin C being available from

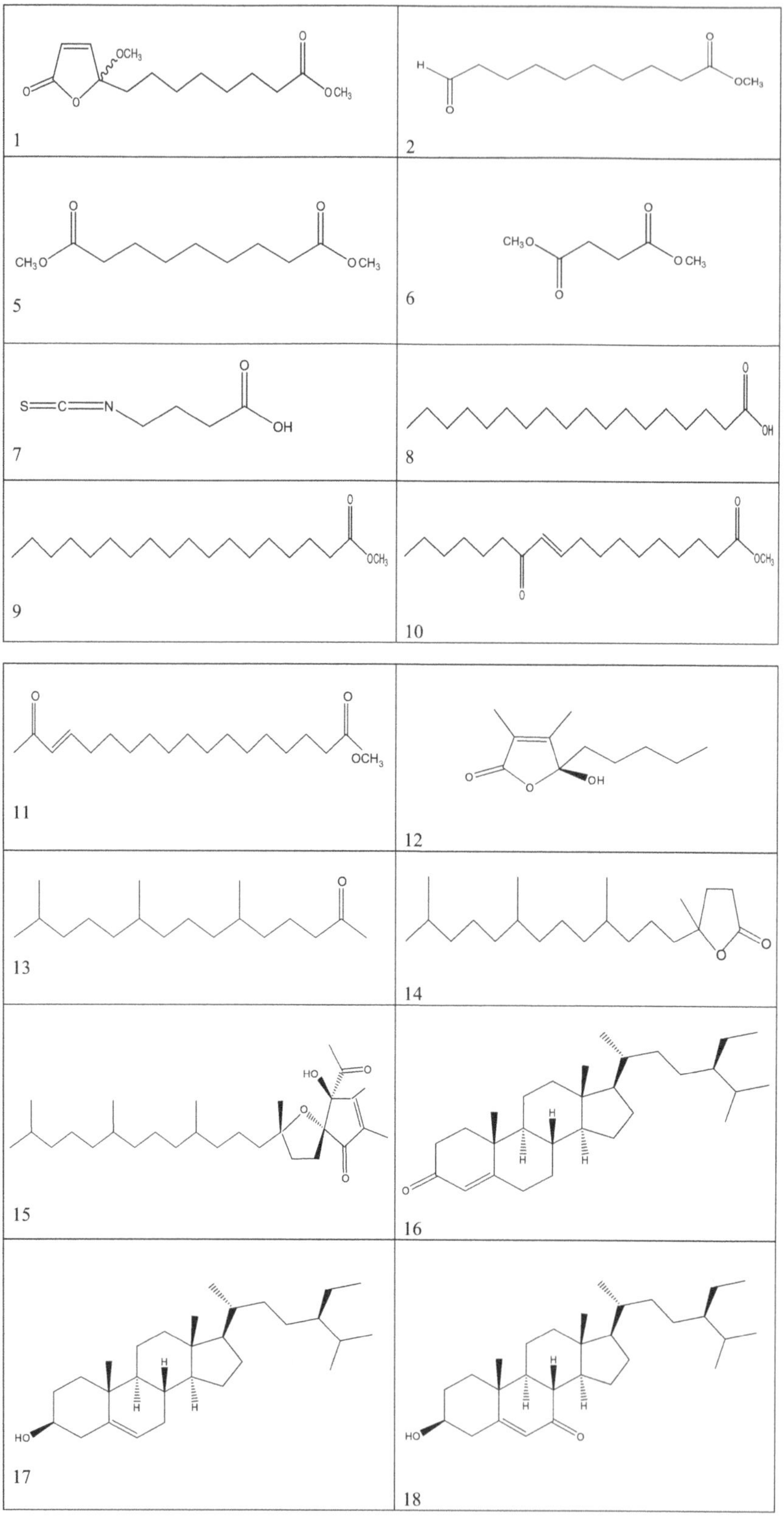

FIGURE 4.3 Phytochemical constituents present in rhizome of *Eutrema japonicum* (Park et al., 2022).

100 g of wasabi. Primarily, however, the pharmacophore of the plant is the isothiocyanates and *p*-hydroxycinnamic acid derivatives (Yamaguchi et al., 2016).

Because it contains high levels of isothiocyanates, which are supposed to possess tremendous medicinal properties, wasabi is found to be anti-hypercholesterolaemic and combats colds and allergies. Additionally, *W. japonica* possesses another ITC, sinigrin, an antibiotic that helps to alleviate inflammation, and thin mucus, reduce congestion, and strengthen the immune system. *E. japonicum*, being fibre-rich, helps to remove toxins, reduce constipation symptoms and regulate movements of the bowel. To boost liver and gallbladder function, it promotes bile (Thomaz et al., 2020). Excess intake may cause diarrhoea, nausea, or acid reflux in some individuals. To keep us healthy, a natural detoxifier like *E. japonicum* can be used which acts by expelling toxins from liver tissues and the digestive system. Anti-ageing benefits of wasabi are due to the presence of a sulphinyl compound, namely 6-MSITC, which aids the body in reducing reactive oxygen species, which cause many aberrant cell divisions and overall body weakness. 6-MSITC also inhibits allergies due to pollen, lowers hypertension, and improves reflexes and circulation. Potassium-rich foods are recommended for the regulation of blood pressure and blood flow and *E. japonicum* is one of them. Potassium deficiency increases the risk of heart disease, cardiac arrest, and hypertension. *E. japonicum* is the appropriate solution since it has low salt and high potassium content, therefore assisting to relax blood vessels, improve blood flow, and safeguard against hypertension. Phytochemicals in *E. japonicum* extracts are capable of killing cells of lung, breast, colon, pancreatic, oral, blood, and skin cancers, as well as bone metastases (Nomura et al., 2005; Fuke et al., 2006; Chen et al., 2014; Lee et al., 2018; Yano et al., 2018, 2019). Extracts have shown inhibitory effects on bacteria that cause food poisoning, respiratory disease, dental caries, and duodenal ulcers also preventing yeasts, fungi and moulds from spreading (Kirlin et al., 1999; Hou et al., 2000; Morimitsu et al., 2000). Wasabi extracts are also found to inhibit platelet aggregation, protecting against heart attack, and preventing a range of inflammatory conditions (Mochid et al., 2008), as well as being a potential weight loss promoter. Comparatively high levels of isosaponarin in *E. japonicum* also enhance collagen synthesis and stimulate hair growth (Nagai et al., 2010). Studies also indicate that wasabi extracts had antiviral activity towards influenza (Nagai et al., 2010). Researchers are working on two major phytochemicals in *E. japonicum* with computational molecular docking data from the SARS-Cov2 genome in an attempt to develop a blended tea for the treatment of long COVID conditions (Ina et al., 1989). The effects of *E. japonicum* in the body are now better understood as a result of research on the plant and traditional wisdom. Because of the high percentage of natural components in *E. japonicum,* many other biomedical applications for wasabi are also suggested. Many such activities are discussed below:

4.3.1 Antimicrobial Effect

Wasabi rhizome paste is reported to show the potential to inhibit the growth of microbes responsible for food poisoning, like *Escherichia coli, Staphylococcus aureus* and *Vibrio parahaemolyticus.* The AIT volatile component also exhibited an antimicrobial effect against bacteria responsible for dental caries, such as *Streptococcus mutans* (Masuda et al., 2000). These findings encourage the use of *E. japonicum* extracts in toothpaste and mouthwashes to prevent cavities. Studies also showed that ITCs hinder the bacterium *Helicobacter pylori* responsible for stomach ulcers and subsequent stomach cancers (Shin et al., 1998; Masuda et al., 2004; Haristoy et al., 2005). Even if the bacterium has invaded the stomach lining, wasabi extracts have been demonstrated to impair its growth, preventing ulcers and stomach cancer.

4.3.2 Anticancer Effect

The ITC constituents of wasabi are also known for their anticarcinogenic properties by inhibiting enzymes that transform pre-carcinogenic compounds into carcinogenic compounds (Phase I enzymes), similar to glutathione S-transferase (Morimitsu et al., 2000); they may also be able to inhibit the development of cancerous growths by stimulating detoxifying Phase II enzymes (Hecht et al., 1999; Kirlin et al., 1999; Gao et al., 2004; Yano et al., 2018, 2019). In a recent study, compounds isolated from the rhizome extract of *E. japonicum* significantly reduced the proliferation of cells of the A549 and SK-MEL-2 cancer cell lines (Park et al., 2022). Further findings revealed that an extract of the rhizome of *E. japonicum* had anti-colon cancer capabilities due to its potential to stimulate apoptosis and autophagy, resulting in its use as a potential therapy for colon cancer (Hsuan et al., 2016). Additionally, wasabi extracts induce cell proliferation or autophagy in cancerous cells. This property was observed in breast cancer cells (Nomura et al., 2005), leukaemia cells (Fimognari et al., 2004; Nakamura et al., 2001), colorectal cancer cells (Lund et al., 2001), and lung cancer cells (Kuang et al., 2004). Another study revealed that antioxidant and/or radical scavenger properties are exhibited by (–)-(R) methylsulphinylalkyl isothiocyanates isolated from wasabi rhizome. Thus, wasabi extracts might be potent inhibitors of human carcinogenesis (Kinae et al., 2000).

4.3.3 Anti-inflammatory Effect

An inflammatory substance is any drug that reduces inflammation. It is well known that wasabi exhibits anti-inflammatory properties. 6-MITC impedes lipoxygenase, cyclooxygenase, and cAMP phosphodiesterases, all of which are implicated in inflammation. Depree et al. (1998) reported that the isothiocyanates of *E. japonicum* have anti-inflammatory properties, whereas Dorsch et al. (1984) reported anti-asthmatic potential. Wasabi extract is also useful for seasonal allergies and asthma. Park et al. (2022) observed that one of the compounds from *W. japonica* rhizomes suppressed nitric oxide (NO) generation in lipopolysaccharide-stimulated BV-2 cells, exhibiting substantial anti-neuroinflammatory potential.

4.3.4 Antidiabetic Effect

In *in-vivo* studies, 50 mg wasabi leaf per kg bodyweight markedly reduced diabetes and other associated implications, such as obesity, hyperlipidaemia, and non-alcoholic fatty liver disease (NAFLD) (Choi et al., 2020). 6-MITC has been identified as useful in type 2 diabetes treatment. It has been suggested that the consumption of functional foods containing *E. japonicum* could benefit people suffering from type 2 diabetes.

4.3.5 Other Health Effects

In addition, *E. japonicum* and its isothiocyanates have other biological activities associated with them, such as antidiarrhoea activity (Nakayama et al., 1998) and a cardiovascular function (Wu et al., 2004). It also has platelet aggregation activity.

4.4 SIDE-EFFECTS

- *Bad for gastritis*: Isothiocyanate which is the main component of *W. japonica* irrithes your mucous lining of the stomach and can lead to gastritis and cause ulcers.
- *Allergies*: Consuming excess amounts of *E. japonicum* may cause a number of allergies.
- *Slows down blood clotting*: Consumption of wasabi slows down coagulum. Isothiocyanates can even destroy sensitive blood vessels.
- *Cirrhosis of the liver*: The presence of compounds responsible for liver damage in *E. japonicum* is not harmful in small doses. However, if consumed in relatively high concentrations, it might cause deterioration of the liver's function.
- *Acid reflux*: Because of its highly flavoured, aromatic profile, wasabi can cause acid reflux. If one has digestive disorders, stomach ulcers, gastritis, or heartburn, wasabi should not be taken at all. It should also not be given to pregnant women.

4.5 CONCLUSION AND FUTURE PERSPECTIVES

E. japonicum rhizomes and leaves have been used as 'Japanese medicine' for a long time and are considered to be associated with people's good health. As an oral medicine, *E. japonicum* is effective for asthma and respiratory problems. It is also used to treat urinary tract stones, breathlessness, and pneumonia, as well as being valued as a 'stomachic' (appetite stimulant), and antiseptic, as well as having antibacterial properties. A total of 18 bioactive compounds have been isolated from wasabi, among which are AITC, BITC, 5-MPITC, 6-MHITC, and 6-MITC. ITCs present in the rhizome of the plant have anticancer, anti-inflammatory, antioxidant, antidiabetic, antiparasitic, and antimicrobial properties. Wasabi also has nutritional value and is found to contain fibre, protein, calcium, magnesium, vitamin B_6, and vitamin C at acceptable concentrations to represent a nutritional supplement. Because of the high concentrations of natural bioactives in *E. japonicum,* many biomedical applications are also suggested. Very few negative side-effects have been associated with its prolonged use, although it is not recommended for people who are suffering from gastritis or stomach ulcer as it has strong stimulating properties. As a consequence of its beneficial effects, the cultivation of wasabi is increasing in many countries, as it is the highest-value legal crop for the food market alone. Since *E. japonicum* is of tremendous medicinal value, research into the mechanisms of action of bioactive constituents isolated from different parts of the plant will expand our insights into the medicinal functions of this plant and its preparations.

REFERENCES

Chadwick, C.I., Lumpkin, T.A. & Elberson, L.R. (1993) The botany uses and production of *Wasabi japonica* (Miq.) (Cruciferae) Matsum. *Economic Botany*, 47(2), 113–135.

Chen, Y.J., Huang, Y.C., Tsai, T.H. & Liao, H.F. (2014) Effect of wasabi component 6-(methylsulfinyl) hexyl isothiocyanate and derivatives on human pancreatic cancer cells. *Evidence-Based Complementary and Alternative Medicine* http://dx.doi.org/10.1155/2014/494739.

Choi, B.R., Kim, H.J., Lee, Y.J. & Ku, S.K. (2020) Anti-diabetic obesity effects of *Wasabi japonica* Matsum leaf extract on 45% kcal high-fat diet-fed mice. *Nutrients*, 12(9), 2837.

Depree, J.A., Howard, T.M. & Savage, G.P. (1998) Flavour and pharmaceutical properties of the volatile sulphur compounds of Wasabi (*Wasabi japonica*). *Food Research International*, 31(5), 329–337.

Dorsch, W., Adam, O., Weber, J. & Ziegeltrum, T. (1984) Antiasthmatic effects of onion extracts-detection of benzyl-and other isothiocyanates (mustard oils) as antiasthmatic compounds of plant origin. *European Journal of Pharmacology*, 107(1), 17–24.

Etoh, H., Nishimura, A., Takasawa, R., Yagi, A., Satio, K., Sakata, K. et al. (1990) ω -Methylsulfinylalkyl Isothiocyanates in Wasabi, *Wasabia japonica* Matsum. *Agricultural and Biological Chemistry*, 54, 1578–1589.

Fimognari, C., Nüsse, M., Berti, F., Iori, R., Cantelli-Forti, G. & Hrelia, P. (2004) A mixture of isothiocyanates induces cyclin B1-and p53-mediated cell-cycle arrest and apoptosis of human T lymphoblastoid cells. *Mutation Research/Fundamental and Molecular Mechanisms of Mutagenesis*, 554(1–2), 205–214.

Fuke, Y., Haga, Y., Ono, H., Nomura, T. & Ryoyama, K. (1997) The anti-carcinogenic activimethyl sulfinyl hexylnylhexyl isothiocyanate-, is an active anti-proliferative principal of wasabi (Eutrema wasabi Maxim.). *Cytotechnology*, 25(1), 197–203.

Fuke, Y., Hishinuma, M., Namikawa, M., Oishi, Y. & Matsuzaki, T. (2014) Wasabi-derived 6-(methylsulfinyl) hexyl isothiocyanate induces apoptosis in human breast cancer by the possible involvement of the NF-κB pathways. *Nutrition and Cancer*, 66(5), 879–887.

Fuke, Y., Shinoda, S., Nagata, I., Sawaki, S., Murata, M., Ryoyama, K. et al. (2006) Preventive effect of oral administration of 6-(methylsulfinyl) hexyl isothiocyanate derived from wasabi (*Wasabi japonica* Matsum) against pulmonary metastasis of B16-BL6 mouse melanoma cells. *Cancer Detection and Prevention*, 30(2), 174–179.

Gao, X. & Talalay, P. (2004) Induction of phase 2 genes by sulforaphane protects retinal pigment epithelial cells against photooxidative damage. *Proceedings of the National Academy of Sciences*, 101(28), 10446–10451.

Haristoy, X., Fahey, J.W., Scholtus, I. & Lozniewski, A. (2005) Evaluation of the antimicrobial effects of several isothiocyanates on helicobacter pylori. *Planta Medica*, 71(04), 326–330.

Hecht, S.S. (1999) Chemoprevention of cancer by isothiocyanates, modifiers of carcinogen metabolism. *The Journal of Nutrition*, 129(3), 768S–774S.

Hou, D.X., Fukuda, M., Fujii, M. & Fuke, Y. (2000) Transcriptional regulation of nicotinamide adenine dinucleotide phosphate: Quinone oxidoreductase in murine hepatoma cemethylsulfonylhylsufinyl) hexyl isothiocyanate, an active principle of wasabi (Eutrema wasabi Maxim). *Cancer Letters*, 161(2), 195–200.

Hsuan, S.W., Chyau, C.C., Hung, H.Y., Chen, J.H. & Chou, F.P. (2016) The induction of apoptosis and autophagy by *Wasabi japonica* extract in colon cancer. *European Journal of Nutrition*, 55(2), 491–503.

Ina, K., Ina, H., Ueda, M., Yagi, A. & Kishima, I. (1989) ω-Methylthioalkyl isothiocyanates in wasabi. *Agricultural and Biological Chemistry*, 53(2), 537–538.

Kim, M.W., Choi, S., Kim, S.Y., Yoon, Y.S., Kang, J.H. & Oh, S.H. (2018) Allyl isothiocyanate ameliorates dextran sodium sulfate-induced colitis in the mouse by enhancing tight junction and mucin expression. *International Journal of Molecular Sciences*, 19(7), 2025.

Kinae, N., Masuda, H., Kaneko, S., Furugori, M. & Shimoi, K. (2001, April) Comparison of the inhibitory effect of Sawa-wasabi (*Wasabi japonica*) on the growth of Helicobacter pylori. In *Abstracts of Papers of the American Chemical Society* (Vol. 221, pp. U36–U36). Washington, DC: American Chemical Society.

Kinae, N., Masuda, H., Shin, I.S., Furugori, M. & Shimoi, K. (2000) Functional properties of wasabi and horseradish. *Biofactors*, 13(1–4), 265–269.

Kirlin, W.G., Cai, J., DeLong, M.J., Patten, E.J. & Jones, D.P. (1999) Dietary compounds that induce cancer preventive phase 2 enzymes activate apoptosis at comparable doses in HT29 colon carcinoma cells. *The Journal of Nutrition*, 129(10), 1827–1835.

Kuang, Y.F. & Chen, Y.H. (2004) Induction of apoptosis in a non-small cell human lung cancer cell line by isothiocyanates is associated with P53 and P21. *Food and Chemical Toxicology*, 42(10), 1711–1718.

Kumagai, H., Kashima, N., Seki, T., Sakurai, H., Ishii, K. & Ariga, T. (1994) Analysis of volatile components in the essential oil of upland wasabi and their inhibitory effects on platelet aggregation. *Bioscience, Biotechnology, and Biochemistry*, 58(12), 2131–2135.

Lee, M.J., Tseng, W.S., Lai, J.C.Y., Shieh, H.R., Chi, C.W. & Chen, Y.J. (2018) Differential pharmacological activities of oxygen numbers on the sulfoxide moiety of wasabi compound 6-(methylsulfinyl) hexyl isothiocyanate in human oral cancer cells. *Molecules*, 23(10), 2427.

Li, L., Lee, W., Lee, W.J., Auh, J.H., Kim, S.S. & Yoon, J. (2010) Extraction of allyl isothiocyanate from wasabi (*Wasabi japonica* Matsum) using supercritical carbon dioxide. *Food Science and Biotechnology*, 19(2), 405–410.

Lund, E.K., Smith, T.K., Clarke, R.G. & Johnson, I.T. (2001) Cell death in the colorectal cancer cell line HT29 in response to glucosinolate metabolites. *Journal of the Science of Food and Agriculture*, 81(9), 959–961.

Manesh, C. & Kuttan, G. (2003) Effect of naturally occurring allyl and phenyl isothiocyanates in the inhibition of experimental pulmonary metastasis induced by B16F-10 melanoma cells. *Fitoterapia*, 74(4), 355–363.

Masuda, H., Kinae, N., Shin, I.S., Furugori, M. & Shimoi, K. (2000) Functional properties of Wasabi and horseradish. BioFactors, 13(1–4), 265–269.

Masuda, H., Shin, I.S. & Naohide, K. (2004) Bactericidal activity of wasabi (Wasabia japonica) against Helicobacter pylori. *International Journal of Food Microbiology*, 94, 255–261.

Masuda, S., Masuda, H., Shimamura, Y., Sugiyama, C. & Takabayashi, F. (2017) Improvement effects of wasabi (*Wasabi japonica*) leave and allyl isothiocyanate on stomach lesions of Mongolian gerbils infected with Helicobacter pylori. *Natural Product Communications*, 12(4), 1934578X1701200431.

Mizuno, K., Kume, T., Muto, C., Takada-Takatori, Y., Izumi, Y., Sugimoto, H. et al. (2011) Glutathione biosynthesis via activation of the nuclear factor E2–related factor 2 (Nrf2)–antioxidant-response element (ARE) pathway is essential for neuroprotective effects of sulforaphane and 6-(methylsulfinyl) hexyl isothiocyanate. *Journal of Pharmacological Sciences*, 1102230521–1102230521.

Mochida, K. & Ogawa, T. (2008) Anti-influenza virus activity of extract of Japanese wasabi leaves discarded in summer. *Journal of the Science of Food and Agriculture*, 88(10), 1704–1708.

Morimitsu, Y., Hayashi, K., Nakagawa, Y., Fujii, H., Horio, F., Uchida, K. et al. (2000) Antiplatelet and anticancer isothiocyanates in Japanese domestic horseradish, wasabi. *Mechanisms of Ageing and Development*, 116(2–3), 125–134.

Nagai, M., Akita, K., Yamada, K. & Okunishi, I. (2010) The effect of isosaponarin isolated from wasabi leaf on collagen synthesis in human fibroblasts and its underlying mechanism. *Journal of Natural Medicines*, 64(3), 305–312.

Nakayama, H., Suzuki, T. & Suzuki, Y. (1998) Effect of wasabi on Ion secretion in guinea pig colon. *Journal of the Agricultural Chemical Society of Japan (Japan)*, 72, 499–507.

Nomura, T., Shinoda, S., Yamori, T., Sawaki, S., Nagata, I., Ryoyama, K. et al. (2005) Selective sensitivity to wasabi-derived 6-(methylsulfinyl) hexyl isothiocyanate of human breast cancer and melanoma cell lines studied in vitro. *Cancer Detection and Prevention*, 29(2), 155–160.

Ono, H., Adachi, K., Fuke, Y. & Shinohara, K. (1996) Purification and structural analysis of substances in wasabi (*Eutrema wasabi* Maxim.) that suppress the growth of MKN-28 human stomach cancer cells. *Nippon Shokuhin Kogyo Gakkai-Shi*, 43(10), 1092–1097.

Park, J.E., Lee, T.H., Ham, S.L., Subedi, L., Hong, S.M., Kim, S.Y. et al. (2022) Anticancer and anti-neuroinflammatory constituents isolated from the roots of *Wasabi japonica*. *Antioxidants*, 11(3), 482.

Shin, I.S. & Lee, J.M. (1998) Study on antimicrobial and antimutagenic activity of horseradish (*Wasabi japonica*) root extracts. *Korean Journal of Fisheries and Aquatic Sciences*, 31(6), 835–841.

Subedi, L., Venkatesan, R. & Kim, S.Y. (2017) Neuroprotective and anti-inflammatory activities of allyl isothiocyanate through attenuation of JNK/NF-κB/TNF-α signaling. *International Journal of Molecular Sciences*, 18(7), 1423.

Sultana, T.A.M.A.N.N.A., Savage, G.P., McNeil, D.L., Porter, N.G. & Martin, R.J. (2000) Flavour components in the rhizome of soil-grown wasabi. Proceedings of the Nutrition Society, 25, 95–106.

Sultana, T., Savage, G.P., McNeil, D.L., Porter, N.G. & Clark, B. (2003) Comparison of flavour compounds in wasabi and horseradish. *Journal of Food, Agriculture and Environment*, 1, 117–121.

Thomaz, F.S., Tomsett, K.I., Panchal, S.K., Worrall, S. & Nitert, M.D. (2020) Wasabi supplementation alters the composition of the gut microbiota of diet-induced obese rats. *Journal of Functional Foods*, 67, 103868.

Wu, L., Ashraf, M.H.N., Facci, M., Wang, R., Paterson, P.G., Ferrie, A. et al. (2004) Dietary approach to attenuate oxidative stress, hypertension, and inflammation in the cardiovascular system. *Proceedings of the National Academy of Sciences*, 101(18), 7094–7099.

Yamaguchi, M. (2016) The botanical molecule p-hydroxycinnamic acid as a new osteogenic agent: Insight into the treatment of cancer bone metastases. *Molecular and Cellular Biochemistry*, 421(1), 193–203.

Yamasaki, M., Ogawa, T., Wang, L., Katsube, T., Yamasaki, Y., Sun, X. et al. (2013) Anti-obesity effects of hot water extract from wasabi (*Wasabi japonica* Matsum.) leaves in mice fed high-fat diets. *Nutrition Research and Practice*, 7(4), 267–272.

Yano, S., Wu, S., Sakao, K. & Hou, D.X. (2018) Wasabi 6-(methylsulfinyl) hexyl isothiocyanate induces apoptosis in human colorectal cancer cells through the 53-independent mitochondrial dysfunction pathway. *Biofactors*, 44(4), 361–368.

Yano, S., Wu, S., Sakao, K. & Hou, D.X. (2019) Involvement of ERK1/2-mediated ELK1/CHOP/DR5 pathway in 6-(methylsulfonyl) hexyl isothiocyanate-induced apoptosis of colorectal cancer cells. *Bioscience, Biotechnology, and Biochemistry*, 83(5), 960–969.

5 *Boesenbergia pandurata* *An Overview of Medicinal Use, Ethnopharmacology, and Phytochemistry*

Addisu Tamir Wassie, Archana Bachheti, Azamal Husen, and Rakesh Kumar Bachheti

5.1 INTRODUCTION

Since ancient times, people have identified and used therapeutic plants, often known as medicinal herbs, for medical purposes (Moges, 2019; Beshah et al., 2020). Plant-based medicines are still widely used all over the world. Most people in the developing world, especially in rural areas, use complementary and alternative medicines, local traditional medicines which generally act as the only source of health care. The bulk of items made from medicinal plants in the traditional way has yet to have their efficacy and safety clinically assessed. Furthermore, the quality of traditional healers' products in developing countries and those sold in industrialized economies varied and occasionally contained hazardous contaminants (Chan, 2015). An estimated 85% of the world's population trusts herbal medicines to maintain their health, whereas pharmaceutical companies rely on them as a source for novel drug research. Studies show that 80% of all synthetic pharmaceuticals are produced, directly or indirectly, from medicinal herbs, making such plants important even today. Based on organoleptic assessments of compatibility and quality, people have recognized and chosen food and medicinal plants for thousands of years. However, the quick transition from utilizing smell, touch, and sight to employing sophisticated technology has only occurred in the past seven decades with the introduction of key analytical methods, such as paper chromatography, gas chromatography and mass spectrometry (Fitzgerald et al., 2020; Abate et al., 2021; Asfaw et al., 2022).

Powerful and reliable medicines have long been derived from plants. Nowadays, medicinal plants are essential to the medical sector and provide ecological, economic, and cultural advantages. In order to defend themselves against herbivorous mammals and insects, and fungal and bacterial pathogens, among other things, plants produce many chemical substances. These chemical substances from medicinal plants are rich sources of components and can be used to make pharmaceutical, non-pharmaceutical, and synthetic remedies (Khan, 2016; Abate et al., 2022). The main approaches to treating illnesses and preventing infections worldwide have their roots in the use of medicinal herbs. The World Health Organization (WHO) estimates that close to 3.5 billion persons in developing nations regularly use plant remedies because they are confident in their efficacy (Moges, 2019). Despite the global population's trust in the therapeutic benefits of conventional medications, culturally relevant practices, and their moderately lower price, medicinal plants are in high demand in developing nations. Due to poor health care facilities, a lack of access to synthetic pharmaceuticals, and a staffing shortage in clinics, people in developing countries, especially those from low-income communities and those who reside in rural regions, may turn to traditional health centres, which raises the need for medicinal plants (Bekele & Reddy, 2015).

 DOI: 10.1201/b22924-5

Boesenbergia pandurata is a perennial herb of the Zingiberaceae, which is used for medicinal purposes (Figure 5.1). Due to its resemblance of its rhizomes in shape to human fingers, this plant is widely known as a fingerroot (Choi et al., 2020). This geophytic shrub is extensively farmed across Indonesia, especially in the teak forests. There are numerous synonyms with *B. pandurata* exist, including *Boesenbergia rotunda* Linn. Mansft, *Curcuma rotunda* L., *Gastrochilus panduratum* Ridl., and *Kaempferia pandurata* Roxb., (Chahyadi et al., 2014). Vietnam, Thailand, Myanmar, Indonesia, and Malaysia are among the Southeast Asian countries that cultivate it (Nguyen et al., 2022). The young rhizome of *B. pandurata*, known as 'Ngai bun' in Vietnam, is particularly popular for seasoning vegetables and is used as a spice (Lee et al., 2022; Nguyen et al., 2017).

5.1.1 Importance of *Boesenbergia pandurata*

Historically, people have eaten uncooked *B. pandurata* roots or used them to flavour drinks or other foods. Sliced rhizomes, when chewed with areca nut (*Areca catechu*), is used as a traditional remedy to alleviate asthma and a dry cough. Bloated stomach in children was treated with *B. pandurata* rhizome and *Pimpinella anisum* combined as a food and used as a diuretic and anti-anthelmintic. In contrast, *B. pandurata* rhizome and coconut milk were combined as a remedy for worm infestation (Chahyadi et al., 2014). *B. pandurata* has been associated with many biological roles, including antioxidant, anti-inflammatory, anti-obesity, and antibacterial properties (Hardi et al., 2016). Panduratin A, a chalcone molecule, has been discovered to be the primary bioactive component of *B. pandurata*. Panduratin A has been shown to have anti-obesity (Kim et al., 2018), anticancer (Karmakar et al., 2017), antimicrobial, and skin-whitening properties. *In-vitro* and *in-vivo* studies indicated that panduratin A has a particular antiphotoageing effect. *B. pandurata* may be effective when used orally to improve skin function (Kim et al., 2017). Using dermal fibroblasts and hairless mice, researchers have also investigated the antiphotoageing properties of *B. pandurata* extract (BPE) (Woo et al., 2015). Additionally, women's uterine irritation and vaginal infections were treated using the rhizome of *B. pandurata* in conjunction with other herbs and spices (Chahyadi et al., 2014). As well as being used to cure dysmenorrhoea, fatigue, and flatulence (Nguyen et al., 2022)., the rhizome of *B. pandurata* is also utilized as folk medicine in Cambodia, Vietnam, the People's Republic of China, and Laos. It also stimulates bile secretion (Nguyen et al., 2017). Functions of *B. pandurata* may be attributed to specific components, including alpinetin, pinostrobin, panduratin A, quercetin,

FIGURE 5.1 *Boesenbergia pandurata* morphology. *B. pandurata* whole plant (A), *B. pandurata* flower (B), and *B. pandurata* rhizome (C).

cardamonin, pinocembrin, kaempferol, and naringin. Enhancing skin hydration, gloss, and suppleness, and preventing wrinkling are all benefits of *B. pandurata* extract (Kim et al., 2017). In light of these findings, *B. pandurata* is an excellent natural antioxidant option, and it could help the food industry meet the increasing demand for environmental, social, and governance management (Lee et al., 2022). The purpose of this chapter is to condense several well-researched pharmacological investigations that have been conducted on the *B. pandurata* plant and to provide information about the chemical compounds that makes this plant bioactive. Another goal is to briefly summarize various potential extraction techniques that might be used to isolate biologically active chemicals from *B. pandurate*, using multiple solvents, and identify significant phytochemicals structure present in *B. pandurata*.

5.2 ETHNOPHARMACOLOGY USE OF *B. PANDURATA*

In several Asian nations, including Malaysia, Thailand, Indonesia, India, and China, *B. pandurata* is a popular food component. Due to its aromatic flavour, which stimulates appetite, it is typically grown on tiny home farms and used as a condiment in foods like curries and soup. Muscle pain, rheumatism, gastrointestinal disorders, fever, gout, flatulence, stomach discomfort, peptic ulcer, and dyspepsia are traditionally treated using *B. pandurata* (Chahyadi et al., 2014; Kim et al., 2017). *B. pandurata* is commonly used in Indonesia to make 'jamu', a well-liked traditional remedy for women after childbirth, a cosmetic aid for adolescent girls, and to prevent leucorrhea. The fresh rhizome of *B. pandurata* is used as a diuretic to treat inflammatory ailments, such as dental caries, dermatitis, dry coughs and colds, teeth and gum diseases, swelling, wounds, diarrhoea, and dysentery. Additionally, it is employed as an anti-scabies agent to ease itching caused by mite bites, an antifungal and antiparasitic agent to treat fungal infections, and as an antihelmintic for roundworm removal for treatment for the human intestine (Kim et al., 2018). Thai people utilize this herb for aphrodisiac purposes. Furthermore, it has been demonstrated that eating its leaves helps treat food poisoning and allergies. In Thailand, AIDS patients have also taken *B. pandurata* for self-medication (Eng-Chong et al., 2012). Cyclohexenyl chalcone, a racemate of panduratin A, was initially isolated from *B. pandurata* in 1984. Some of the beneficial biological properties of panduratin A include anti-inflammatory, anticancer, anti-angiogenic, antibacterial, anti-metalloproteinase-9, antioxidant, anti-HIV, and competitive dengue-2 virus NS3 protease inhibitory action (Kim et al., 2017, 2018; Thy et al., 2018).

5.3 PHYTOCHEMISTRY OF *B. PANDURATA*

It has been determined that *B.pandurata* contains a variety of essential oils (EOs) and several flavonoid compounds with numerous biologically active properties. Most of its flavonoids are distinctive due to the inclusion of certain prenyl substituents in their primary structures. Table 5.1 shows the different isolated compounds, various extraction methods, and uses of the rhizome of *B. pandurata*.

5.4 APPLICATION OF *B. PANDURATA*

The extracts of *B. pandurata* rhizomes contain many flavonoid compounds and essential oils presenting various intriguing pharmacological activities, including antifungal, antibacterial, and antioxidant activities. It exhibits potent *in-vitro* action against *Streptococcus pyogenes*. The extract of *B. pandurata* can lyse bacterial cells and raises the possibility of cell wall and cell membrane disruption as the mode of action (Chahyadi et al., 2014; Hardi et al., 2016). *B.pandurata* is the principal source of panduratin A, which is well-known for its numerous potential health benefits (Ghosh & Parida, 2020). Numerous studies have investigated the biological activity of *B. pandurata* in relation to its traditional medicinal uses. Both the extract and the chemicals from this plant exhibit a

TABLE 5.1
Compounds Isolated from the Rhizome of *Boesenbergia pandurata*

Name of Isolated Compound	Method of Extraction	Class of Compound	Uses	References
Boesenberol I Boesenberol J Boesenberol K	Maceration with methanol	Terpenes	Tumour necrosis factor-related apoptosis-induced ligand (TRAIL)-resistance-overcoming activity	(Karmakar et al., 2017)
Panduratin A	Maceration with ethanol or water	Flavonoid (chalcone)	Antibacterial, antioxidant, anti-inflammatory, anticancer, anti-obesity, activities and skin hydration	(Kim et al., 2018; Taheri et al., 2022; Woo et al., 2015)
Panduratins B1 and panduratins B2	Maceration with ethanol	Flavonoid	Fighting fatigue, enhancing exercise performance capability, and promoting muscle mass growth	(Win, Awale, Esumi et al., 2008)
Panduratin C	Maceration with ethanol	Flavonoid	Preventing and treating/relieving a metabolic disease, protecting against cisplatin-induced apoptosis of renal proximal tubular cells, and activity against the virus SARS-CoV-2	(Hwang & Kim, 2014)
Panduratin D Panduratin E Panduratin H Panduratin I Panduratin F Panduratin G	Maceration with methanol	Flavonoid Flavonoid		(Gurung et al., 2022) (Vergoten & Bailly, 2022)
(-)-Alpinetin (-)-7,4'-Dihydroxy-5-methoxyflavanone	Refluxed with methanol	Flavanone derivative	Anti-austerity activity	(Nguyen et al., 2022)
(-)-Pinostrobin (-)-Pinocembrin	Refluxed with methanol	Flavanone derivative	Anti-austerity activity	(Okuno et al., 2019)
Cardamonin Pinocembrin Helichrysetin 2',6'-Dihydroxy-4'-methoxychalcone Flavokawain C	Solvent (chloroform, ether, benzene, and ethanol) extraction	Chalcone	Treatment of infection, anticancer, inflammation, and liver cirrhosis	(Adhikari et al., 2020; Chahyadi et al., 2014; Eng-Chong et al., 2012)
2',4',6'-Trihydroxychalcone Uvangoletin		Hydrochalcones	Treatment of hypertension, diabetes, malaria, yellow fever, breast cancer, dengue fever, gastroenteric, and inflammatory disorders.	(Stompor et al., 2019)

(*Continued*)

TABLE 5.1 (CONTINUED)
Compounds Isolated from the Rhizome of *Boesenbergia pandurata*

Name of Isolated Compound	Method of Extraction	Class of Compound	Uses	References
5,6-Dehydrokawain	Hydro-distillation, solvent extractions (petroleum ether and ethanol)	Terpene	Anti-inflammatory and anti-allergic activities	(Adhikari et al., 2020; Charisma et al., 2018; Donadu et al., 2020; Oktavianawati et al., 2018)
Prenylated phenylbenzoic acid		Terpene	Anticancer and anti-inflammatory activities	
2',4'-Dihydroxy-6'-phenethylbenzoic acid methyl ester		Terpene	Antibacterial and wound-healing activities	
Geranyl-2',4'-dihydroxy-6'-phenethylbenzoate		Terpene	Antioxidant and antiulcer activities	
Dihydro-5,6'-dehydrokawain		Terpene	Antiulcer and wound- healing activities	
γ-Terpinene, geranial		Terpene	Antibacterial effect	(Huong et al., 2021)
1,8-Cineole		Terpene	Anticancer and anti-inflammatory activities	(Kar et al., 2015)
Terpineol		Terpene	Antiulcer activities and antioxidant	(Oktavianawati et al., 2018)
β-Ocimene		Terpene	Wound-healing and antibacterial activities	(Oktavianawati et al., 2018)
Neral		Terpene	Wound-healing, and antibacterial activities	(Oktavianawati et al., 2018)
Geraniol		Terpene	Antiulcer and wound-healing activities	(Jitviriyanon et al., 2016)
Camphor		Terpene	Antibacterial effect	(Oktavianawati et al., 2018)
Myrcene		Terpene	Antibacterial and anti-inflammatory activities	(Sukandar et al., 2016)
Borneol		Terpene	Antibacterial and antiulcer activities	
Camphene		Terpene	Antioxidant and antiallergic activities	
Methyl cinnamate		Terpene	Antioxidant and antiulcer activities	

variety of intriguing behaviours, especially the panduratin derivatives that showed potent antibacterial activity (Chahyadi et al., 2014).

5.4.1 Medicinal and Pharmaceutical Properties of *B. pandurata*

Researchers have effectively isolated various bioactive chemicals from *B. pandurata* over time, employing different methods and technologies (Table 5.1) (Mohammed et al., 2019; Saah et al., 2021). Nearly one hundred different compounds were discovered and structurally elucidated, including terpenoids, esters, kawains, chalcone derivatives, and derivatives of flavonoid compounds (see Figure 5.2) (Ongwisespaiboon & Jiraungkoorskul, 2017; Youn & Jun, 2019).

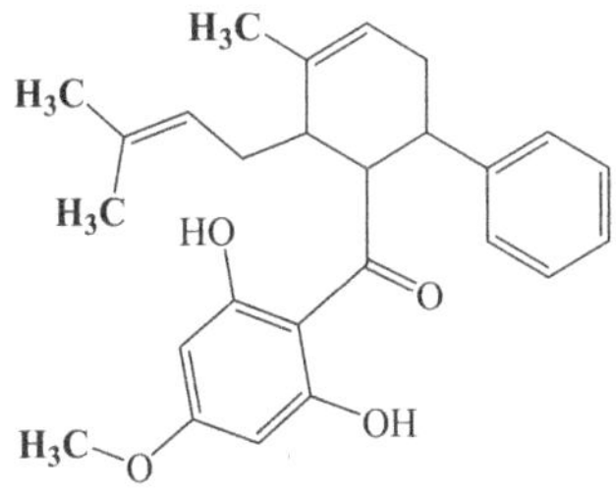
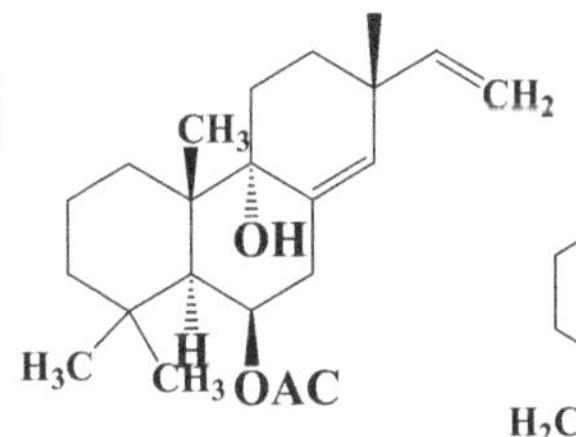
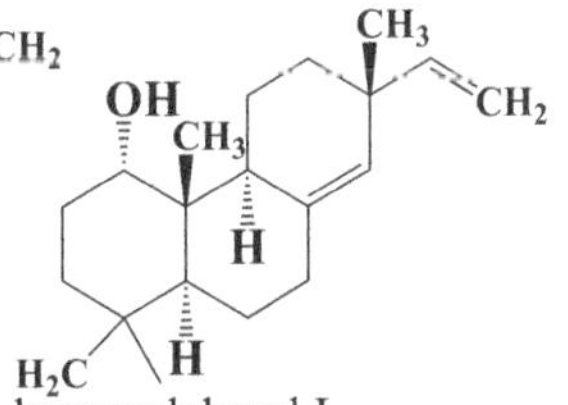

Panduratin A | kaempulchraol K | kaempulchraol I

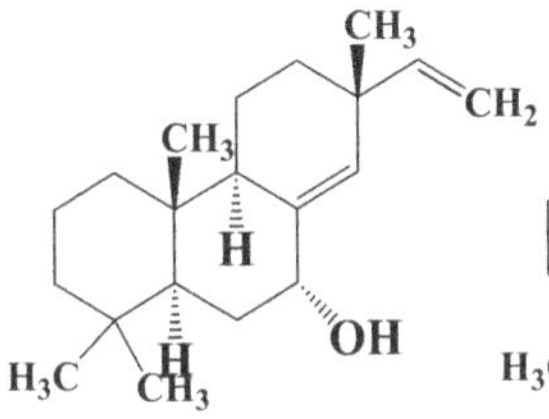
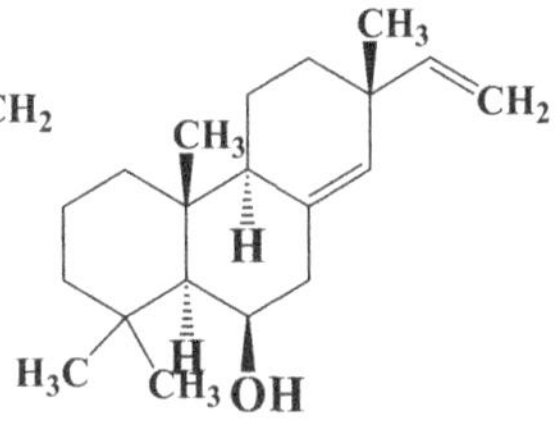
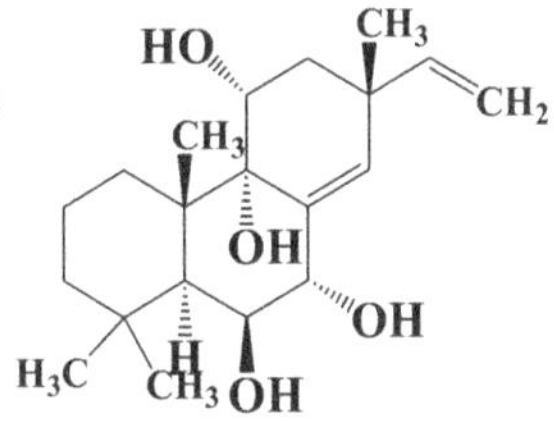

Boesenberol I | boesenberols J | boesenberols K

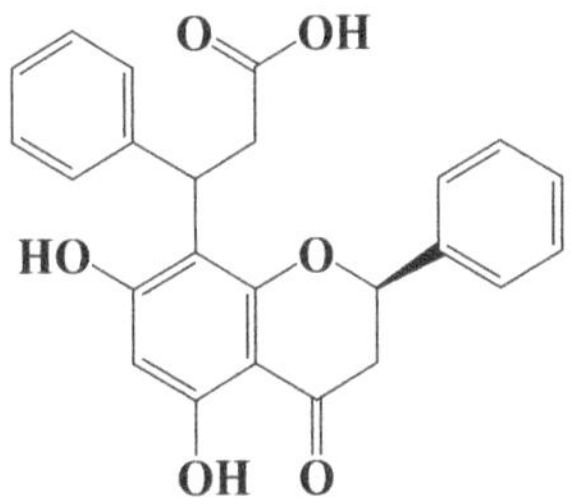
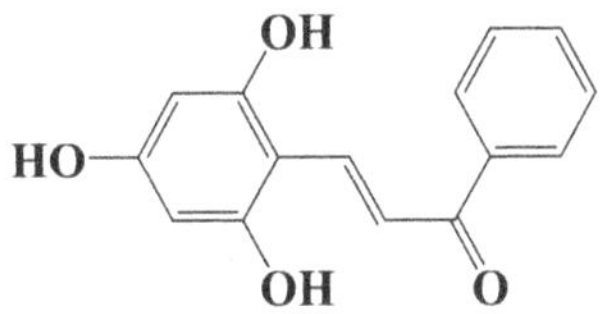

(2R,7S)-8-(1-phenyl-2-carboxyethyl)pinocembrin | 2,4,6-trihydroxychalcone

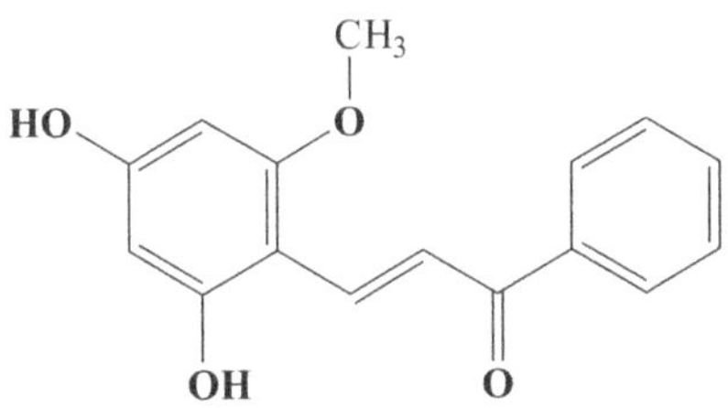
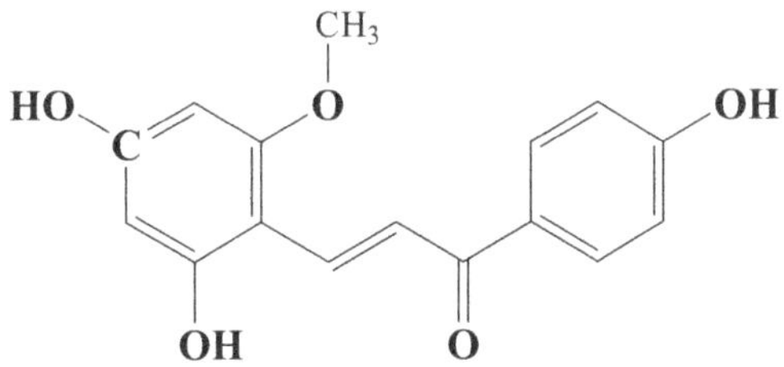

2,4-dihydroxy-6-methoxy chalcone | 2,4,4-trihydroxy-6-methoxychalcone

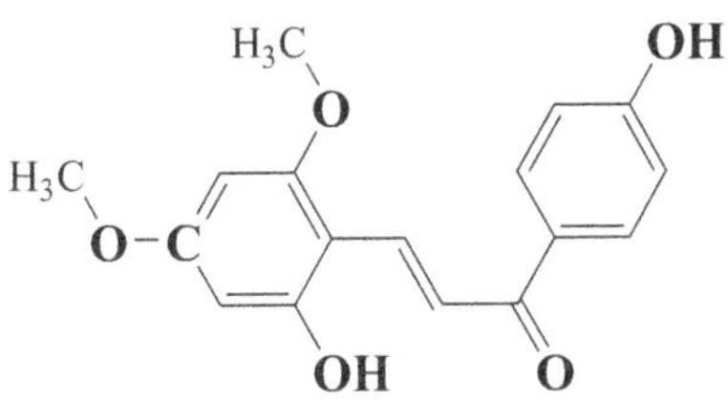
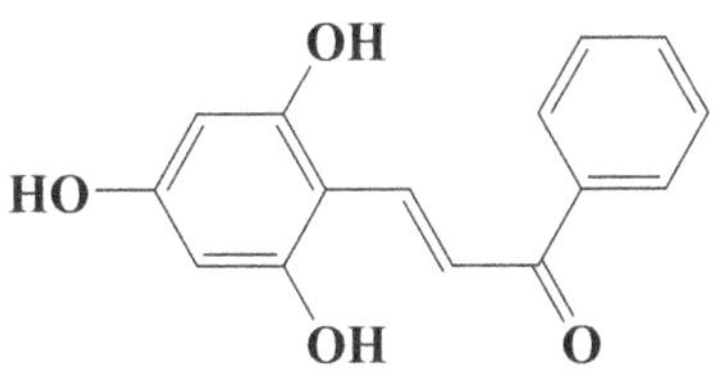

2,4-dihydroxy-4,6-dimethoxychalcone | 2,4,6-trihydroxychalcon

FIGURE 5.2 Structures of significant phytochemicals present in *Boesenbergia pandurata*.

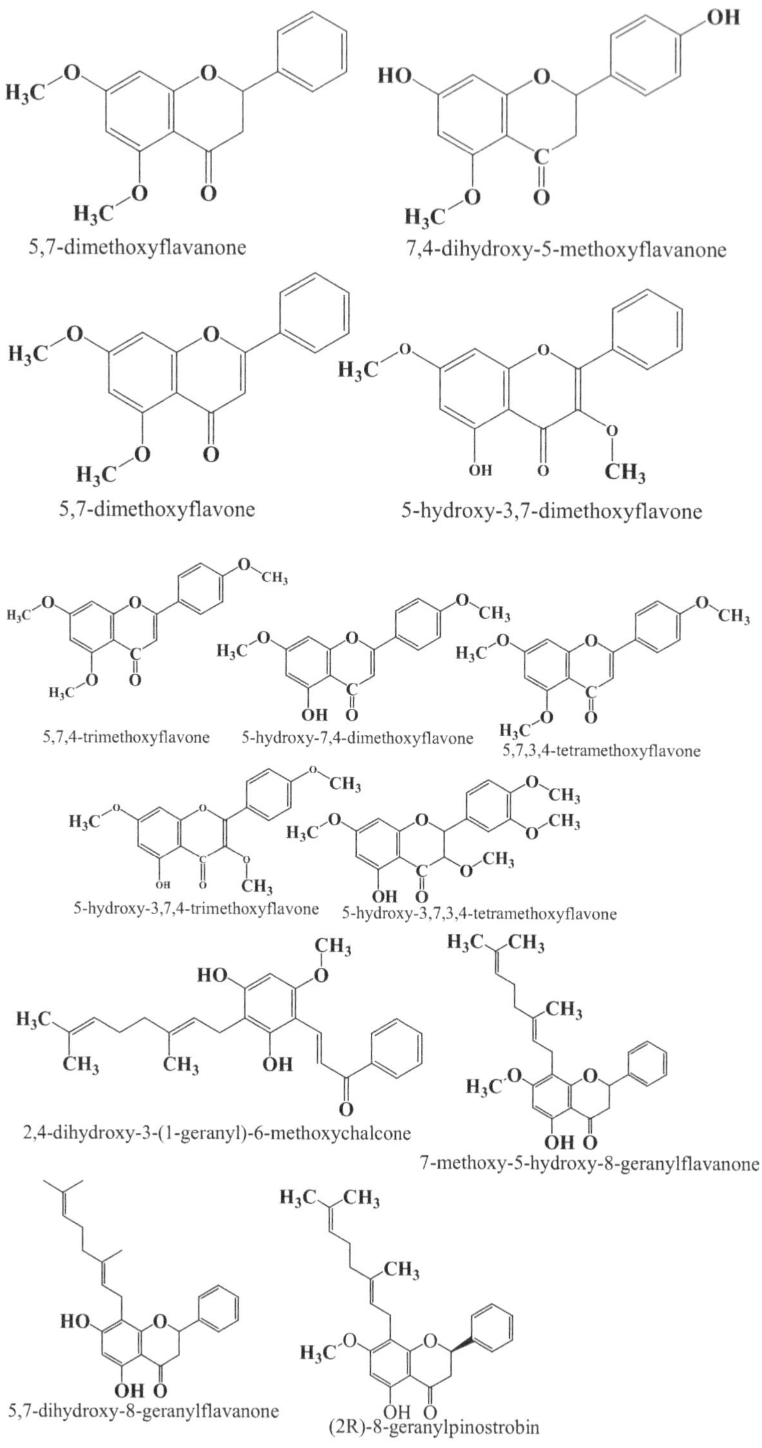

FIGURE 5.2 (CONTINUED) Structures of significant phytochemicals present in *Boesenbergia pandurata*.

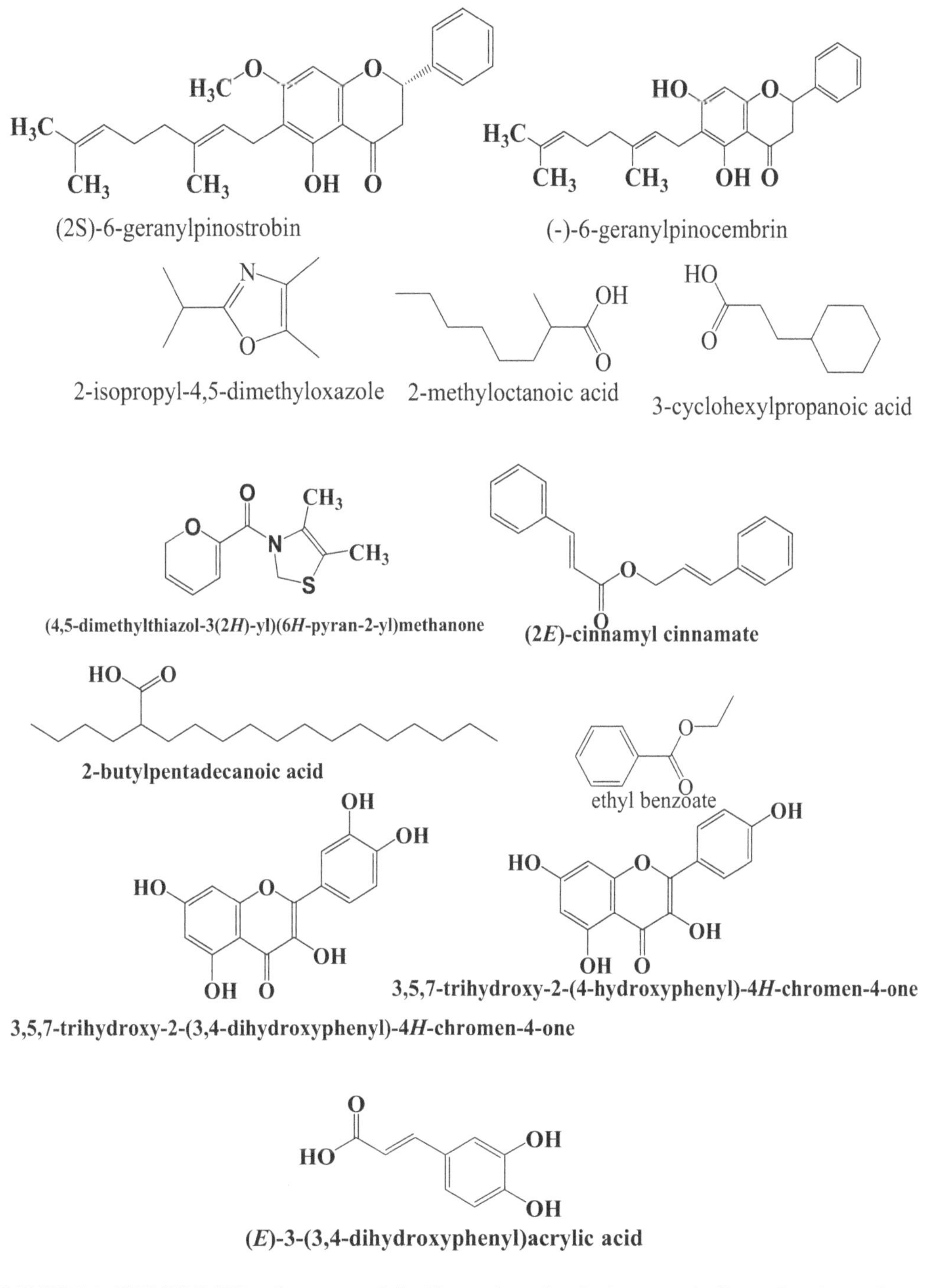

FIGURE 5.2 (CONTINUED) Structures of significant phytochemicals present in *Boesenbergia pandurata*.

5.4.1.1 Potent Antibacterial Activity of *B. pandurata* Extracts

Numerous reports have been made on the antibacterial capabilities of *B. pandurata*. Crude extracts, essential oils (EOs), and purified individual components of *B. pandurata* have all been extensively researched for their antibacterial capabilities. Interestingly, even at low concentrations, they all showed remarkable effectiveness against different bacterial strains (Mohammed et al., 2019). The activity of *B. pandurata* EOs was examined against *Listeria monocytogenes*, *Staphylococcus aureus*, *Bacillus cereus*, *Escherichia coli*, and other food-borne pathogens. For the three Gram-positive bacterial species, *B. cereus*, *S. aureus*, and *L. monocytogenes*, the minimum inhibitory concentration (MIC) values were 12.5, 12.5, and 6.25 mg/mL, respectively, whereas the MIC values for the Gram-negative bacterium, *E. coli*, was 50 mg/mL. The hydrodistillated EOs showed the greatest effectiveness against these three bacterial species. Because of the ability of *B. pandurata* to alter the salt tolerance and penetrability of the cell, a mode of action for EOs towards bacterial cells was proposed. These findings demonstrated that *B. pandurata* extract and EOs are effective disease controllers and natural food preservatives in food systems (Chahyadi et al., 2014).

Panduratin A is one of the main flavonoid components of *B. pandurata* and has the potential to be a candidate for curing *Helicobacter pylori* infection. Compared with other antibacterial medications, panduratin A is more effective and exhibits antibacterial properties against various microorganisms. Additionally, it has been shown to be an antibacterial agent with potent *in vitro* action against clinical enterococci, including microorganisms resistant to other antibiotics. Furthermore, it has been found that panduratin A exhibits antibacterial properties against the cariogenic *Streptococcus mutans*, *Prevotella intermedia*, and *Prevoltella loescheii*. It has been discovered that panduratin A and isopanduratin A have strong antibacterial properties against *Cutibacterium (Propionibacterium). acnes* and other skin bacteria. It has also been found that panduratin A exhibits antistaphylococcal properties *in vitro*. These results imply that the *B. pandurata* extract containing panduratin A can be used as an antibacterial mediator to prevent the development of different pathogenic bacteria (Ghosh & Parida, 2020).

5.4.1.2 Anti-parasitic Activity of *B. pandurata*

Sawangjaroen and colleagues have previously shown that *B. pandurata* has an inhibitory effect against *Giardia lamblia*, a protozoan parasite that causes giardiasis. Giardiasis is a small-intestine infection that causes nutritional deficits, while diarrhoea can also lead to persistent diarrhoea in HIV patients (Hardi et al., 2021; Stompor, 2020). *B. pandurata* extracts in methanol or chloroform yielded IC_{50} values ranging from 20 g/mL to 100 g/mL. The effect of these plant extracts was not as strong as that of the commercial antibiotic metronidazole, which has a 0.48 g/mL IC_{50} value. On the other hand, these extracts are thought to contain bioactive chemicals that may be able to stop giardiasis (Eng-Chong et al., 2012). Additionally, methanol and chloroform extracts of *B. pandurata* demonstrated anti-amoebic action against *Entamoeba hitolytica*, with 45.8 and 57.6 g/mL IC_{50} values, respectively (Chahyadi et al., 2014).

5.4.1.3 Antioxidant Activity of *B. pandurata*

In a rat brain homogenate model, the *B. pandurata* rhizome extract possesses neuroprotective and strong antioxidant activity (Ghosh & Parida, 2020; Ramli et al., 2020). According to various studies, the flavonoid panduratin A has reducing potential and acts predominantly through an antioxidant mechanism, exhibiting hepatoprotective efficacy against thioacetamide-induced oxidative stress-mediated injuries. In a dose-dependent way, panduratin A has been reported to alleviate the depletion of reduced glutathione as well as to lower the degree of lipid peroxidation (Ghosh & Parida, 2020; Sundram et al., 2019).

The oxidative breakdown of lipids, known as lipid peroxidation, is harmful because the production of lipid peroxidation promotes the propagation reactions of free radicals. Pathological problems, including atherosclerosis and myocardial infarction, are known to be caused by free radicals (Chandra & Lo, 2021; Eng-Chong et al., 2012). Eng-Chong and colleagues demonstrated the *B. pandurata* rhizome antioxidant activity. They investigated the free radical scavenging activities of six bioactive compounds from *B. pandurata* rhizome extract, including (-)-4-hydroxypanduratin A, (-)-panduratin A, 2′, 6′-dihydroxy- 4′-methoxychalcone, 5-hydroxy-7-methoxyflavanone, 5,7-dihydroxyflavanone and 2′,4′-dihydroxy-6′-methoxychalcone (Charisma et al., 2018; Eng-Chong et al., 2012; Lee et al., 2022).

5.4.1.4 Anti-ulcer Effect of *B. pandurata*

People in Indonesia and Thailand have used *B. pandurata* preparations to treat ulcers. Rosdianto and coworkers have investigated the antiulcer properties of the methanol extract and its phytoconstituents. According to reports, rats with ulcers showed cytoprotective benefits from the *B. pandurata* extract. Additionally, this plant extract dramatically reduced submucosal oedema and leukocyte infiltration (Bailly 2022; Eng-Chong et al., 2012; Ghosh & Parida, 2020; Rosdianto et al., 2020).

5.4.1.5 An Antiviral Activity of *B. pandurata*

Antiviral medications are a subclass of antibiotics that can eradicate viruses. Just as individual antibiotic medications kill specific bacteria, so antiviral drugs are only effective against specific viruses. Even so, several broad-spectrum antiviral medications work well and are used for various viral infections. The ability of panduratin A to inhibit dengue viruses is well documented. According to studies reveal that Dengue viral protein is the highest affinity target for panduratin A (Atta-ur-Rahman & Atta-ur-Rahman, 2019; Ghosh & Parida, 2020).

Additionally, research indicates that variants of panduratin A have stronger antiviral action against the dengue *flavivirus* than the original compound. Designing different panduratin A derivatives and assessing and predicting their potential as inhibitors of the primary dengue proteins have relied heavily on computational analyses. To further understand the mechanism, molecular dynamics, and the kinetics underlying panduratin A's inhibitory action on the interaction between dengue proteins and human nuclear proteins, research has also been done. In addition to being the key participant in the polyprotein's post-translational proteolytic processing, the catalytic triad is a crucial element in viral propagation. The catalytic triad has been discovered to be blocked by panduratin A and its derivatives. Computational docking investigations reveal that panduratin A has been discovered to have binding energy superior to other traditional antiviral medications against dengue, especially mycophenolic acid and ribavirin (Ghosh & Parida, 2020; Kaushik et al., 2018).

5.4.1.5.1 Anti-HIV-1 Protease Activity of B. pandurata

A purified methanolic extract of *B. pandurata* rhizomes produced chalcone derivatives like 2,4,6-trihydroxyhydrochalcone, uvangoletin, and helichrysetin, as well as the cyclohexenyl chalcones panduratin A, panduratin C, and hydroxypanduratin A, which are responsible for the rhizomes' anti-HIV protease activity. Their findings demonstrated that, in contrast to other bioactive compounds that showed lesser inhibition, panduratin A and hydroxypanduratin A exhibited significant inhibition at IC_{50} values of 5.6 M and 18.7 M, respectively. According to research on the structure-activity relationship (SAR), chalcones' hydroxylation and prenylation are connected to the HIV-1 protease inhibitory activity (Sudsai et al., 2017; Zubair et al., 2020). Sudsai and coworkers also looked into the anti-HIV protease activity in methanol, chloroform, or water extracts of various indigenous Thai herbs that the locals use as AIDS self-medication. Compared with other plant species, *B. panburata* chloroform extract showed the strongest inhibition against HIV-1 protease (64.92±4.75%), followed by the methanolic extract with 51.92±0.22% inhibition (Chahyadi et al., 2014; Sudsai et al., 2017). The ethanolic extract of *B. pandurata* rhizomes was used in another investigation to isolate four flavonoids, namely pinostrobin, cardamonin, pinocembrin, and alpinetin. Cardamonin exhibited an IC_{50} value of 31.0 g/mL and the antiviral protease assay revealed that it caused a maximum inhibition of 75.11 ±1.44%. Possible therapeutic targets to suppress the HIV-1 protease activity include cardamonin, panduratin A, and hydroxypanduratin A (Eng-Chong et al., 2012; Zubair et al., 2020).

5.4.1.6 Antifungal Activity of *B. pandurata*

The antifungal activity of the chloroform extract of *B. pandurata* rhizomes was evaluated using clinical isolates of *Candida albicans, Cryptococcus neoformans*, and *Microsporum gypsum.* The chloroform extract showed high activity against *C. albicans* (MIC > 512 g/mL), but showed weak activity against *M. gypseum* and *C. neoformans*, according to the MIC values (Chahyadi et al., 2014; Sawatdikarn 2016). The fungi listed above commonly cause infections in HIV patients, so the study evidence suggests using *B. pandurata* extracts as a potential therapy for opportunistic fungal infections in AIDS patients. The *B. pandurata* extract was examined for its possible application as a fungicide in farming due to its antifungal activity (Chahyadi et al., 2014; Hardi et al., 2016). A soil-borne pathogen that causes epidemic disease in several crops, *Phytopthora capsici,* exhibited inhibition of mycelial development in response to the ethyl acetate extract of *B. pandurata. B. pandurata* has traditionally been used to treat fungal conditions, including irritation and dental caries associated with mouth and dental issues. Since *C. albicans* is a fungus found in the oral cavity, it is frequently linked to denture stomatitis, adheres to oral surfaces, and ultimately plays a role in the growth of mouth and tooth infections. The effect of the *B. pandurata* ethanol extract on the adhesion of *Candida albicans* to the acrylic surface of dentures was examined (Chahyadi et al., 2014; Khewkhom & Shangchote 2009). The adherence of the *C. albicans* clinical strain was observed to be reduced by the ethanol extract in a dose-dependent manner. The amount of adhering fungus was reduced by around 75% by pretreatment with 100 mg/mL extract. In contrast, the positive control of 0.2% chlorhexidine gluconate resulted in a significant inhibitory effect of more than 90%. Therefore, using *B. pandurata* to treat *Candida*-related denture stomatitis and assistance to avoid infection reappearance may be beneficial (Chahyadi et al., 2014; Sukandar et al., 2016; Zakuan et al., 2018)

5.4.1.7 The Anti-ageing Activity of *B. pandurata*

The compounds panduratin A and 4-hydroxypanduratin A and the extract of *B. pandurata* showed a protective advantage and an anti-ageing impact on human skin fibroblast cells exposed to ultraviolet (UV) radiation (Chandra & Lo, 2021). By increasing the activity of matrix metalloproteinases (MMPs) and reducing collagen production, UV radiation causes photoageing. It's interesting to note that panduratin A, at concentrations between 0.001 and 0.1M, could reduce the impact of UV radiation on fibroblast cells by decreasing MMP-1 expression and increasing type-1 procollagen expression. At 0.01 to 0.5 g/mL, the *B. pandurata* extract demonstrated a potent anti-ageing inhibitory effect with a related mechanism. These findings supported the authors' hypothesis that the chemicals panduratin A and 4-hydroxypanduratin A, as well as the extract, have the potential to become new natural anti-ageing agents (Chahyadi et al., 2014; Chandra & Lo, 2021; Kim et al., 2017; Listyawati, 2015; Mukherjee et al., 2011; Won et al., 2021).

5.4.1.8 Anti-obesity Activity of *B. pandurata*

Due to the obese body's excessive quantity of fat and compromised lipid metabolism, obesity is regarded as a pathological disease. Obesity is a serious condition because it causes many associated difficulties and health issues. Numerous factors, such as genetic predisposition, eating habits, lifestyle factors, and food composition, may contribute to obesity (Eng-Chong et al., 2012). According to research studies, *B. pandurata* extract may reduce the risk of obesity from a high-fat diet by modulating fat breakdown and stimulating protein kinase (AMPK). Through the activation of genes associated with fatty acid oxidation, AMPK activation will boost fatty acid oxidation (Chandra & Lo, 2021). Reduced sterol regulatory element-binding protein and PPAR phosphorylation will stop lipid production due to this procedure. Although there has not yet been any concrete proof in the study, *B. pandurata* is a rich source of panduratin A; hence, panduratin A may have played a role in the anti-obesity properties of the *B. pandurata* extract (Chahyadi et al., 2014; Eng-Chong et al., 2012; Ghosh & Parida, 2020). Without reducing food consumption, panduratin A compound (at a dose of 50 mg/kg bodyweight/day) or *B. pandurata* plant extract (at a dose of 200 mg/kg/day) can dramatically lower weight gain and fat mass. It was discovered that the panduratin A substance might control lipid metabolism (Chandra & Lo, 2021).To validate panduratin A's effectiveness against obesity, more research is needed (Chandra & Lo, 2021; Ghosh & Parida, 2020)

5.4.1.9 Anticancer Activity

Different cancer cells have been used to test the anticancer effects of *B. pandurata* extracts and isolated chemicals. The outcomes seem to support the long-standing usage of *B. pandurata* as a natural cancer preventative. Through an experiment using human B-lymphoblastoid (Raji) cells and tumour promoter-induced Epstein-Barr virus (EBV) stimulation, it was discovered that the *B. pandurata* methanol extract inhibits the growth of tumours. According to cell viability values (CV≤ 30%), the extract at 20 μg/mL was highly toxic and significantly inhibited the teleocidin (20 μg)-induced EBV activation with an inhibitory effect (IE) value of ≥ 70% (Chahyadi et al., 2014; Eng-Chong et al., 2012). With good cell viability (CV>90%), the anticipated bioactive substance, cardamonin, was discovered to totally inhibit EBV activation at a concentration of 25 μM. The activity of cardamonin (IC_{50} = 3.1 M) was equivalent to 1'-acetoxychavicol acetate (ACA; IC_{50} = 1.5 M), a highly effective EBV activation inhibitor. Cardamonin is one of the most potent chalcone-type inhibitors against EBV activation, in the opinion of scientists (Nurrachma et al., 2018). Treatment with food containing 4% (v/w) ethanolic extracts of *B. pandurata* showed a decrease in the generation of ACFs in the azoxymethane-induced aberrant crypt foci (ACF, colon cancer) model in rats, but the change was not statistically significant. Furthermore, at 10 g/mL, the *B. pandurata* chloroform extract showed 100% preferential cytotoxicity (PC100) against human pancreatic cancer (PANC-1) cells and cytotoxic action against human promyelocytic (HL-60) cancer cells (Chahyadi et al., 2014; Listyawati et al., 2016).

5.4.1.10 Toxicity Effect of *B. pandurata*

Because of the high consumption of this plant and the deficiency of precise confirmation supporting its safety, recent research has concentrated on acute toxicity examining various *B. pandurata* extracts via *in-vivo* methods to ascertain the safety of *B. pandurata* rhizome for consumption (Charoensin et al., 2010). Ethanolic extract of *B. pandurata* proved safe to consume because *in vivo* tests on fed rats revealed no appreciable changes in their bodyweight (Isa et al., 2012). Additionally, no unfavourable effects were detected in any haematological or histological markers used to assess toxicity. The safety of consumption was further demonstrated by the fact that two bioactive components from *B. pandurata*, pinocembrin and pinostrobin, had no mutagenic or harmful effects on Wistar rats (Eng-Chong et al., 2012; Fattepur et al., 2018; Kim et al., 2017).

5.4.1.11 Wound Healing Activity of *B. pandurata*

Wound healing in rats has been demonstrated to be accelerated by exposure to ethanol extracts of the *B. pandurata* rhizome. Visually, it was revealed that wound dressings containing extract of rhizomes and Intrasite gel settled considerably more quickly than those using vehicle-based treatments (Jitvaropas et al., 2012). Histological analysis of wounds dressed with rhizome extract demonstrated much smaller scar breadth at wound closure and healed wounds contained more collagen and angiogenesis compared with treated wounds dressed with the vehicle alone. According to the previously discussed empirical findings, *B. pandurata* may also contain bioactive substances with various therapeutic effects that can be used to treat, alleviate, or prevent several disorders (Abdalbasit et al., 2010; Won et al., 2021).

5.5 FUTURE PROSPECTS OF *B. PANDURATA*

The ethnomedicinal applications of *B. pandurata*, backed up by compelling scientific indications of its potential therapeutic effects, amply demonstrate that this herbal deserves to be advanced to the next stage of drug discovery research. To further define the processes and pathways contributing to its desirable features, investigations should concentrate on issues of genomics, metabolomics, proteomics, transcriptomics, and bioinformatics issues. Nanotechnology is used in new drug discovery investigations involving plant source chemicals, drug-DNA/protein interactions, polypharmacology, solubility, definite drug constancy, and transport to the targeted organ. The wide range of diseases that *B. pandurata* inhibits, including cancer and infection with bacteria, viruses, and parasites, requires further study.

5.6 CONCLUSION

B. pandurata is a plant utilized as a food condiment in many nations worldwide, including Thailand, Malaysia, India, Indonesia, and China. In these countries, *B. pandurata* is used as a traditional medicine to cure various ailments, as a traditional tonic, particularly for childbirth, as a cosmetic service for adolescent females, and as a preventative treatment for leukorrhoea in women. Along with being used as an antiparasitic, antifungal, and aphrodisiac, fresh *B. pandurata* rhizomes are again used to treat illnesses. Locals utilize its leaves to treat food allergies and sicknesses. Additionally, AIDS patients use *B. pandurata* as self-medication to treat the disease. Due to technological advancements, it is now possible to use *in-vivo* as well as *in-vitro* studies to demonstrate the functions of the plant extracts to explain their ethnomedicinal applications. The bioactive chemicals extracted from *B. pandurata* offer great potential for treating many ailments. The development of more active and potent drugs to treat diseases in the future may be sped up as a result of molecular advancements that further induce researchers to explore the best path for the synthesis of biologically active *B. pandurata* compounds. In the overview provided in this chapter, we attempted to explore several pharmacological activities, the methods of phytochemical extraction, and the major active compounds and their chemical structures in *B. pandurata*.

REFERENCES

Abdalbasit, A.M., Siddig, I.A., Salmah, I. & Fouad, A.-B. (2010) Potential activity of ethanolic extract of Boesenbergia rotunda (L.) rhizomes extract in accelerating wound healing in rats. *Journal of Medicinal Plants Research*, 4(15), 1570–1576.

Abate, L., Tadesse, M.G., Bachheti, A. & Bachheti, R.K. (2022) Traditional and phytochemical bases of herbs, shrubs, climbers, and trees from Ethiopia for their anticancer response. *BioMed Research International*. https://doi.org/10.1155/2022/1589877

Abate, L., Bachheti, A., Bachheti, R.K. & Husen, A. (2021) Antibacterial properties of medicinal plants: Recent trends, progress, and challenges. In *Traditional herbal therapy for the human immune system*, CRC Press, pp. 13–54.

Adhikari, D., Gong, D., Oh, S.H., Sung, E.H., Lee, S.O., Kim, D.W. et al. (2020) Vasorelaxant effect of Boesenbergia rotunda and its active ingredients on an isolated coronary artery. *Plants*, 9(12), 1688.

Asfaw, T.B., Esho, T.B., Bachheti, A., Bachheti, R.K., Pandey, D.P. & Husen, A. (2022) Exploring important herbs, shrubs, and trees for their traditional knowledge, chemical derivatives, and potential benefits. In *Herbs, shrubs, and trees of potential medicinal benefits*, CRC Press, pp. 1–26.

Atta-ur-Rahman, F. (2019) *Studies in natural products chemistry*. Elsevier.

Beshah, F., Hunde, Y., Getachew, M., Bachheti, R.K., Husen, A. & Bachheti, A. (2020) Ethnopharmacological, phytochemistry and other potential applications of *Dodonaea* genus: A comprehensive review. *Current Research in Biotechnology*, 2, 103–119.

Bailly, C. (2022) Toward the use of Boesenbergia rotunda extracts and the chalcone panduratin A to treat periodontitis. *Journal of Oral Biosciences*, 64, 183–192.

Bekele, G. & Reddy, P.R. (2015) Ethnobotanical study of medicinal plants used to treat human ailments by Guji Oromo tribes in Abaya District, Borana, Oromia, Ethiopia. *Universal Journal of Plant Science*, 3(1), 1–8.

Chahyadi, A., Hartati, R. & Wirasutisna, K.R. (2014) Boesenbergia pandurata Roxb., an Indonesian medicinal plant: Phytochemistry, biological activity, plant biotechnology. *Procedia Chemistry*, 13, 13–37.

Chan, M. (2015) WHO director-general addresses the sixty-sixth World Health Assembly. 2013.

Chandra, S., Lo, D. (2021) *Functional Properties of Boesenbergia Pandurata and Murraya Paniculata: A Review*. Paper presented at the IOP Conference Series: Earth and Environmental Science.

Charisma, S.L., Rahayu, W.S. & Wahyuingrum, R. (2018) Determination of sun protection factor and antioxidant properties of cream formulation of kencur (Kaempferia galanga L) and temu kunci (Boesenbergia pandurata (Roxb.) Schlecht) rhizomes extract. *Pharmaciana*, 8(2), 321–330.

Charoensin, S., Punvittayagul, C., Pompimon, W., Mevatee, U. & Wongpoomchai, R. (2010) Toxicological and clastogenic evaluation of pinocembrin and pinostrobin isolated from Boesenbergia pandurata in Wistar rats. *Thai Journal of Toxicology*, 25(1), 29–29.

Choi, S., Kim, C., Son, H., Hwang, J.K. & Kang, W. (2020) Estimation of an appropriate human dose of Boesenbergia pandurata extracts based on allometric scaling data of panduratin A in mice, rats, and dogs. *Journal of Medicinal Food*, 23(4), 453–458.

Donadu, M.G., Trong Le, N., Viet Ho, D., Quoc Doan, T., Tuan Le, A., Raal, A. & Madeddu, S. (2020) Phytochemical compositions and biological activities of essential oils from the leaves, rhizomes and whole plant of Hornstedtia bella Škorničk. *Antibiotics*, 9(6), 334.

Eng-Chong, T., Yean-Kee, L., Chin-Fei, C., Choon-Han, H., Sher-Ming, W., Li-Ping, C.T. & Karsani, S.A. (2012) Boesenbergia rotunda: from ethnomedicine to drug discovery. *Evidence-Based Complementary and Alternative Medicine*, 2012, 473637.

Fattepur, S., Nilugal, K.C., Rajendran, R., Asmani, F. & Yusuf, E. (2018) Anti-hyperlipidemic activity of methanolic extract of boesenbergia pandurata (finger root) in experimental induced hypercholestrolemic Sprague Dawley rats. *Asian Journal of Pharmaceutical and Clinical Research*, 11(3), 8–12.

Fitzgerald, M., Heinrich, M. & Booker, A. (2020) Medicinal plant analysis: A historical and regional discussion of emergent complex techniques. *Frontiers in Pharmacology*, 10, 1480.

Ghosh, D. & Parida, P. (2020) Multipotential therapeutic bioactive compound: Panduratin A. *Everyman's Science*,1, 24–29.

Gurung, A.B., Ali, M.A., Al-Hemaid, F., El-Zaidy, M. & Lee, J. (2022) In silico analyses of major active constituents of fingerroot (Boesenbergia rotunda) unveils inhibitory activities against SARS-CoV-2 main protease enzyme. *Saudi Journal of Biological Sciences*, 29(1), 65–74.

Hardi, E.H., Kusuma, I.W., Suwinarti, W., Agustina, A. & Nugroho, R.A. (2016) Antibacterial activity of Boesenbergia pandurata, Zingiber zerumbet and Solanum ferox extracts against Aeromonas hydrophila and Pseudomonas sp. *Nusantara Bioscience*, 8(1, 18–21).

Hardi, E.H., Nugroho, R.A., Abinawanto, A., Fujaya, Y., Murtiyoso, B., Winanto, T. et al. (2021) Antiparasitic Activity of Single Solanum ferox extract and concoction with Zingiber zerumbet and Boesenbergia pandurata extracts to control Argulus sp. on goldfish (Cyprinus carpio), 1, 513.

Huong, L.T., Sam, L.N., Dai, D.N. & Ogunwande, I.A. (2021) Investigation into the chemical compositions and antimicrobial activity of essential oil from the rhizomes of Boesenbergia quangngaiensis NS Lý from Vietnam. *Journal of Essential Oil Bearing Plants*, 24(5), 1125–1133.

Hwang, J.-K. & Kim, D.U. (2014) Use of panduratin derivative or Boesenbergia pandurata extract: Google patents, 8, 143.

Isa, N., Abdelwahab, S., Mohan, S., Abdul, A., Sukari, M., Taha, M. et al. (2012) In vitro anti-inflammatory, cytotoxic and antioxidant activities of boesenbergin A, a chalcone isolated from Boesenbergia rotunda (L.)(fingerroot). *Brazilian Journal of Medical and Biological Research*, 45, 524–530.

Jitvaropas, R., Saenthaweesuk, S., Somparn, N., Thuppia, A., Sireeratawong, S. & Phoolcharoen, W. (2012) Antioxidant, antimicrobial and wound healing activities of Boesenbergia rotunda. *Natural Product Communications*, 7(7), 1934578X1200700727.

Jitviriyanon, S., Phanthong, P., Lomarat, P., Bunyapraphatsara, N., Porntrakulpipat, S. & Paraksa, N. (2016) In vitro study of anti-coccidial activity of essential oils from indigenous plants against Eimeria tenella. *Veterinary Parasitology*, 228, 96–102.

Kar, B., Panda, P.C., Sahoo, S., Dash, B. & Nayak, S. (2015) GC-MS analysis of rhizome essential oil of Bosenbergia longiflora (Wall) Kuntze rich in longipnocarvone, a sesquiterpenoid. *International Journal of Pharmacy and Pharmaceutical Sciences*, 7, 568–570.

Karmakar, U.K., Arai, M.A., Koyano, T., Kowithayakorn, T. & Ishibashi, M. (2017) Boesenberols I–K, new isopimarane diterpenes from Boesenbergia pandurata with TRAIL-resistance overcoming activity. *Tetrahedron Letters*, 58(40), 3838–3841.

Kaushik, S., Kaushik, S., Sharma, V. & Yadav, J. (2018) Antiviral and therapeutic uses of medicinal plants and their derivatives against dengue viruses. *Pharmacognosy Reviews*, 12(24), 77–85.

Khan, M.A. (2016) Introduction and importance of medicinal plants and herbs. *Unani*. Zahid, India.

Khewkhom, N., Shangchote, S. (2009) Postharvest antifungal activity of extracts and compounds from Cinnamomum zeylanicum, Boesenbergia pandurata and Syzygium aromaticum against Colletotrichum gloeosporioides and Botryodiplodia theobromae. *Asian Journal of Food and Agro-Industry*, 2(Special Issue), 125–132.

Kim, D.U., Chung, H.C., Kim, C. & Hwang, J.K. (2017) Oral intake of Boesenbergia pandurata extract improves skin hydration, gloss, and wrinkling: A randomized, double-blind, and placebo-controlled study. *Journal of Cosmetic Dermatology*, 16(4), 512–519.

Kim, H., Kim, C., Kook, K.E., Choi, S., Kang, W. & Hwang, J.K. (2018) Inhibitory effects of standardized boesenbergia pandurata extract and its active compound panduratin A on lipopolysaccharide-induced periodontal inflammation and alveolar bone loss in rats. *Journal of Medicinal Food*, 21(10), 961–970.

Lee, J.-J., Choi, J. & Ha, J.-H. (2022) Physicochemical and storage characteristics of pork Tteokgalbi treated with Boesenbergia pandurata (Roxb.) powder. *Applied Sciences*, 12(5), 2425.

Listyawati, S. (2015) Potensi ekstrak terkuantifikasi rimpang boesenbergia pandurata (roxb.) Schlecht. Sebagai agen kemoprevensi kanker kulit. Universitas Gadjah Mada.

Listyawati, S., Sismindari, S., Mubarika, S., Murti, Y.B. & Ikawati, M. (2016) Anti-proliferative activity and apoptosis induction of an ethanolic extract of boesenbergia pandurata (Roxb.) schlecht. Against HeLa and vero cell lines. *Asian Pacific Journal of Cancer Prevention*, 17(1), 183–187.

Moges, A. & Moges, Y. (2019) Ethiopian common medicinal plants: Their parts and uses in traditional medicine-ecology and quality control. *Plant Science-Structure, Anatomy and Physiology in Plants Cultured in Vivo and in Vitro*, 21, 78–101.

Mohammed, I.A., Akhtar, M.N., Biau, F.J., Tor, Y.S., Zareen, S., Binti Shahabudin, S. et al. (2019) Isolation of cardamonin and pinostrobin chalcone from the rhizomes of Boesenbergia rotunda (L.) Mansf. and their cytotoxic effects on H-29 and MDA-MB-231 cancer cell lines. *The Natural Products Journal*, 9(4), 341–348.

Mukherjee, P.K., Maity, N., Nema, N.K. & Sarkar, B.K. (2011) Bioactive compounds from natural resources against skin aging. *Phytomedicine*, 19(1), 64–73.

Nguyen, M.T.T., Nguyen, H.X., Le, T.H., Do, T.N.V., Dang, P.H., Pham, T.V. et al. (2022) A new flavanone derivative from the rhizomes of Boesenbergia pandurata. *Natural Product Research*, 36(8), 1959–1965.

Nguyen, N.T., Nguyen, M.T.T., Nguyen, H.X., Dang, P.H., Dibwe, D.F., Esumi, H. et al. (2017) Constituents of the rhizomes of Boesenbergia pandurata and their antiausterity activities against the PANC-1 human pancreatic cancer line. *Journal of Natural Products*, 80(1), 141–148.

Nurrachma, M.Y., Fadliyah, H. & Meiyanto, E. (2018) Fingerroot (Boesenbergia pandurata): A prospective anticancer therapy. *Indonesian Journal of Cancer Chemoprevention*, 9(2), 102–109.

Oktavianawati, I., Kurniati, H.I., Maghfiroh, K., Hanifah, N., Handayani, W. & Winata, I.N.A. (2018) *Essential Oils from Rhizhomes of Five Zingiberaceae Species in Meru Betiri National Park*. Paper presented at the AIP Conference Proceedings.

Okuno, Y., Marumoto, S. & Miyazawa, M. (2019) Antimutagenic activity of flavonoids from Sozuku. *Natural Product Research*, 33(6), 862–865.

Ongwisespaiboon, O. & Jiraungkoorskul, W. (2017) Fingerroot, Boesenbergia rotunda and its aphrodisiac activity. *Pharmacognosy Reviews*, 11(21), 27.

Ramli, S., Radu, S., Shaari, K. & Rukayadi, Y. (2020) Toxicity analysis of Syzygium polyanthum (Wight) Walp. leaves extract and its stability against different pH and temperature. *Emirates Journal of Food and Agriculture*, 32, 461–468.

Rosdianto, A.M., Puspitasari, I.M., Lesmana, R. & Levita, J. (2020) Bioactive compounds of Boesenbergia sp. and their anti-inflammatory mechanism: A review. *Journal of Applied Pharmaceutical Science*, 10(7), 116–126.

Saah, S., Siriwan, D. & Trisonthi, P. (2021) Biological activities of Boesenbergia rotunda parts and extracting solvents in promoting osteogenic differentiation of pre-osteoblasts. *Food Bioscience*, 41, 101011.

Sawatdikarn, S. (2016) Antifungal activity of crude extracts of some medicinal plants against Fusarium sp., the pathogen of dirty panicle disease in rice. *Journal of Medicinal Plants Research*, 10(19), 248–255.

Stompor, M. (2020) A review on sources and pharmacological aspects of sakuranetin. *Nutrients*, 12(2), 513.

Stompor, M., Broda, D. & Bajek-Bil, A. (2019) Dihydrochalcones: Methods of acquisition and pharmacological properties—A first systematic review. *Molecules*, 24(24), 4468.

Sudsai, T., Leajae, S., Dangmanee, N., Chatgat, W., Chaniad, P. & Tewtrakul, S. (2017) Antibacterial and anti-HIV-1 integrase properties of isolated compounds from Boesenbergia kingii. *Songklanakarin Journal of Science and Technology*, 39(1), 131–135.

Sukandar, E.Y., Kurniati, N.F., Wikaningtyas, P. & Agprikani, D. (2016) Antibacterial interaction of combination of ethanolic extract of Zingiber officinale var rubrum rhizome, Boesenbergia pandurata rhizome, and Stevia rebaudiana leaves with certain antibiotics against infectious mouth microbial. *Asian Journal of Pharmaceutical and Clinical Research*, 9(1), 332–335.

Sundram, T., Zakaria, M.H.B. & Nasir, M. (2019) Antioxidant and cytotoxic effects of Curcuma Mangga and Bosenbergia Rotunda Ethanolic extracts on Mcf-7 cancer cell lines. *Science*, 3(2), 10–14.

Taheri, S., Teo, C.H., Heslop-Harrison, J.S., Schwarzacher, T., Tan, Y.S., Wee, W.Y. et al. (2022) Genome assembly and analysis of the flavonoid and phenylpropanoid biosynthetic pathways in Fingerroot ginger (Boesenbergia rotunda). *International Journal of Molecular Sciences*, 23(13), 7269.

Thy, C.K., Murthy, S., Lee, Y.K., Yaeghoobi, M., Rahman, N.A. & Chee, C.F. (2018) A simple aluminum bromide-promoted diastereoselective Synthesis of panduratin A derivatives. *Synlett*, 29(10), 1358–1361.

Vergoten, G. & Bailly, C. (2022) Interaction of panduratin A and derivatives with the SARS-CoV-2 main protease (mpro): A molecular docking study. *Journal of Biomolecular Structure and Dynamics*, 1–11.

Win, N.N., Awale, S., Esumi, H., Tezuka, Y. & Kadota, S. (2008) Panduratins D—I, novel secondary metabolites from rhizomes of Boesenbergia pandurata. *Chemical and Pharmaceutical Bulletin*, 56(4), 491–496.

Won, J., Noh, K., Hwang, J.-K., Shin, B.S. & Kang, W. (2021) Pharmacokinetics of panduratin A following oral administration of a Boesenbergia pandurata extract to rats. *Journal of Food & Drug Analysis*, 29(4), 676–683.

Woo, S.W., Rhim, D.-B., Kim, C. & Hwang, J.-K. (2015) Effect of standardized Boesenbergia pandurata extract and its active compound panduratin A on skin hydration and barrier function in human epidermal keratinocytes. *Preventive Nutrition and Food Science*, 20(1), 15.

Youn, K. & Jun, M. (2019) Biological evaluation and docking analysis of potent BACE1 inhibitors from Boesenbergia rotunda. *Nutrients*, 11(3), 662.

Zakuan, Z., Mustapa, S., Sukor, R. & Rukayadi, Y. (2018) Antifungal activity of Boesenbergia rotunda (temukunci) extract against filamentous spoilage fungi from vegetables. *International Food Research Journal*, 25(1, 433–438).

Zubair, M.S., Maulana, S., Widodo, A., Mukaddas, A. & Pitopang, R. (2020) Docking study on anti-HIV-1 activity of secondary metabolites from Zingiberaceae plants. *Journal of Pharmacy & Bioallied Sciences*, 12(Suppl 2), S763.

6 Phylogeny, Phytochemistry, Traditional Uses and Pharmaceutical Properties of *Thapsia* spp. Roots

Khaled Taïbi, Leila Aït Abderrahim, Mohamed Boussaid, Fadhila Taïbi, Kada Souana, Mohamed Achir, and Abdelkader Tadj

6.1 INTRODUCTION

The Apiaceae family contains several plant species which are widely distributed throughout the Mediterranean region. Members of the Apiaceae are very popular due to their commercial usage as drugs and spices. They are rich in valuable secondary metabolites, such as sesquiterpene lactones, coumarins, and essential oils (Alilou & Akssira, 2021).

Thapsia has recently gained a great deal of interest among the genera of the Apiaceae family. This small genus of herbaceous perennial species is commonly known as the deadly carrots, Derias in Arabic. The generic name is derived from the ancient Greek θαψία (thapsía); the Greeks believe that *Thapsia* plants originated from the ancient Sicilian village Thapsos, so they called them thapsia (Casiglia et al., 2015). The centre of diversity of the genus *Thapsia* is around the western Mediterranean, North Africa, and the Iberian Peninsula, extending to the Atlantic coasts (Andersen et al., 2015). However, its geographic limits of distribution extend to Greece and Turkey in the eastern Mediterranean region (Weitzel et al., 2014).

6.2 DESCRIPTION OF THE GENUS *THAPSIA*

Thapsia is composed of herbaceous perennial plants. The main characteristics of the genus are the bright yellow flowers, four-winged seeds with flat endosperm, large umbels, and stems that can reach a height of up to two meters. The inflorescences are large with regularly distributed umbels. However, the elliptic fruits have two membranous wings, a feature that is the main characteristic of this genus very peculiar (Pujades-Salvà et al., 2003). The flowering period extends from June to August. *Thapsia* plants are self-fertile, with hermaphrodite flowers pollinated usually by insects. *Thapsia* plants grow better in dry or moist soil but cannot grow in the shade.

6.3 PHYLOGENY, DISTRIBUTION, AND IMPORTANCE OF *THAPSIA*

Several errors and conflicting views regarding the identification of members of the genus *Thapsia* make it poorly defined even at present (Anderson et al., 2015). Misidentifying and using the wrong plants in traditional medicines might lead to intoxication and numerous health problems. Several cases of diarrhea and vomiting have been reported in Morocco due to women using roots of plants

DOI: 10.1201/b22924-6

different from but morphologically similar to what should be *Thapsia villosa* L. in traditional medicine, to treat sterility or weight gain (Bellakhdar, 1997).

Numerous species aggregates are recognized within the *Thapsia* genus, including diploid, tetraploid, and hexaploid forms, which are usually morphologically similar. Nevertheless, they can be distinguished based on chemotaxonomy and nuclear ribosomal DNA (nrDNA) internal transcribed spacer (ITS) sequences, and some of the identified chemotypes have been awarded taxonomic recognition (Banasiak et al., 2016).

In fact, according to the *Flora Europaea* (Tutin et al., 1968), the genus *Thapsia* comprises three species: *Thapsia garganica* L., *Thapsia maxima* Miller, and *Thapsia villosa* (Figure 6.1). However, chemotaxonomic studies that have been carried out to investigate the distribution of specific

bioactive compounds, mainly polyoxygenated guaianolides, have revealed significant new intra- and interspecies relationships, indicating a more complex deviation.

Recent phylogenetic studies have reported 14 species of *Thapsia* (Weitzel et al., 2014), nine species of which are distributed in North Africa and the Iberian Peninsula (http://ww2.bgbm.org/euroPlusMed/).

It has been reported that *Thapsia garganica* L. and *Thapsia transtagana* Brot. (Figure 6.1) are distinct species (Smitt et al., 1995), whereas *Thapsia maxima* Miller comprises two morphotypes (Avato et al., 1992). However, the most marked heterogeneity was observed among plants recognized as *T. villosa*; it comprises at least five diverse types, some of which are polyploids (Smitt et al., 1990; Smitt, 1995), for example, *Thapsia minor* Hoffgg. and Link. Moreover, *Thapsia laciniata* Rouy was reported to be different from *Thapsia villosa*. The new species, *Thapsia gymnesica* Rossell & Pujadas, was recorded by Pujadas et al. (1991). In addition, *Thapsia smittii* Simonsen & al., a segregate of *Thapsia maxima* Mill., was designated as a distinct new species by Weitzel et al. (2014).

Thapsia garganica is considered the most extensively distributed species, growing in Algeria, Tunisia, Libya, Spain, Italy, and Greece. *Thapsia transtagana* is distributed in Portugal, Spain, and Morocco, *Thapsia maxima* in the Iberian Peninsula, and *Thapsia villosa* in Northwest Africa, the Iberian Peninsula and France. In contrast, *Thapsia gymnesica* grows exclusively in Minorca, Mallorca, and the Balearic Islands.

To date, thapsigargins, guaianolide sesquiterpenes commonly used for therapeutic purposes to study cell activation, promotion of tumors, and homeostasis of calcium, are exclusively present in *Thapsia garganica*, *Thapsia transtagana*, *Thapsia gymnesica*, and two other polyploid types of *Thapsia villosa* (with 44 or 66 chromosomes).

6.4 TRADITIONAL USES

Thapsia species have been employed as traditional medicines in the Mediterranean region for thousands of years. Some historians believe that *Thapsia garganica*, formerly known as silphion, was exported by the Greek colony of Cyrene as a medicinal plant for use as a purgative and an emetic (Greive, 1996).

Bertherand (1857) reported using a medical bandage containing resin extracted from *Thapsia garganica* roots to cure lung diseases, rheumatic pains, and catarrh in Europe. The resin of the root, *Resin Thapsiae*, and *Radix Thapsiae*, have been included in the 1937 edition of the French Pharmacopoeia (Smitt et al., 1995). Moreover, Bellakhdar (1997) and Ouarghidi et al. (2013) have also reported using resin extracted from *Thapsia garganica* roots to treat pulmonary diseases, pneumonia, fever, catarrh, rheumatic pains, and female sterility. In addition, Tlili et al. (2015) reported using plants of the *Thapsia* genus as an anti-inflammatory remedy to manage various conditions, such as eczema. The properties of the resin have been previously described by Theophrastus (372–287 BC), Dioscorides (approximately AD 50), and Plinius (AD 24–79). Hippocrates and Theophrastus described the skin-irritating effects and medicinal uses of *Thapsia* species around 300–400 BC.

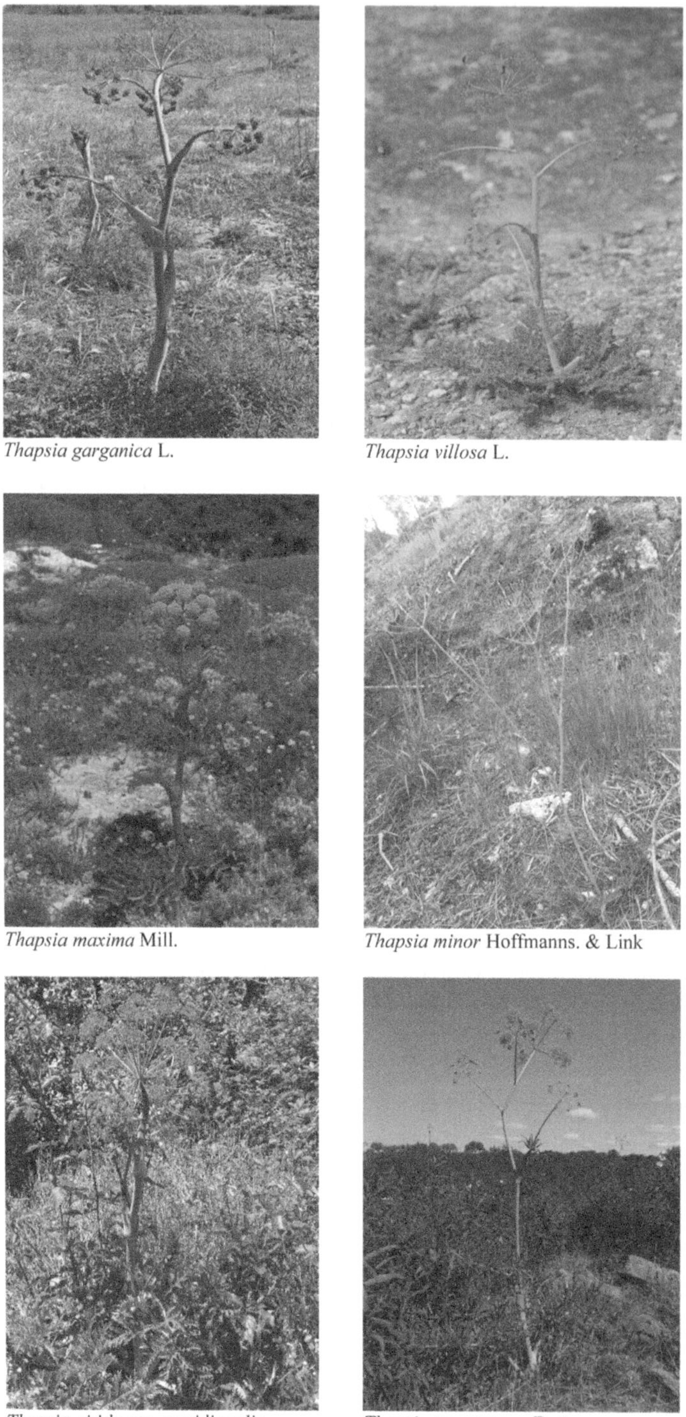

Thapsia garganica L.

Thapsia villosa L.

Thapsia maxima Mill.

Thapsia minor Hoffmanns. & Link

Thapsia nitida var. *meridionalis*

Thapsia transtagana Brot.

FIGURE 6.1 Some species of the genus *Thapsia*.

The beneficial effects of skin-irritation have been employed in traditional Arabian medicines for millennia (Perrot, 1943). These deadly plants to camels have been used as pain relievers in Algeria (Casiglia et al., 2015). However, for the time being, *Thapsia* spp. are still employed in traditional North African medicines; the previous studies of Boudjelal et al. (2013) and Benderradji et al. (2014) have stated their uses to manage respiratory diseases, rheumatic pain, eczema, and bronchitis. However, Boughrara and Belgacem (2016) and Meddour and Meddour (2016) have reported their uses as an antipyretic to recover from weakness and to manage weight loss.

Taïbi et al. (2020) have even reported using *Thapsia garganica* root decoctions to treat prostate and skin cancers. Furthermore, in another study by Taïbi et al. (2021), *Thapsia garganica* roots infusion has been prescribed to treat bladder infections, kidney inflammation, and urinary tract cancers. Furthermore, Djahafi et al. (2021) has reported the use of this species, locally known as Derias, to treat anorexia and wounds in the region of Tiaret (Northwest Algeria).

6.5 PHYTOCHEMICAL COMPOSITION

Phytochemical research on *Thapsia* species revealed the presence of certain unique metabolites, represented mainly by several hexaoxygenated guaianolides (thapsigargicin, thapsigargin, thapsitranstagin, thapsivillosins (A–E), thapsivillosins (G–K), and 2-acetoxytrilobolide (Table 6.1)) (Christensen et al., 1980; Christensen et al., 1997; Harmatha et al., 2013).

In addition, some pentaoxygenated guaianolides (nortrilobolide, trilobolide, and thapsivillosin F) have also been reported in plants belonging to the genus along with other uncommon metabolites like thapsanes (Christensen et al., 1997), tethered lipids (Liu et al., 2004) and C19 terpenoids (Appendino et al., 2005). It should be noted that hexaoxygenated and pentaoxygenated guaianolides have only been found within species belonging to the genus *Thapsia* except for *Laser trilobum* L. (Borkh).

Despite sesquiterpenoids being prevalent within the genus *Thapsia*, thapsigargin and similar compounds, the guaianolides, are the most important. Thapsigargin is the name of a family of guaianolides characterized by its richness with a highly oxygenated tricyclic skeleton, bearing seven or eight stereogenic centres and numerous oxygenated positions esterified with acyl groups.

Thapsigargin is a highly oxygenated sesquiterpene lactone, with three or four ester groups, with a specific stereochemistry, with a *cis*-annulated lactone ring that has only been reported in *Thapsia garganica* (Smitt et al., 1995) and *Thapsia gymnesica* (Christensen et al., 1997). The unique structure elements of thapsigargins are a *cis*-annulation of the lactone ring and two non-esterified hydroxy groups at C-7 and C-11. All thapsigargins are esterified with angelic acid and acetic acid at C-2 and C-10, respectively, whereas different acids may be esterified to the hydroxy groups at C-1 and C-8.

The procurement of thapsigargin and its analogues is only possible by isolation from the roots and fruits of wild-growing plants distributed in the Mediterranean region, North Africa, and the Iberian Peninsula. Their concentration varies depending on the environmental conditions: wild plants of *Thapsia garganica* have thapsigargin concentration of 0.2%–1.2% of root dry weight and 0.7%–1.5% of the dry weight of the ripe fruits, whereas the concentration in the dried stems and leaves is 0.1%–0.5% and 0.1% respectively (Smitt et al., 1995).

Phytochemical studies focused on *Thapsia garganica* have reported that, alongside its richness with respect to several sesquiterpenoids, it also contains numerous flavonoids, terpenoids, alkaloids, tannins, saponins, coumarins, and anthraquinones exhibiting remarkable biological properties (Drew et al., 2009; Alghazeer et al., 2012; Andersen et al., 2015). Flavonoids have been detected in *Thapsia garganica* plants collected from Algeria (Chibani et al., 2014). However, coumarins have been identified in the roots of a population collected in Lybia (Larsen & Sandberg, 1970). Aici and Benmehdi (2021) reported in their phytochemical analysis of *Thapsia garganica* roots that the major chemical constituents of the extract were flavonoids, tannins, volatile oils, and saponosides.

In addition, root oil composition has been studied in depth by Avato and Smith (2000); the main components found in *Thapsia maxima* types I and II and the diploid and polyploid types of

TABLE 6.1
Hexaoxygenated Guaianolides of the Genus *Thapsia*

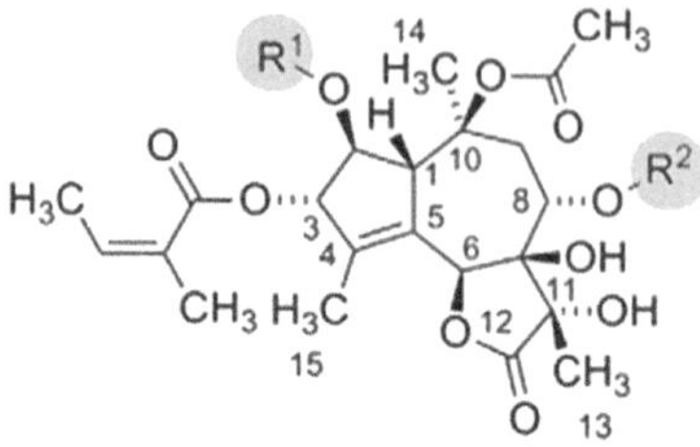

Compound	R1	R2
Thapsigargin	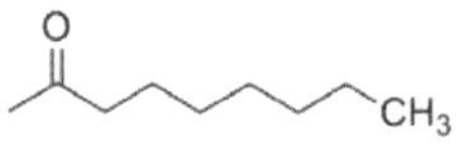	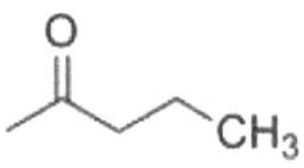
Thapsigargicin	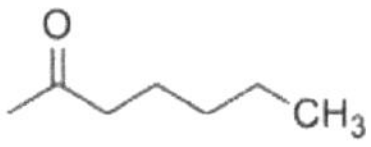	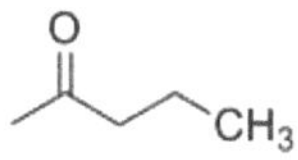
Thapsitranstagin	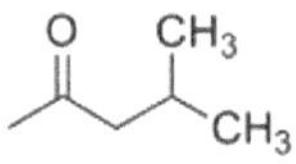	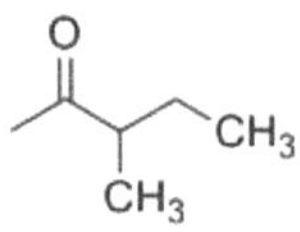
Thapsivillosin A	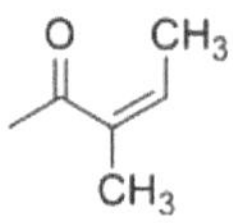	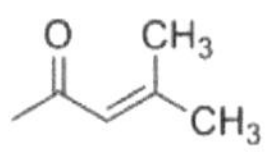
Thapsivillosin B		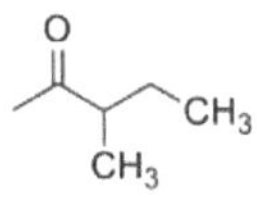
Thapsivillosin C	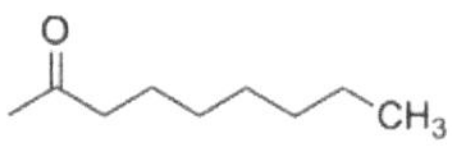	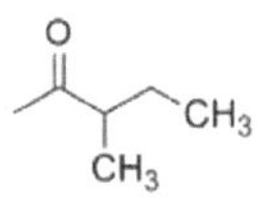
Thapsivillosin D	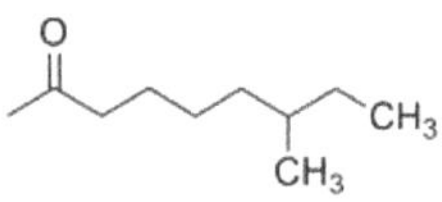	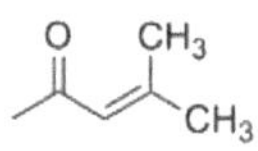
Thapsivillosin E		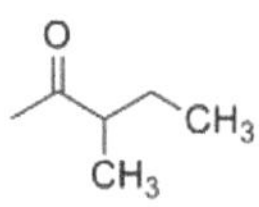
Thapsivillosin G	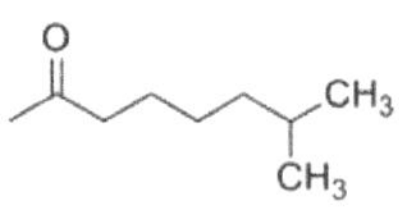	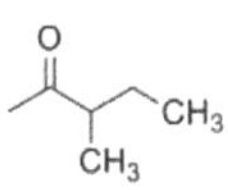
Thapsivillosin H	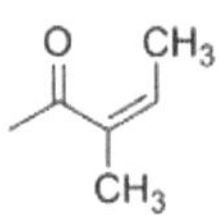	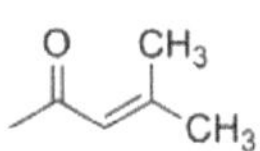
Thapsivillosin I		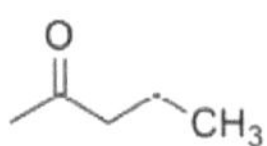
Thapsivillosin J	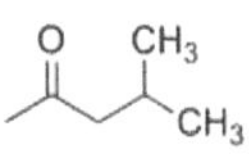	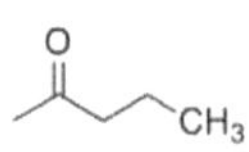

(*Continued*)

TABLE 6.1 (CONTINUED)
Hexaoxygenated Guaianolides of the Genus *Thapsia*

Compound	R1	R2
Thapsivillosin K	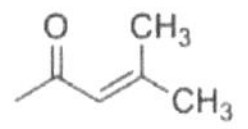	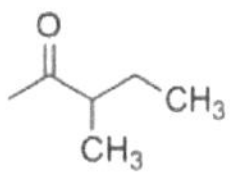
2-Acetoxytrilobolide	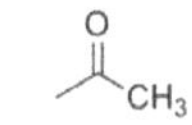	

R1 and R2 : Hexaoxygenated guaianolides.

Thapsia villosa were sesquiterpenoids. These authors have also demonstrated that the root oil of both *Thapsia maxima* I and II has a comparable composition of caryophyllene oxide, spathulenol, and elemicin. Nevertheless, limonene was absent from the root oil, whereas methyl eugenol was detected only in *Thapsia maxima* type II. In addition, a higher concentration of humulene oxide was detected in *Thapsia maxima* type I.

Several papers have been published regarding the essential oil composition of the different organs of *Thapsia garganica* (Avato, 1991; Avato & Rosito, 2002; Ladjel et al., 2011; Drew et al., 2012; Evergetis & Haroutounian, 2014; Hassen et al., 2015). It has been found that oils distilled from *Thapsia garganica* roots do not resemble those of other *Thapsia* species. They are mainly composed of phenylpropane derivatives. Elemicin and latifolone are the two primary compounds in all the analyzed root extracts from *Thapsia garganica*. However, monoterpenes and myristicin were also detected as minor components.

6.6 PHARMACOLOGICAL PROPERTIES

Thapsia species have been reported to exhibit several interesting biological activities, i.e., anticancer (Jakobsen et al., 2001), cytotoxic (Liu et al., 2006), antioxidant (Djeridane et al., 2006), antifungal, and anti-inflammatory activities (Gonçalves et al., 2012). This is why these species are still used in traditional medicines (Ouarghidi et al., 2012, 2013; Taïbi et al., 2020, 2021; Djahafi et al., 2021).

Thapsigargin and thapsigargicin are the two main active compounds in *Thapsia garganica* (Christensen et al., 1997). Particularly, the biological activity of thapsigargin has been well researched. It has been demonstrated that thapsigargin acts as a potent histamine liberator on rat mast cells with an EC_{50} value of 15 nM (Rasmussen et al., 1978). In addition to mast cells, all tested cells of the immune system were found to be activated by thapsigargin (Norup et al., 1986).

The most important pharmacological activity of thapsigargin, like other sesquiterpenoid lactones, is its effective ability to alter and inhibit the function of the sarco–endoplasmic reticulum Ca^{2+}-ATPase (SERCA) pumps, involved in the active transport of calcium ions from the cell cytosol to the sarco- and endoplasmic reticulum, which induce an increase in the cytoplasm calcium levels of mammalian cells and, consequently, lead eventually to apoptosis or the death of the cell (Thastrup et al., 1990; Christensen et al., 1992, 2009; Guerra et al., 2011; Dubois et al., 2013; Andersen et al., 2015). There is a potential role for this discovery in treating cancer diseases, mainly prostate cancer (Jakobsen et al., 2001).

Huang et al. (2018) suggested that thapsigargin significantly decreased cell proliferation and increased the apoptosis rate and the enzymatic activities of caspase-9 and -3 in human prostate cancer (PC3 cell line). Caspases are activated once cells suffer from endoplasmic reticulum stress because of the failure to fold proteins correctly during the early stage of apoptosis, which induces an increase in mitochondrial outer membrane permeability (Cella et al., 2015). However, the cytochrome C liberated in the cytoplasm promotes the formation of an apoptosis complex, activates effector caspases, and induces apoptosis (Kwegyir-Afful et al., 2015). In addition, inhibition of the Akt-mTOR pathway and modulation of the F-actin/cofilin-1/paxillin pathway by thapsigargin may suppress the growth of prostate cancer cells (Kato et al., 2016).

Previous *in-vitro* studies of the antimicrobial potency of *Thapsia* spp., carried out by Casiglia et al. (2016) and Aici and Benmehdi (2021), showed that the methanolic root extract was active against *Enterococcus faecalis* and the fungus *Penicillium purpurogenum*. Furthermore, the flavonoid extract of *Thapsia* spp. also exerted potential effects against the bacteria *Bacillus cereus*, *E. faecalis* and *S. aureus*. In addition, the saponin and flavonoid extracts of both leaves and roots showed effective antifungal activity against *Aspergillus flavus*. These activities have been attributed to the high levels of total phenolic content, flavonoids, and condensed tannins in these plant extracts. Riccobono (2016) also reported that the essential oil of *T. garganica* displayed significant antimicrobial activity against the bacterium *Bacillus subtilis* and the fungus *Candida albicans* and moderate activity against *S. aureus* and *Staphylococcus epidermidis*.

Gonçalves et al. (2012) provided evidence that *Thapsia minor* essential oils possess effective anti-fungal activity against the dermatophyte species *Candida neoformans* and *C. guillermondii*, and anti-inflammatory properties at concentrations that did not affect the viability of mammalian cells, such as macrophages and keratinocytes. These findings highlight its great potential clinical use in treating skin disorders and infections alike in dermatomycoses that can be severe, particularly in immunocompromised patients.

Methanolic leaf extracts of *Thapsia garganica* (2 g/kg) reduced histological damage and increased the survival time of mice injected with lethal doses of *Buthus occitanus* (common yellow scorpion) venom. In addition, Bouimeja et al. (2018) presented strong evidence supporting the traditional use of *Thapsia garganica* as an anti-scorpion therapy against one of the most toxic venoms in North Africa. Such a discovery would help develop complementary medicines for scorpion envenomation treatments.

6.7 CONCLUSION

The present chapter highlights the main phytochemical and pharmacological properties of *Thapsia* roots, which represent a rich source of secondary metabolites presenting several healing potentialities with a wide range of biological activities. Phytochemical studies revealed the richness of root extracts in various phenolic compounds, flavonoids, tannins, steroids, sesquiterpenoids, and phenylpropanoids. The main compounds of interest are hexaoxygenated guaianolides, particularly thapsigargins, widely used for therapeutic purposes. The wide-spectrum biological activities displayed by these molecules confirm their potential as new drug agents against several antimicrobial and anti-inflammatory diseases. In addition, the ability of the thapsigargins to induce apoptosis suggests their potential as novel promising therapeutic agents in cancer therapy. However, further studies are needed to confirm the activities of new *Thapsia* compounds along with identification of potential pharmacological targets, permitting the development of clinically suitable therapeutics.

REFERENCES

Aici, D. & Benmehdi, H. (2021) Phytochemical content, antioxidant and antimicrobial effects of *Thapsia garganica* L. leaves and roots grown wild in North-West Algeria. *Indian Journal of Agricultural Research*, 55(5), 519–526.

Alghazeer, R., El-Saltani, H., Saleh, N., Al-Najjar, A. & Hebail, F. (2012) Antioxidant and antimicrobial properties of five medicinal Libyan plants extracts. *Natural Sciences*, 4(5), 324–335. http://dx.doi.org/10.4236/ns.2012.45045

Alilou, H., Akssira, M. (2021) Chemical composition, antibacterial, antioxidant and insecticidal activities of Moroccan *Thapsia transtagana* essential oil. *Saudi Journal of Biological Sciences*, 28(12), 6756–6764. https://doi.org/10.1016/j.sjbs.2021.07.052

Andersen, T.B., López, C.Q., Mancza, T., Martinez, K. & Simonsen, H.T. (2015) Thapsigargin-from Thapsia L. to mipsagargin. *Molecules (Basel, Switzerland)*, 20(4), 6113–6127. https://doi.org/10.3390/molecules20046113

Appendino, G., Prosperini, S., Valdivia, C., Ballero, M., Colombano, G., Billington, R.A. et al. (2005) SERCA-inhibiting activity of C-19 terpenolides from *Thapsia garganica* and their possible biogenesis. *Journal of Natural Products*, 68, 1213–1217.

Avato, P., Jacobsen, N. & Smitt, U.W. (1992) Chemotaxonomy of *Thapsia maxima* Miller. Constituents of the essential oil of the fruits. *Journal of Essential Oil Research*, 4, 467–473.

Avato, P. & Rosito, I. (2002) Essential oils from the roots of *Thapsia garganica. Journal of Essential Oil Research*, 14, 20–22.

Avato, P. & Smitt, U.W. (2000) Composition of the essential oils from the roots of *Thapsia maxima* Miller and *T. villosa* L. *Journal of Essential Oil Research*, 12, 303–309.

Avato, P. (1991) Essential oil of *Thapsia garganica. Planta Medica*, 57, 585–586.

Banasiak, L., Wojewódzka, A., Baczyński, J., Reduron, J.P., Piwczyński, M., Kurzyna-Młynik, R. et al. (2016) Phylogeny of Apiaceae subtribe Daucinae and the taxonomic delineation of its genera. *Taxon*, 65, 563–585. http://dx.doi.org/10.12705/653.8

Bellakhdar, J. (1997) La Pharmacopée Marocaine Traditionnelle: Médecine Arabe Ancienne et Savoirs Populaires. Ed. Ibis Press: Paris, France, 1370.

Benderradji, L., Rebbas, K., Ghadbane, M., Bounar, R., Brini, F. & Bouzerzour, H. (2014) Ethnobotanical study of medicinal plants in Djebel Messaad region (M'sila, Algeria). *Global Journal of Research on Medicinal Plants & Indigenous Medicine*, 3(12), 445–459.

Bertherand, E.L. (1857) Notice sur la résine de *Thapsia garganica* et son emploi comme emplatre révulsif. *Journal of Méditerranean Chirurgy and Pharmacolgy*, 24, 273–274.

Boudjelal, A., Henchiri, C., Sari, M., Sarri, D., Hendel, N., Benkhaled, A. & Ruberto, G. (2013) Herbalists and wild medicinal plants in M'Sila (North Algeria): An ethnopharmacology survey. *Journal of Ethnopharmacology*, 148(2), 395–402.

Boughrara, B. & Belgacem, L. (2016) Ethnobotanical study close to the population of the extreme north east of Algeria: The municipalities of El Kala National Park (EKNP). *Industrial Crops and Products*, 88, 2–7.

Bouimeja, B., Abdelmonaim El Hidan, M., Touloun, O., Ait Laaradia, M., Ait Dra, L., El Khoudri, N. et al. (2018) Anti-scorpion venom activity of *Thapsia garganica* methanolic extract: Histopathological and biochemical evidence. *Journal of Ethnopharmacology*, 211, 340–347. https://doi.org/10.1016/j.jep.2017.10.003

Casiglia, S., Ben Jemia, M., Riccobono, L., Bruno, M., Scandolera, E. & Senatore, F. (2015) Chemical composition of the essential oil of *Moluccella spinosa* L, (Lamiaceae) collected wild in Sicily and its activity on microorganisms affecting historical textiles. *Natural Product Research*, 29(13), 1201–1206. https://doi.org/10.1080/14786419.2014.995654

Casiglia, S., Riccobono, L., Bruno, M., Rosselli, S., Senatore, F. & Senatore, F. (2016) Chemical composition of the essential oil from *Thapsia garganica* L. (Apiaceae) grown wild in Sicily and its antimicrobial activity. *Natural Product Research*, 30, 1042–1052.

Cella, D., Ivanescu, C., Holmstrom, S., Bui, C.N., Spalding, J. & Fizazi, K. (2015) Impact of enzalutamide on quality of life in men with metastatic castration-resistant prostate cancer after chemotherapy: Additional analyses from the AFFIRM randomized clinical trial. *Annal of Oncology*, 26, 179–185.

Chibani, S., Al-Dabbas, M., Abuhamdah, S., Aburjai, T., Bencheraiet, R., Kabouche, A. et al. (2014) Flavonoids and antioxidant activity of *Thapsia garganica* from Algeria. *Chemistry of Natural Compounds*, 50, 357–359.

Christensen, S.B., Andersen, A., Lauridsen, A., Moldt, P., Smitt, U.W. & Thastrup, O. (1992) Thapsigargin: A lead to design of drugs with the calcium pump as target. In: Krogsgaard-Larsen, P., Christensen, S.B., Kofod, H. (Eds.) *New leads and targets in drug research.* Alfred Benzon Symp 33, Munksgaard, Copenhagen, pp. 243–252.

Christensen, S.B., Andersen, A. & Smitt, U.W. (1997) Sesquiterpenoids from thapsia species and medicinal chemistry of the thapsigargins. *Chemistry of Natural Products*, 71, 131–167.

Christensen, S.B., Rasmussen, U. & Christophersen, C. (1980) Thapsigargin, constitution of a sesquiterpene lactone histamine liberator from *Thapsia garganica. Tetrahedron Letters*, 21, 3829–3830.

Christensen, S.B., Skytte, D.M., Denmeade, S.R., Dionne, C., Møller, J.V., Nissen, P. et al. (2009) A trojan horse in drug development: Targeting of thapsigargins towards prostate cancer cells. *Anticancer Agents in Medicinal Chemistry*, 9, 276–294.

Djahafi, A., Taïbi, K., Aït Abderrahim, L. (2021) Aromatic and medicinal plants used in traditional medicine in the region of Tiaret, North West of Algeria. *Mediterranean Botany*, 42, e71465. https://dx.doi.org/10.5209/mbot.71465

Djeridane, A., Yousfi, M., Nadjemi, B., Boutassouna, D., Stocker, P. & Vidal, N. (2006) Antioxidant activity of some Algerian medicinal plants extracts containing phenolic compounds. *Food Chemistry*, 97, 654–660.

Drew, D.P., Krichau, N., Reichwald, K. & Simonsen, H.T. (2009) Guaianolides in Apiaceae: Perspectives on pharmacology and biosynthesis. *Phytochemistry Review*, 8, 581–599.

Drew, D.P., Rasmussen, S.K., Avato, P. & Simonsen, H.T. (2012) A comparison of headspace solid-phase microextraction and classic hydrodistillation for the identification of volatile constituents from *Thapsia* spp. Provides insights into guaianolide biosynthesis in Apiaceae. *Phytochemical Analysis*, 23, 44–51.

Dubois, C., Abeele, F.V., Sehgal, P., Olesen, C., Junker, S., Christensen, S.B. et al. (2013) Differential effects of thapsigargin analogues on apoptosis of prostate cancer cells. *FEBS Journal*, 280, 5430–2540.

Evergetis, E. & Haroutounian, S.A. (2014) Exploitation of Apiaceae family plants as valuable renewable source of essential oils containing crops for the production of fine chemicals. *Industrial Crops and Products*, 54, 70–77.

Gonçalves, M.J., Cruz, M.T., Tavares, A.C., Cavaleiro, C., Lopes, M.C., Canhoto, J. et al. (2012) Composition and biological activity of the essential oil from *Thapsia minor*, a new source of geranyl acetate. *Industrial Crops and Products*, 35, 166–171.

Greive, M. (1996) A modern herbal. Available from: http://botanical.com/botanical/mgmh/mgmh.html

Guerra, F.M., Moreno-Dorado, F.J., Jorge, Z.D. & Massanet, G.M. (2011) Recent advances in the synthesis of sesquiterpenolides from Umbelliferae. *Natural Product Communications*, 6(4), 491–496.

Harmatha, J., Budesinsky, M., Vokac, K., Kostecka, P., Kmonickova, E. & Zidek, Z. (2013) Trilobolide and related sesquiterpene lactones from *Laser trilobum* possessing immunobiological properties. *Fitoterapia*, 89C, 157–166.

Larsen, P.K. & Sandberg, F. (1970) Constituents of umbelliferous plants. XV. Coumarins from Thapsia garganica L. The structure of a new coumarin. *Acta Chem Scand.*, 24, 1113–1114.

Hassen, I., M'Rabet, Y., Belgacem, C., Kesraoui, O., Casabianca, H. & Hosni, K. (2015) Chemodiversity of volatile oils in *Thapsia garganica* L. (Apiaceae). *Chemical Biodiversity*, 12, 637–651.

Huang, F., Wang, P. & Wang, X. (2018) Thapsigargin induces apoptosis of prostate cancer through cofilin-1 and paxillin. *Oncology Letters*, 16(2), 1975–1980. https://doi.org/10.3892/ol.2018.8833

Jakobsen, C.M., Denmeade, S.R., Isaacs, J.T., Gady, A., Olsen, C.E. & Christensen, S.B. (2001) Design, synthesis, and pharmacological evaluation of thapsigargin analogues for targeting apoptosis to prostatic cancer cells. *Journal of Medical Chemistry*, 44, 4696–4703.

Kato, M., Banuelos, C.A., Imamura, Y., Leung, J.K., Caley, D.P., Wang, J. et al. (2016) Cotargeting androgen receptor splice variants and mTOR signaling pathway for the treatment of castration-resistant prostate cancer. *Clinical Cancer Research*, 2, 2744–2754.

Kwegyir-Afful, A.K., Ramalingam, S., Purushottamachar, P., Ramamurthy, V.P. & Njar, V.C. (2015) Galeterone and VNPT55 induce proteasomal degradation of AR/AR-V7, induce significant apoptosis via cytochrome C release and suppress growth of castration resistant prostate cancer xenografts *in vivo*. *Oncotarget*. 6, 27440–27460.

Ladjel, S., Zellagui, A. & Gherraf, N. (2011) Reinvestigation of essential oil content of *Thapsia garganica* grown in the east of Algeria. *Revue des Sciences Fondamentales et Appliquées*, 3, 30–34.

Liu, H., Jensen, K.G., Tran, L.M., Chen, M., Zhai, L., Olsen, C.E. et al. (2006) Cytotoxic phenylpropanoids and an additional thapsigargin analogue isolated from *Thapsia garganica*. *Phytochemistry*, 67, 2651–2658.

Liu, H., Olsen, C.E. & Christensen, S.B. (2004) Tethered lipids from *Thapsia garganica*. *Journal of Natural Products*, 67, 1339–1340.

Meddour, R. & Meddour-Sahar, O. (2016) Medicinal plants and their traditional uses in Kabylia (Tizi Ouzou, Algeria). *Arab Journal of Medicinal and Aromatic Plants*, 1(2), 137–151.

Norup, E., Smitt, U.W. & Christensen, S.B. (1986) The potencies of thapsigargins and analogues as activators of rat peritoneal mast cell. *Planta Medica*, 4, 251–255.

Ouarghidi, A., Gary, J.M. & Abbad, A. (2013) Botanical identification and ethnomedicinal uses of some underground part of medicinal plants collected and traded in Marrakech region. *Journal of Medicinal Plants Research*, 7, 2165–2169.

Ouarghidi, A., Powell, B., Martin, G., de Boer, H. & Abbad, A. (2012) Species substitution in medicinal roots and possible implications for toxicity of herbal remedies in Morocco. *Economic Botany*, 66, 370–382.

Perrot, E. (1943) *Matières premières usuelles du règne végétal*. Masson, Paris, 1630–1632.

Pujadas, A., Rossel, J.A. & Barcelo, P. (1991) De flora baleárica adnotationes. 10. *Thapsia gymnesica*. *Candollea*, 46, 65–74.

Pujades-Salvà, A.J. & Rosselló, J.A. (2003) Thapsia L. Ed. Flora Ibérica (Vol. X): Araliaceae—Umbelliferae. Real Jardín Botánico, C.S.I.C: Madrid, Spain, 10, 401–410.

Rasmussen, U., Christensen, S.B. & Sandberg, F. (1978) Thapsigargine and thapsigargicine, two new histamine liberators from *Thapsia garganica* L. *Acta Pharmaceutica Suecica*, 15, 133–140.

Riccobono, L. (2016) Metabolites from Mediterranean plants: Characterization and transformation. Chemotaxonomic assessment and biological activity. Doctorate thesis, Università degli Studi di Palermo, p. 168.

Smitt, U.W., Avato, P. & Christensen, S.B. (1990) New pro azulene guaianolides from *Thapsiavillosa* L. *Journal of Natural Products*, 53, 1479–1484.

Smitt, U.W., Jäger, A.K., Adsersen, A. & Gudiksen, L. (1995) Comparative studies in phytochemistry and fruit anatomy of *Thapsia garganica* and *T. transtagana*, Apiaceae (Umbelliferae). *Botanical Journal of the Linnean Society*, 117, 281–292.

Smitt, U.W. (1995) A chemotaxonomic investigation of *Thapsiavillosa* L., Apiaceae (Umbelliferae). *Botanical Journal of the Linnean Society*, 119(4), 367–377.

Taïbi, K., Aït Abderrahim, L., Boussaid, M., Taïbi, F., Achir, M., Souana, K. et al. (2021) Unraveling the ethnopharmacological potential of medicinal plants used in Algerian traditional medicine for urinary diseases. *European Journal of Integrative Medicine*, 44, 101339.

Taïbi, K., Aït Abderrahim, L., Ferhat, K., Betta, S., Taïbi, F., Bouraada, F. et al. (2020) Ethnopharmacological study of natural products used for traditional cancer therapy in Algeria. *Saudi Pharmaceutical Journal*, 28(11), 1451–1465.

Thastrup, O., Cullen, P.J., Drobak, B.K., Hanley, M.R. & Dawson, A.P. (1990) Thapsigargin, a tumor promoter, discharges intracellular Ca2+ stores by specific inhibition of the endoplasmic reticulum Ca2+-ATPase. *Proceedings of the National Academy of Sciences USA*, 87, 2466–2470.

Tlili, N., Mejri, H., Lajnef, H.B., Feriani, A., Khaldi, A. & Nasri, N. (2015) Unexploited *Thapsiagarganica*, *Orlayamaritima* and *Retama raetam* seeds: Potential sources of unsaturated fatty acid and natural antioxidants. *Journal of the American Oil Chemists' Society*, 92, 1175–1181.

Tutin, T.G., Heywood, V.H., Burges, N.A., Moore, D.M., Valentine, D.H., Walters, S.M. et al. (1968) Flora Europaea, vol 2. University Press, Cambridge, 370.

Weitzel, C., Rønsted, N., Spalik, K. & Simonsen, H.T. (2014) Resurrecting deadly carrots: Towards a revision of Thapsia (Apiaceae) based on phylogenetic analysis of nrITS sequences and chemical profiles. *Botanical Journal of the Linnean Society*, 174, 620–636.

7 Secondary Metabolites, Ethnopharmacology, and Commercial Application of *Glycyrrhiza glabra*

Noureddine Chaachouay, Abdelhamid Azeroual, Bouchaib Bencharki, and Lahcen Zidane

7.1 INTRODUCTION

Nature is an appealing source of novel therapeutic candidate chemicals since millions of animals, marine creatures, microbes, and plants contain a considerable variety of chemicals which could be possible medical agents (Chaachouay et al., 2023; Sharma et al., 2018). For millennia, people have used plant derivatives to treat various ailments. The information about medicinal plant qualities was learned by early civilizations and passed down from generation to generation to the present day (Al-Snafi, 2018; Abate et al., 2021, 2022; Asfaw, 2022).

Glycyrrhiza glabra L. is among the most extensively studied economically beneficial medicinal herbs (Fenwick et al., 1990; Öztürk et al., 2018). The term '*Glycyrrhiza*' is derived from the ancient Greek word *Glykos* (sweet) and *Rhiza* (root). *G. glabra*, often known as liquorice (UK), licorice (US), and 'sweet wood', as well as 'Arq Sûs' in North Africa (Arabic). It is a member of the Fabaceae (pea) family, and its relatives are now extensively utilized as food and feed. Liquorice is a hardy plant or undershrub that grows to about 2 m in height and is endemic to the Mediterranean region, India, Central and Southwest Asia, and Southeast Asia (Dwivedi, 1999).

Approximately 30 species comprise the genus *Glycyrrhiza*, including *G. glabra*, *Glycyrrhiza aspera* Pall., *Glycyrrhiza inflata* Batalin, *Glycyrrhiza korshinskyi* Grig., *Glycyrrhiza eurycarpa* P.C. Li., and *Glycyrrhiza uralensis* Fisch. (The Plant List, 2013). *G. glabra* is an herbaceous perennial that grows to 1–2 m in height and has pinnate leaves that are 7–15 cm long and have 9—17 leaflets. The blooms are violet to pale bluish-white and grouped in a hermaphrodite inflorescence (Huxley & Griffiths, 1992; Öztürk et al., 2018). This plant is cultivated in the United Kingdom, the United States of America, Italy, Spain, Germany, France, Russia, China, and northern India. Large-scale farming is viable in England, Spain, and Sicily (Chaachouay et al., 2019; Anagha et al., 2012; Armanini et al., 2005; Fenwick et al., 1990; Nomura et al., 2002). *G. glabra* thrives in warm temperate and subtropical climates. It prefers well-limed, well-drained, loose, friable, deep soil rich in organic matter in full sun (Öztürk et al., 2018; Zhao et al., 2005) and clay soil beside a creek or river with enough water (Öztürk et al., 2018).

In ancient medical history, *G. glabra* was a commonly employed medicinal herb species as a folk therapy, traditional medicine, and flavouring herb. The earliest historical references to this plant are related to its wide-ranging therapeutic benefits (Fenwick et al., 1990). The Egyptian, Babylonian, Assyrian, Phoenician, Aryan, Persian, Greek, Roman, Chinese, and Indian civilizations all used this plant. The first reference to its therapeutic use is recorded in the '*Code of Hammurabi*' (2100

 DOI: 10.1201/b22924-7

BC). It was also among the valuable plants mentioned in Assyrian herbals (2000 BC) (Anagha et al., 2012). Hippocrates (400 BC) described it as a treatment for the quenching of thirst and ulcers. Dioscorides and Theophrastus (322BC) documented *G. glabra* as a medicinal herb and described its therapeutic properties (Kaur et al., 2013). In the classic Arabic medical system, *G. glabra* is used internally for treatment of asthma, arthritis, bronchitis, infertility, and peptic ulcers (Al-Snafi, 2018; Chaachouay et al., 2021b, 2021a; Chaachouay & Zidane, 2021; Sayed, 1980). This medicinal herb appears to have a consistent bioactive composition profile, traditional applications, and pharmacological effects. This chapter provides a complete overview with up-to-date information on the ethnobotany, phytochemistry, and pharmacological activities of *G. glabra*.

7.2 POSITION OF THE TAXON IN THE SCIENTIFIC CLASSIFICATION

7.2.1 Domain

G. glabra fits under the Eukaryota domain because it has a membrane-bound cell that contains a nucleus. It fits under the Plantae kingdom, Tracheophyta phylum, Magnoliopsida class, Fabales order, Fabaceae family, and *Glycyrrhiza* genus (IT IS, 2022).

Synonyms: **(1)** *Glycyrrhiza glabra* var. *caduca* XY Li, **(2)** *Glycyrrhiza glabra* var. *glabra,* **(3)** *Glycyrrhiza glabra* subsp. *glandulifera* (Waldst. & Kit.) Ponert, **(4)** *Glycyrrhiza glabra* var. *glandulifera* (Waldst. & Kit.) Regel & Herder, **(5)** *Glycyrrhiza glabra* var. *glandulifera* (Waldst. & Kit.) Boiss., **(6)** *Glycyrrhiza glabra* var. *glandulosa* XY Li, **(7)** *Glycyrrhiza glabra* var. *laxifoliolata* XY Li, **(8)** *Glycyrrhiza glabra* var. *typica* L., **(9)** *Glycyrrhiza glabra* var. *violacea* (Boiss. & Noe) Boiss (The Plant List, 2013).

7.2.2 Vernacular Names

Arabic: Arq As-Sus, Irqu As-Sus, Irqu Al-Sus, Sous, Sus; *Afrikaans:* Drop; *Albanian:* Gliciriza E Shogët, Glicirizë; *Armenian:* Madoodag, Matutak; *Azeri:* Biyanlıq; Basque: Erregaliz, Gotxerro, Makilgoxo; *Chinese:* Gān Cǎo; *Czech:* Lékořice, Lékořice Lysá, Sladký Dřevo; *Danish:* Glat Lakrids, Lakrids, Lakrids Rod, Lakridsrod, Lakridsplante; Dutch: Zoethout; *English:* Liquorice (UK), Licorice (US); *Farsi:* Shirin Bayan; *French:* Réglisse; *German:* Gemeines Süssholz; *Greek:* Glikoriza; *India:* Jesthimadhu; *Italian:* Liquirizia; *Portuguese:* Alcaçuz; *Russian:* Koren Solodki; *Spanish:* Alcazuz; *Swahili:* Susu; *Turkish:* Biyam (Akbar, 2020; Bellakhdar, 1997; Chaachouay et al., 2023a; Chaachouay et al., 2020).

7.3 BOTANY

Leaves: imparipinnate, 7–15 cm long, with 9–17 elliptic leaflets, oblong-lanceolate to ovate-oblong (1.7–4.0 by 0.8–2.0 cm), abaxially thickly lamellate glandular punctate and pubescent on veins, adaxially glabrescent or pilose (Öztürk et al., 2018). Stipules are linear, caducous, 1–2 mm. The flowers range from purple to a light bluish blue, borne in axillary spikes and grouped in a hermaphrodite inflorescence. The calyx is small, campanulate, with trapezoidal ends, and produces capillary hair (Zeng et al., 1988). The fruit is a pod or compressed legume up to 1.5 cm long, flat, erect, glabrous, slightly pitted, and typically contains 35 brown reniform seeds with lanceolate ends, and glandular hairs (Huxley & Griffiths, 1992; Öztürk et al., 2018).

7.4 DISTRIBUTION

G. glabra is native to Morocco, Iraq, Syria, Palestine, Saudi Arabia, Turkey, Lebanon, Afghanistan, Albania, Bulgaria, Central European Russia, North-Central China, Cyprus, East Aegean Islands, East European Russia, Greece, Iran, Italy, Kazakhstan, Kirgizstan, Krym, Mongolia, North

Caucasus, Pakistan, Romania, Sardinia, Sicily, South European Russia, Tadzhikistan, Transcaucasus, Turkmenistan, Ukraine, Uzbekistan, West Siberia, Xinjiang, Serbia, Croatia, Montenegro, and Bosnia-Herzegovina (Fenwick et al., 1990; Fiore et al., 2005; Lange 1998; Olukoga & Donaldson, 1998; Öztürk et al., 2018) (Figure 7.1).

It was introduced into Algeria, Egypt, Portugal, Austria, Bangladesh, Cape Provinces, Czechoslovakia, Hungary, France, New South Wales, South Australia, the Maldives, Spain, Switzerland, and Victoria. Afghanistan, Iraq, Azerbaijan, Pakistan, Iran, the People's Republic of China, Turkmenistan, Turkey, and Uzbekistan are the countries producing *G. glabra* commercially. Three botanical varieties of *G. glabra* L. have been documented; the Italian and Spanish liquorice, assigned to *Glycyrrhiza glabra* var. *typica* L.; Russian licorice to *Glycyrrhiza glabra* var. *glandulifera* (Waldst. & Kit.) Boiss.; and Turkish and Persian liquorice to *Glycyrrhiza glabra* var. *violacea* (Boiss. & Noe) Boiss (Nomura et al., 2002).

7.5 ECOLOGY

Liquorice is not affected by freezing since it is inactive in winter and benefits from the hard times during the winter, which causes the transfer of resources to the underground rhizomes (Öztürk et al., 2018; Zhao et al., 2005). The rhizomes are thick enough to be extracted only after three years of development.

7.6 SECONDARY METABOLITES

In recent years, secondary metabolites of *G. glabra* have been the subject of several investigations (Hamad et al., 2020; Hatano et al., 2000; Husain et al., 2015; Karkanis et al., 2018; Öztürk et al., 2018; Thakur et al., 2016). Several active ingredients have been isolated from the rhizomes of *G. glabra*, such as alkaloids, coumarins, essential oils, flavonoids, phenolic compounds, saponins, tannins, triterpenes, asparagines, gums, bitters, lipids, mineral salts, vitamins, mucilage, pectins, resins, volatile oils, starches, the female sex hormone oestrogen, and other molecules (Amani et al., 2005; Hamad et al., 2020; Hayashi et al., 2003; Husain et al., 2015; Wang et al., 2015). The yellow colour of the inside of *G. glabra* rhizomes results from the flavonoid components. Chalcones, flavones, flavanonols, flavanones, isoflavones, isoflavanones, isoflavans, isoliquiritin, liquiritin, and liquiritigenin, are among the flavonoids discovered (Fu et al., 2005; Fu et al., 2004; Fukai et al., 2002; Kitagawa et al., 1994; Yahara & Nishioka, 1984). *G. glabra* root includes phenolic ingredients such as glycerin, licocoumarone, licoarylcoumarin, and glycycoumarin (Coruh et al., 2007; Hamad et al., 2020; Hatano et al., 1989; Zhang et al., 2005). The important terpene constituents of *G. glabra* were documented as glabrolide, liquiritic acid, 11-deoxoglycyrrhetinic acid, isoglabrolide, 11-deoxoglabrolide, glypallidifloric acid, and 3-β-hydroxy-ll (Kitagawa et al., 1993; Kiuchi et al., 1990; Nomura et al., 2002). Major minerals identified in the roots were Al, As, Ca, Cu, Fe, K, Mg, Mn, Na, P, S, Si, Sr, Ti, and Zn (Coruh et al., 2007). The essential oil extracted from *G. glabra* is rich in benzoic acid, octanoic acid, furfuryl formate, γ-nonalactone, 2,3-butanediol, 1-methyl-2-formylpyrrole, *N*-methyl-2-pyrrolidone, propionic acid, furfuraldehyde, trimethyl pyrazine, and maltol (Chouitah et al., 2011; Preedy, 2015; Quirós-Sauceda et al., 2016; Tamir et al., 2001; Zahedi et al., 2021). Additionally, several volatile compounds, including geraniol, α -terpineol, hexanol, terpinen-4-ol, and pentanol are found in the root of *G. glabra*, imparting a distinctive odour (Farag & Wessjohann, 2012; Frattini et al., 1977). Table 7.1 details the essential secondary metabolites of *G. glabra*.

7.7 ETHNOPHARMACOLOGY

7.7.1 Traditional Uses

Since prehistoric times, Indigenous communities have used *G. glabra* root, which is well established in written form throughout the antique civilizations. Historically, the Arabs used *G. glabra* to alleviate

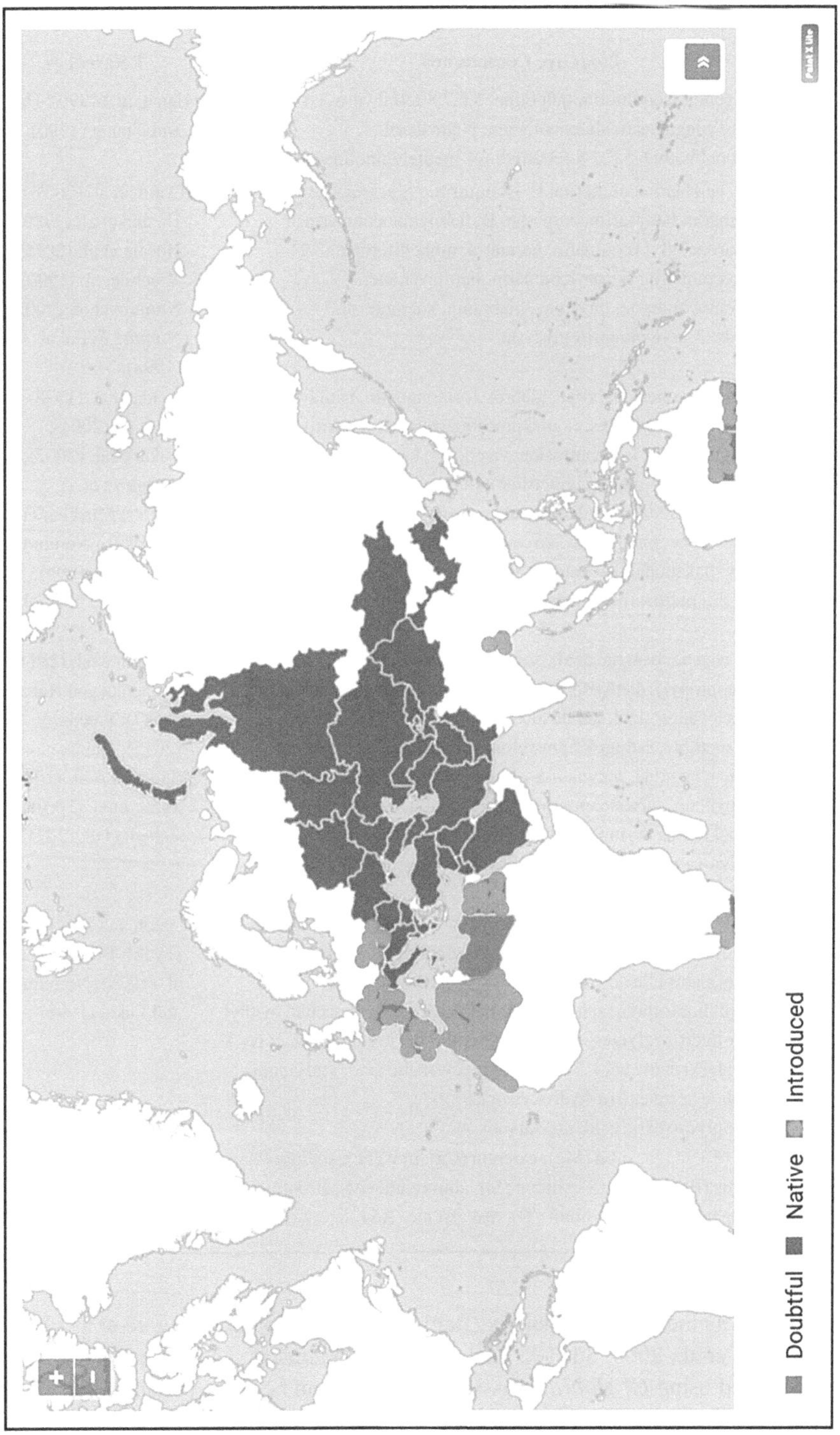

FIGURE 7.1 Geographical distribution of *Glycyrrhiza glabra* L.

TABLE 7.1
Different Secondary Metabolites of *Glycyrrhiza glabra* L.

Secondary Metabolites	Bioactive Compounds	References
Alkaloids	Isoliquiritigenin, isoquinolin, quinoline, 5,6,7,8-tetrahydro-2,4-dimethy lquinoline, yunganoside, liconeolignan, β-sitosterol, tetrahydropalmatine, 5,6,7,8-tetrahydro-4-methylquinoline	Han et al. (1990); Han and Chung (1990)
Phenols and coumarins	Astragalin, hedysarimcoumestan E, isoliquiritin, glycerol, glycocoumarin, hedysarimcoumestan B, licopyranocoumarin, licocoumarone, glycycoumarin, herniarin, umbelliferone, licopyranocoumarin, licoarylcoumarin, licoftavanone, isolicopyranocoumarin, chalcone, glabranin, kaempferol-3-*O*-diglucoside, 3-arylcoumarin-glycyrin	(Coruh et al. (2007-; Hatano et al. (1989); Husain et al. (2015); Kiuchi et al. (1990); Nomura et al. (2002); Nomura & Fukai (1998)
Flavonoids (flavanonols, flavanones, isoflavones, and hispaglabridin A and B)	Glycyrdione, flavones, glabrene, glabrol, formononetin, isoflavanones, shinpterocarpin, isoflavans, liquiritin, shinflavanone, liquiritigenin, prenyllicoflavone A, isoschaftoside, vicenin-2, kanzonol Y, afrormosin-7-*O*-(6- malonylglucoside), licoagrodin, licoagrochalcone, licoagrocarpin, licoagrochalcone, calycosin 7-*O*-glucoside, licoagrochalcone, ononin, licoagroaurone, licochalcone, wistin, 7,4′ dihydroxy–flavanone, β-sitosterol, stigmasterol, glucoliquiritin apioside, shinpterocarpin, neoisoliquiritin	(Asada et al. (1998); Fu et al. (2004); Fukai et al. (2002); Kitagawa et al. (1994); Mitscher et al. (1980); Nomura and Fukai (1998); Yahara and Nishioka (1984)
Essential Oils	Linalool, benzene, benzoic acid, isoniazid, hexanol, octanoic acid, 4-(2-aminopropyl), diethyltoluamide, iodoquinol, phenol, hexanoic acid, hexadecanoic acid, cumic alcohol, γ-nonalactone, indole, eugenol, anethole, estragole, prasterone, decanal, (*E*)-2-heptenal, himachalene epoxide, *p*-cymen-8-ol, methyl eugenol, α-terpineol, β-caryophyllene, carvone, piperitone, lavandulol, terpinen-4-ol, (*E*)-linalool oxide, thymol, furfural, 3-methyl-6-propryl phenol, 1-methoxy4-isopropyl benzene, β-dihydro-ionone, isobutyl adipate, 5-pentylpyran-2-one	Chouitah et al. (2011); Kameoka and Nakai (1987); Preedy (2015); Quirós-Sauceda et al. (2016); Tamir et al. (2001); Zahedi et al. (2021)
Triterpenoid saponins	Glycyrrhizic acid, glabrolide, aglycone, soyasaponins I and II, liquoric acid, liquiritic acid, glabranin-A, II-deoxoglabrolide, isoglabrolide, 11-desoxo-glabrolide acetate, 11-deoxoglycyrrhetinic acid, glycyrretol, liquiridiolic acid, 24-hydroxyliquiritic acid, acetate methyl ester, 18α-hydroxyglycyrrhetic acid, glypallidifloric acid, 24-acetoxy-l 1-desoxoglycyrrhetic acid, 24-hydroxyglycyrrhetic acid, uralsaponin B, glabrolide acetate, 21α-hydroxyisoglabrolide, 11-desoxoglycyrrhetic acid, uralsaponin A	Beaton and Spring (1956); Bogatkina et al. (1975); Nomura and Fukai (1998)
Mineral elements	Potassium (K), magnesium (Mg), calcium (Ca), iron (Fe), sodium (Na), phosphorus (P), sulphur (S), silicon (Si), aluminium (Al), strontium (Sr), manganese (Mn), titanium (Ti), and arsenic (AS)	Lim (2016)

constipation and coughs via its medicinal properties (Benlarbi et al., 2023; Benkhnigue et al., 2023; Orch et al., 2021; Armanini et al., 2005; Saad & Said, 2011). Theophrastus, an antique Greek botanist (371–286 BCE), reported using *G. glabra* to assist with asthma and coughs (Chaachouay et al., 2019b; Chaachouay et al., 2019a; Saad & Said, 2011; Zeng et al., 1988). Since 25 CE, the Chinese and Tibetans have used the plant to treat and prevent coughs as an expectorant and adjuvant in numerous herbal formulations (Foster & Chongxi, 1992; Zeng et al., 1988). *G. glabra* has a long history as an Ayurvedic therapy due to its expectorant, anti-inflammatory, and laxative effects (Onstad, 1996;

Zadeh et al., 2013). *G. glabra* was mentioned in the *De Materia Medica* as an expectorant, diuretic, bronchitis, emollient, haemoptysis, demulcent, and laxative, and to treat laryngitis, etc. (Dioscorides, 1829). Classically, *G. glabra* has been used to treat various ailments, including Addison's disease, skin diseases, kidney diseases, haemorrhagic diseases, asthma, arthritis, acidity, hyperdipsia, sore throat, tonsillitis, flatulence, shingles, heartburn, rheumatism, swellings, eczema, bleeding, stomach ulcers, epilepsy, fever, colic, herpes, sexual debility, paralysis, leucorrhoea, peptic ulcer, jaundice, and coughs (Chaachouay et al., 2022b, 2022a; Chaachouay et al., 2022; Chaachouay et al., 2021; Chaachouay et al., 2021c; Chaachouay et al., 2021a; Anagha et al., 2012; Armanini et al., 2005; Braun, 2007; Damle, 2014; Fenwick et al., 1990; Hayashi et al., 2003; Karkanis et al., 2018; Kaur et al., 2013; Khare 2004; Kitagawa et al., 1994; Lim 2016; Tewari et al., 2017; Zadeh et al., 2013).

7.7.2 Pharmacological Properties

G. glabra is an excellent herbal treatment which has been used for a long time. Many pharmacological investigations into its bioactivities have shown that it might be the basis for novel medications and therapies for treating various illnesses and disorders. Numerous pharmacological effects include antioxidants, anti-inflammatory, antimicrobial, antiviral, anti-ulcer, anticarcinogenic, antitumour, antiallergic, antidiabetic, anxiolytic, cytotoxic, and dermatitis, eczema, hypoglycaemic, hypocholesterolaemic, hepatoprotective, and to treat skin eruptions, have been documented for the roots of *G. glabra*. Furthermore, *G. glabra* may be employed to treat Alzheimer's disease, dementia, learning disabilities, and other neurodegenerative ailments. A summary of the various biological activities of *G. glabra* is given in Table 7.2.

7.8 COMMERCIAL APPLICATIONS

G. glabra is an important medicinal plant in many nations' commercial and economic circles, being primarily imported. *G. glabra* has been employed as a cosmetic, food additive, pharmaceutical product, sweetener, confectionery food, soft drink, flavour additive for tobacco, and as a condiment. Flavonoids extracted from *G. glabra* are widely used in cosmetics (Bown, 1995). The leading exporters and producers of *G. glabra* are as follows: China, Russia, Iran, and Turkey. The major exporters of the liquorice extract are as follows: France, Iran, Italy, Iraq, the United States, and Japan, as well as Turkey.

7.8.1 Use of *G. glabra* Products as Food Additives

G. glabra is a flavouring ingredient and sweetener in low-calorie and non-cariogenic meals. It provides aroma and sparkle to beer and sweets, respectively (Fenwick et al., 1990; Nieman, 1957; Tamura, 2017). Moreover, *G. glabra* extracts are used in spirits and beer for foam stability and flavour. *G. glabra* extracts and glycyrrhizin are utilized as food additives in various foods, including quick noodles, sauces, marmalades, chewing gum, ice creams, diet cola, chocolate, brown colouring matter, jams, sherbets, and sweet snacks (Chaachouay et al., 2019c; Fenwick et al., 1990; Liu et al., 1985; Nieman, 1957). Glycyrrhizin has been reported to have salt-softening properties, reducing the perception of saltiness in food. This property can be helpful in specific applications where a reduction in sodium content is desired. It is also used to reduce the saltiness of salty dishes, such as sausages, soy sauce, soy-cooked fish, boiling fish paste, savoury snacks, other spices, and Japanese pickles (Hayashi et al., 2003; Hayashi & Sudo 2009; Liu et al., 2000).

7.8.2 Application of *G. glabra* Extracts for Tobacco

G. glabra extracts are widely employed in tobacco (Housemna & Lacey, 1929; Nieman 1957). It is most often found in chewing tobacco, although it is commonly used in pipes, cigarettes, cigars, and

TABLE 7.2
Biological Properties Documented from *Glycyrrhiza glabra* L.

Pharmacological Activities	Extract/Plant Part	References
Antioxidant	Methanol-water extract	Martins et al. (2015); Wang et al. (2015)
	Methanol extract	Chopra et al. (2013); Fu et al. (2013); Karahan et al. (2016); Morteza-Semnani et al. (2003); Shinde et al. (2016); Sultana et al. (2010)
	Ethanol root extract	(Frattaruolo et al. (2019); Karami et al. (2013)
	Root and bark lyophilized aqueous extracts/ethanol extracts	(Tohma and Gulçin (2010)
	Infusion preparation	Cheel et al. (2010)
	Lipid extract of the leaves	Biondi et al. (2003)
	Aqueous and ethanolic extracts	Visavadiya et al. (2009)
	Methanol, water, and ethanol extracts	Thakur et al. (2016)
	Chloroform extract	Chandra and Gunasekaran (2017)
	Methanol-water root extract	Varsha et al. (2013)
	Ethyl acetate extract	Siracusa et al. (2011)
Antibacterial	Ethanol root extract	Karami et al. (2013); Nair and Geetha (2014); Soulef and Yahia (1970)
	Methanol extract	Chopra et al. (2013); Karahan et al. (2016); Rahnama et al. (2019)
	Methanol, water, and ethanol extracts	Masad et al. (2019); Thakur et al. (2016)
	Chloroform extract	Chandra et al. (2017)
	Methanol-water root extract	Nirmala and Selvaraj (2011); Varsha et al. (2013)
Antiproliferative	Methanol extract	Sahin Yaglıoglu et al. (2016); Shinde et al. (2016); Tamir et al. (2001)
Antiviral	Alcohol-water extract	Badam (1997)
Anti-inflammatory	Ethanol extract and other procedures	Frattaruolo et al. (2019); Račková et al. (2007); Yu et al. (2015)
	Alcohol-water extract	Nirmala and Selvaraj (2011)
	Ethyl acetate extract	Siracusa et al. (2011)
	Methanol extract	Fu et al. (2013)
	Alcohol-water plant extracts	Herold et al. (2003)
Anti-atherogenic	Aqueous and ethanol extracts	Visavadiya et al. (2009)
Immunostimulatatory	Infusion preparation	Cheel et al. (2010)
Anti-biofilm	Methanol extract	Rahnama et al. (2019)
	Methanol and water extracts	Masad et al. (2019)
Anti-genotoxic	Ethyl acetate extract	Siracusa et al. (2011)
	Methanol root extract	Yavuz Kocaman and Güzelkokar (2018)
Genotoxic	Methanol root extract	Yavuz Kocaman and Güzelkokar (2018)
Immunomodulatory	Ethanol root extract	Ayeka et al. (2017)
	Extract	Guo et al. (2015)
Antimycotic	Ethanol extract	Geetha and Anitha (2013)
Oestrogenic	Food-grade ethanol extract	Hajirahimkhan et al. (2013); Somjen et al. (2004); Tamir et al. (2001)
Antiplatelet	Butanol extract	Sajid et al. (1991)
Antitumour	Aqueous extract	Yang et al. (2014)
Renoprotective	Ethanolic extract	Malekinejad et al. (2011)

(*Continued*)

TABLE 7.2 (CONTINUED)
Biological Properties Documented from *Glycyrrhiza glabra* L.

Pharmacological Activities	Extract/Plant Part	References
Antitussive	Licorice, ethanol-water, n-butanol, and ethyl acetate extracts	Kuang et al. (2018)
Choleretic	Dissolved in polyethylene glycol and distilled water	Kim et al. (2009); Raggi et al. (1995)
Anti-convulsant	Ethanol-water extract	Chowdhury et al. (2013)
Radioprotective	Methanol extract	Kovalenko et al. (1989); Shetty et al. (2002)
Antiparasitic	Alcohol extract (licochalcone A)	Chen et al. (1993); Ziegler et al. (2004)
Expectorant	Licorice, ethanol-water, n-butanol, and ethyl acetate extracts	Kuang et al. (2018)

other tobacco products (Fenwick et al., 1990; Hayashi & Sudo, 2009). The casing liquid, sprayed onto the leaf tobacco either by dipping or spraying, contains *G. glabra* extract as a component. *G. glabra* serves three purposes: as a flavouring ingredient, a sweetening ingredient, and as a blending agent to make the taste of the tobacco milder. It can assist in keeping the tobacco wet (Fenwick et al., 1990; Nieman, 1957).

7.8.3 Application of *G. glabra* L. Extracts for Cosmetics

G. glabra preparations and numerous individual glycyrrhizin compounds are commonly used in cosmetics manufacture. Glycyrrhizin, powdered *G. glabra* roots, glycyrrhetinic acid, stearyl glycyrrhizinate, pyridoxine glycyrrhizin, carbenoxolone, and glycyrrhizinate are anti-inflammatory ingredients used in cosmetics. Because it is soluble in water, glycyrrhizin and its salts are utilized in lotions, eye drops, and tonics. Furthermore, glycyrrhizic acid and its derivatives are used in sun oils, creams, and milky lotions (Hayashi & Sudo, 2009). Moreover, glabridin-containing glycyrrhiza flavonoids extracted from *G. glabra* are employed in cosmetic manufacture due to their anti-inflammatory, anti-sensitizing, and skin-whitening activities (Numata et al., 2015; Tamura, 2017).

7.8.4 Application of *G. glabra* Extracts for Flavouring Confectionery Products

G. glabra extract has been used in confectionery since the eighteenth century, when a British chemist called Dunhill mixed *G. glabra* extract, sugar, flour, molasses, and wheat in a recipe for Pontefract cakes (Johnson & Peterson, 1974; Tamura, 2017). *G. glabra* confectionery has had a bad reputation in history due mainly to the frequent substitution of anise oils for the authentic *G. glabra* taste (Housemna & Lacey, 1929). Nowadays, *G. glabra* sweets are widely accessible all over the world, and enormous amounts of *G. glabra* are used in the confectionery sector (Fenwick et al., 1990; Lees 2012; Minifie & Chocolate, 1989). *G. glabra*-flavoured sweets are consumed in various forms worldwide, particularly in Asia and Africa.

7.9 CONCLUSION

This chapter provided a complete overview of the phytochemistry, ethnopharmacology, and commercial uses for *G. glabra*. It is one of the most extensively studied and commercially significant medicinal herbs. This medicinal plant has a long history of use in popular medicine and as a food industry component, notably as a flavouring and sweetening agent. The roots of *G. glabra* have been employed to treat and prevent various diseases, including asthma, arthritis, acidity, hyperdipsia, sore throat, tonsillitis, shingles, heartburn, rheumatism, swellings, eczema, bleeding, flatulence, stomach ulcers, and epilepsy. Glycyrrhizin, glycerol, glabrin A and B, isoliquiritin, flavanones, liquiritic acid, 18-glycyrrhetinic acid, and glabrolide have all been identified and related to the pharmacological effects mentioned, such as antibacterial, anti-inflammatory, antiproliferative, anticancer, antiviral, anti-atherogenic, antioxidant, anti-genotoxic, antimycotic, antitumour, antitussive, or anti-convulsant activities as well as a renoprotective property. In light of this, further experiments assessing the safety and effectiveness of *G. glabra* bioactive compounds in treating various pathological disorders are necessary in the future.

REFERENCES

Abate, L., Tadesse, M. G., Bachheti, A. & Bachheti, R. K. (2022) Traditional and phytochemical bases of herbs, shrubs, climbers, and trees from Ethiopia for their anticancer response. *BioMed Research International*. https://doi.org/10.1155/2022/1589877

Abate, L., Bachheti, A., Bachheti, R. K. & Husen, A. (2021) Antibacterial properties of medicinal plants: Recent trends, progress, and challenges. In *Traditional Herbal Therapy for the Human Immune System*, Edited ByAzamal Husen, CRC Press, pp. 13–54.

Akbar, S. (2020) Handbook of 200 medicinal plants : A comprehensive review of their traditional medical uses and scientific justifications. *Springer Nature Switzerland AG*. https://doi.org/10.1007/978-3-030-16807-0

Asfaw, T. B., Esho, T. B., Bachheti, A., Bachheti, R. K., Pandey, D. P. & Husen, A. (2022) Exploring important herbs, shrubs, and trees for their traditional knowledge, chemical derivatives, and potential benefits. In *Herbs, Shrubs, and Trees of Potential Medicinal Benefits*, Edited By Azamal Husen, CRC Press, pp. 1–26.

Al-Snafi, A. E. 2018. *Glycyrrhiza glabra*: A phytochemical and pharmacological review. *IOSR Journal of Pharmacy*, 8(6), 1–17.

Amani, M., Sotudeh, G. R. & Mostaoufi, N. (2005) *Optimal Extraction of Glycyrrhetinic Acid from Licorice Root*, *Journal of Food Technology*, 3(4), 576–580.

Anagha, K., Manasi, D., Priya, L. & Meera, M. (2012) Comprehensive review on historical aspect of yashtimadhu-*Glycyrrhiza glabra* L. *Global Journal of Research on Medicinal Plants & Indigenous Medicine*, 1(12), 687.

Armanini, D., Fiore, C., Bielenberg, J. & Ragazzi, E. (2005) Licorice (*Glycyrrhiza glabra*). In *Encyclopedia of Dietary Supplements* (pp. 391–399), Coates, P. (ed.). Marcel Dekker Inc.

Asada, Y., Li, W. & Yoshikawa, T. (1998) Isoprenylated flavonoids from hairy root cultures of *Glycyrrhiza glabra*. *Phytochemistry*, 47(3), 389–392.

Ayeka, P. A., Bian, Y., Githaiga, P. M. & Zhao, Y. (2017) The immunomodulatory activities of licorice polysaccharides (Glycyrrhiza uralensis Fisch.) in CT 26 tumor-bearing mice. *BMC Complementary and Alternative Medicine*, 17(1), 1–9.

Badam, L. 1997. In vitro antiviral activity of indigenous glycyrrhizin, licorice and glycyrrhizic acid (Sigma) on Japanese encephalitis virus. *The Journal of Communicable Diseases*, 29(2), 91–99.

Beaton, J. M. & Spring, F. S. (1956) 468Triterpenoids. Part LI. The isolation and characterization of glabric acid, a new triterpenoid acid from liquorice root. *Journal of the Chemical Society* (Resumed), Issue 0, 2417–2419.

Bellakhdar, J. (1997) Contribution à l'étude de la pharmacopée traditionnelle au Maroc : La situation actuelle, les produits, les sources du savoir (enquête ethnopharmacologique de terrain réalisée de 1969 à 1992) [PhD Thesis]. Université Paul Verlaine-Metz.

Benlarbi, F., Mimoune, N., Chaachouay, N., Souttou, K., Saidi, R., Mokhtar, M. R. et al. (2023) Ethnobotanical survey of the traditional antiparasitic use of medicinal plants in humans and animals in Laghouat (Southern Algeria). *Veterinary World*, 16(2), 357–368. https://doi.org/10.14202/vetworld.2023.357-368

Benkhnigue, O., Chaachouay, N., Khamar, H., El Azzouzi, F., Douira, A. & Zidane, L. (2022) Ethnobotanical and ethnopharmacological study of medicinal plants used in the treatment of anemia in the region of Haouz-Rehamna (Morocco). *Journal of Pharmacy & Pharmacognosy Research*, 10(2), 279–302.

Biondi, D. M., Rocco, C. & Ruberto, G. (2003) New dihydrostilbene derivatives from the leaves of *Glycyrrhiza glabra* and evaluation of their antioxidant activity. *Journal of Natural Products*, 66(4), 477–480.

Bogatkina, V. F., Murav'ev, I. A., Stepanova, E. F. & Kir'yalov, N. P. (1975) Triterpene compounds from the epigeal mass of *Glycyrrhiza glabra*. *Chemistry of Natural Compounds*. 11, 114–115.

Bown, D. (1995) *The Royal Horticultural Society Encyclopedia of Herbs & Their Uses*. Dorling Kindersley Limited.

Braun, L. (2007) Liquorice *Glycyrrhiza Glabra*. *Journal of Complementary Medicine: CM, The*, 6(4), 59–61.

Chaachouay, N. & Zidane, L. (2021) Neurological Phytotherapy by Indigenous People of Rif, Morocco. In: Bernardo-Filho, M., Da Cunha De Sá-Caputo, D., Seixas, A. & Taiar, R. (Eds.) *Therapy Approaches in Neurological Disorders*. IntechOpen. 122–137.

Chaachouay, N., Azeroual, A., Bencharki, B., Douira, A. & Zidane, L. (2022a) Ethnoveterinary medicines plants for animal therapy in the Rif, North of Morocco. *South African Journal of Botany*, 147, 176–191.

Chaachouay, N., Azeroual, A., Douira, A. & Zidane, L. (2022b) Ethnoveterinary practices of medicinal plants among the Zemmour and Zayane tribes, Middle Atlas, Morocco. *South African Journal of Botany*, 151, 826–840.

Chaachouay, N., Azeroual, A., Ansari, M. K. A. & Zidane, L. (2023) Use of plants as medicines and aromatics by indigenous communities of morocco: Pharmacognosy, ecology and conservation. In: Ansari, M. K. A., Unal, B. T., Ozturk, M. & Owens, G. (Eds.) *Plants as Medicine and Aromatics*, CRC Press (pp. 33–44).

Chaachouay, N., Belhaj, S., El Khomsi, M., Benkhnigue, O. & Zidane, L. (2023a) Herbal remedies used to treat digestive system ailments by indigenous communities in the Rif region of northern Morocco. *Vegetos* 36(1), 1–18. https://doi.org/10.1007/s42535-023-00606-4

Chaachouay, N., Benkhnigue, O., Douira, A. & Zidane, L. (2021) Poisonous medicinal plants used in the popular pharmacopoeia of the Rif, northern Morocco. *Toxicon*, 189, 24–32.

Chaachouay, N., Benkhnigue, O., El Ibaoui, H., El Ayadi, R. & Zidane, L. (2019) Medicinal plants used for diabetic problems in the Rif, Morocco. *Ethnobotany Research and Applications*, 18(21), 1–19.

Chaachouay, N., Benkhnigue, O., Fadli, M., El Ayadi, R. & Zidane, L. (2019) Ethnobotanical study of medicinal plants used to treat osteoarticular diseases in the Moroccan Rif, Morocco. *Journal of Pharmacy & Pharmacognosy Research*, 7(6), 454–470.

Chaachouay, N., Benkhnigue, O., Fadli, M., El Ibaoui, H., El Ayadi, R. & Zidane, L. (2019a) Ethnobotanical and ethnopharmacological study of medicinal and aromatic plants used in the treatment of respiratory system disorders in the Moroccan Rif. *Ethnobotany Research and Applications*, 18, 1–16.

Chaachouay, N., Benkhnigue, O., Fadli, M., El Ibaoui, H. & Zidane, L. (2019b) Ethnobotanical and ethnopharmacological studies of medicinal and aromatic plants used in the treatment of metabolic diseases in the Moroccan Rif. *Heliyon*, 5(10), e02191.

Chaachouay, N., Benkhnigue, O., Fadli, M., El Ayadi, R. & Zidane, L. (2019c) Ethnobotanical study of medicinal plants used to treat osteoarticular diseases in the Moroccan Rif, Morocco. *Journal of Pharmacy & Pharmacognosy Research*, 7(6), 454–470.

Chaachouay, N., Benkhnigue, O. & Zidane, L. (2020) Ethnobotanical study aimed at investigating the use of medicinal plants to treat nervous system diseases in the Rif of Morocco. *Journal of Chiropractic Medicine*, 19(1), 70–81.

Chaachouay, N., Benkhnigue, O. & Zidane, L. (2022) Ethnomedicinal study of medicinal and aromatic plants used against dermatological diseases by the People of Rif, Morocco. *Journal of Herbal Medicine*, 32, 100542.

Chaachouay, N., Douira, A., Hassikou, R., Brhadda, N., Dahmani, J., Belahbib, N. et al. (2020a) Etude floristique et ethnomédicinale des plantes aromatiques et médicinales dans le Rif (Nord du Maroc) [Phdthesis, Département de Biologie - Université Ibn Tofail - Kénitra]. https://tel.archives-ouvertes.fr/tel-03376377.

Chaachouay, N., Douira, A. & Zidane, L. (2021a) COVID-19, prevention and treatment with herbal medicine in the herbal markets of Salé Prefecture, North-Western Morocco. *European Journal of Integrative Medicine*, 42, 101285.

Chaachouay, N., Douira, A. & Zidane, L. (2021b) Herbal Medicine Used in the Treatment of Human Diseases in the Rif, Northern Morocco. *Arabian Journal for Science and Engineering*, 47, 131–153.

Chaachouay, N., Orch, H. & Zidane, L. (2021c) Cystitis treatment with phytotherapy within the Rif, Northern Morocco. *Future Journal of Pharmaceutical Sciences*, 7, 1–9.

Chandra, J. H. & Gunasekaran, H. (2017) Screening of phytochemical, antimicrobial and antioxidant activity of *Glycyrrhiza glabra* root extract. *Journal of Environmental Biology*, 38(1), 161.

Cheel, J., Van Antwerpen, P., Tuumová, L., Onofre, G., Vokurková, D., Zouaoui-Boudjeltia, K., Vanhaeverbeek, M. & Nève, J. (2010) Free radical-scavenging, antioxidant and immunostimulating effects of a licorice infusion (*Glycyrrhiza glabra* L.). *Food Chemistry*, 122(3), 508–517.

Chen, M., Christensen, S. B., Blom, J., Lemmich, E., Nadelmann, L., Fich, K., Theander, T. G. & Kharazmi, A. (1993) Licochalcone A, a novel antiparasitic agent with potent activity against human pathogenic protozoan species of Leishmania. *Antimicrobial Agents and Chemotherapy*, 37(12), 2550–2556.

Chopra, P., Saraf, B. D., Inam, F. & Deo, S. S. (2013) Antimicrobial and antioxidant activities of methanol extract roots of *Glycyrrhiza glabra* and HPLC analysis. *International Journal of Pharmacy and Pharmaceutical Sciences*, 5(2), 157–160.

Chouitah, O., Meddah, B., Aoues, A. & Sonnet, P. (2011) Chemical composition and antimicrobial activities of the essential oil from *Glycyrrhiza glabra* leaves. *Journal of Essential Oil-Bearing Plants*, 14(3), 284–288.

Chowdhury, B., Bhattamisra, S. K. & Das, M. C. (2013) Anti-convulsant action and amelioration of oxidative stress by *Glycyrrhiza glabra* root extract in pentylenetetrazole-induced seizure in albino rats. *Indian Journal of Pharmacology*, 45(1), 40.

Coruh, I., Gormez, A., Ercisli, S. & Bilen, S. (2007) Total phenolics, mineral elements, antioxidant and antibacterial activities of some edible wild plants in Turkey, *Asian Journal of Chemistry*,19(7), 5755-5762.

Damle, M. (2014) *Glycyrrhiza glabra* (Liquorice)-a potent medicinal herb. *International Journal of Herbal Medicine*, 2(2), 132–136.

Dioscorides, P. (1829) De materia medica (Vol. 1). Knobloch.

Dwivedi, R. S. (1999) Unnurtured and untapped super sweet nonsacchariferous plant species in India. *Current Science*, 76(11), 1454–1461.

Farag, M. A. & Wessjohann, L. A. (2012) Volatiles profiling in medicinal licorice roots using steam distillation and solid-phase microextraction (SPME) coupled to chemometrics. *Journal of Food Science*, 77(11), C1179–C1184.

Fenwick, G. R., Lutomski, J. & Nieman, C. (1990) Liquorice, *Glycyrrhiza glabra* L.—Composition, uses and analysis. *Food Chemistry*, 38(2), 119–143.

Fiore, C., Eisenhut, M., Ragazzi, E., Zanchin, G. & Armanini, D. (2005) A history of the therapeutic use of liquorice in Europe. *Journal of Ethnopharmacology*, 99(3), 317–324.

Foster, S. & Chongxi, Y. (1992) *Herbal Emissaries: Bringing Chinese Herbs to the West: A Guide to Gardening, Herbal Wisdom, and Well-Being*. Inner Traditions/Bear & Co.

Frattaruolo, L., Carullo, G., Brindisi, M., Mazzotta, S., Bellissimo, L., Rago, V., Curcio, R., Dolce, V., Aiello, F. & Cappello, A. R. (2019) Antioxidant and anti-inflammatory activities of flavanones from *Glycyrrhiza glabra* L.(licorice) leaf Phyto complexes: Identification of licoflavanone as a modulator of NF-kB/MAPK pathway. *Antioxidants*, 8(6), 186.

Frattini, C., Bicchi, C., Barettini, C. & Nano, G. M. (1977) Volatile flavor components of licorice. *Journal of Agricultural and Food Chemistry*, 25(6), 1238–1241.

Fu, B., Liu, J., Li, H., Li, L., Lee, F. S. & Wang, X. (2005) The application of macroporous resins in the separation of licorice flavonoids and glycyrrhizic acid. *Journal of Chromatography A*, 1089(1–2), 18–24.

Fu, Y., Chen, J., Li, Y.-J., Zheng, Y.-F. & Li, P. (2013) Antioxidant and anti-inflammatory activities of six flavonoids separated from licorice. *Food Chemistry*, 141(2), 1063–1071.

Fu, Y., Hsieh, T., Guo, J., Kunicki, J., Lee, M. Y., Darzynkiewicz, Z. & Wu, J. M. (2004) Licochalcone-A, a novel flavonoid isolated from licorice root (*Glycyrrhiza glabra*), causes G2 and late-G1 arrests in androgen-independent PC-3 prostate cancer cells. *Biochemical and Biophysical Research Communications*, 322(1), 263–270.

Fukai, T., Marumo, A., Kaitou, K., Kanda, T., Terada, S. & Nomura, T. (2002) Antimicrobial activity of licorice flavonoids against methicillin-resistant Staphylococcus aureus. *Fitoterapia*, 73(6), 536–539.

Geetha, R. V. & Anitha, R. O. Y. (2013) In vitro evaluation of antimycotic activity of ethanolic extract of Glycyrrhiza glabra. *Asian Journal of Pharmaceutical and Clinical Research*, 6(3), 205–206.

Guo, A., He, D., Xu, H.-B., Geng, C.-A. & Zhao, J. (2015) Promotion of regulatory T cell induction by immunomodulatory herbal medicine licorice and its two constituents. *Scientific Reports*, 5(1), 1–12.

Hajirahimkhan, A., Simmler, C., Yuan, Y., Anderson, J. R., Chen, S.-N., Nikolić, D., Dietz, B. M., Pauli, G. F., van Breemen, R. B. & Bolton, J. L. (2013) Evaluation of estrogenic activity of licorice species in comparison with hops used in botanicals for menopausal symptoms. *PloS One*, 8(7), e67947.

Hamad, G. M., Abd Elaziz, A. I., Hassan, S. A., Shalaby, M. A. & Azim Mohdaly, A. A. A. (2020) Chemical Composition, Antioxidant, Antimicrobial and Anticancer Activities of Licorice (*Glycyrrhiza glabra* L.) Root and Its Application in Functional Yoghurt. *Journal of Food and Nutrition Research*, 8(12), 707–715.

Han, Y. N. & Chung, M. S. (1990) A pyrrolo-pyrimidine alkaloid from *Glycyrrhiza uralensis*. *Archives of Pharmacal Research*, 13(1), 103–104.

Han, Y. N., Chung, M. S., Kim, T. H. & Han, B. H. (1990) Two tetrahydroquinoline alkaloids from *Glycyrrhiza uralensis*. *Archives of Pharmacal Research*, 13(1), 101–102.

Hatano, T., Aga, Y., Shintani, Y., Ito, H., Okuda, T. & Yoshida, T. (2000) Minor flavonoids from licorice. *Phytochemistry*, 55(8), 959–963.

Hatano, T., Yasuhara, T., Fukuda, T., Noro, T., & Okuda, T. (1989) Phenolic constituents of licorice II: Structures of licopyranocoumarin, licoarylcoumarin and glisoflavone, and inhibitory effects of licorice Phenolics on xanthine oxidase. *Chemical and Pharmaceutical Bulletin*, 37(11), 3005–3009.

Hayashi, H., Hattori, S., Inoue, K., Khodzhimatov, O., Ashurmetov, O., Ito, M. & Honda, G. (2003) Field survey of *Glycyrrhiza* plants in Central Asia (3). Chemical characterization of G. glabra collected in Uzbekistan. *Chemical and Pharmaceutical Bulletin*, 51(11), 1338–1340.

Hayashi, H. & Sudo, H. (2009) Economic importance of licorice. *Plant Biotechnology*, 26(1), 101–104.

Herold, A., Cremer, L., Călugaru, A., Tamaş, V., Ionescu, F., Manea, S. & Szegli, G. (2003) Hydroalcoholic plant extracts with anti-inflammatory activity. *Roumanian Archives of Microbiology and Immunology*, 62(1–2), 117–129.

Housemna, P. A. & Lacey, H. T. (1929) The licorice root in industry. *Industrial & Engineering Chemistry*, 21(10), 915–917.

Husain, A., Ahmad, A., Mujeeb, M., Khan, S. A., Alghamdi, A. G. & Anwar, F. (2015) Quantitative analysis of total phenolic, flavonoid contents and HPTLC fingerprinting for standardization of *Glycyrrhiza glabra* Linn. Roots. *Herbal Medicine*, 1(1), 1–9.

Huxley, A. J. & Griffiths, M. (1992) *Dictionary of Gardening*. Stockton Press.

ITIS. (2022) Integrated Taxonomic Information System (ITIS) on-line database, www.itis.gov, CC0 https://doi.org/10.5066/F7KH0KBK. http://itis-web-prod.snafu.cr.usgs.gov

Johnson, A. & Peterson, M. (1974) *Sunflower Seed: Encyclopedia of Food Technology*. The Avi Publishing.

Kameoka, H. & Nakai, K. (1987) Components of essential oil from the root of *Glycyrrhiza glabra*. *Nippon Nogeikagaku Kaishi*, 61(9), 1119–1121.

Karahan, F., Avsar, C., Ozyigit, I. I. & Berber, I. (2016) Antimicrobial and antioxidant activities of medicinal plant *Glycyrrhiza glabra* var. *Glandulifera* from different habitats. *Biotechnology & Biotechnological Equipment*, 30(4), 797–804.

Karami, Z., Mirzaei, H., Emam-Djomeh, Z., Mahoonak, A. S. & Khomeiri, M. (2013) Effect of harvest time on antioxidant activity of *Glycyrrhiza glabra* root extract and evaluation of its antibacterial activity. *International Food Research Journal*, 20(5), 2951.

Karkanis, A., Martins, N., Petropoulos, S. A. & Ferreira, I. C. (2018) Phytochemical composition, health effects, and crop management of liquorice (*Glycyrrhiza glabra* L.): A medicinal plant. *Food Reviews International*, 34(2), 182–203.

Kaur, R., Kaur, H. & Dhindsa, A. S. (2013) *Glycyrrhiza glabra*: A phytopharmacological review. *International Journal of Pharmaceutical Sciences and Research*, 4(7), 2470.

Khare, C. P. (2004) *Encyclopedia of Indian Medicinal Plants: Rational Western Therapy, Ayurvedic and Other Traditional Usage, Botany*. Springer.

Kim, Y. W., Kang, H. E., Lee, M. G., Hwang, S. J., Kim, S. C., Lee, C. H. & Kim, S. G. (2009) Liquiritigenin, a flavonoid aglycone from licorice, has a choleretic effect and the ability to induce hepatic transporters and phase-II enzymes. *American Journal of Physiology-Gastrointestinal and Liver Physiology*, 296(2), G372–G381.

Kitagawa, I., Chen, W.-Z., Hori, K., Harada, E., Yasuda, N., Yoshikawa, M. & Ren, J. (1994) Chemical studies of Chinese licorice-roots. I. Elucidation of five new flavonoid constituents from the roots of *Glycyrrhiza glabra* L. collected in Xinjiang. *Chemical and Pharmaceutical Bulletin*, 42(5), 1056–1062.

Kitagawa, I., Hori, K., Uchida, E., Chen, W.-Z., Yoshikawa, M. & Ren, J. (1993) Saponin and sapogenol. L. On the constituents of the roots of *Glycyrrhiza uralensis* Fischer from Xinjiang, China. Chemical structures of licorice-saponin L3 and isoliquiritin apioside. *Chemical and Pharmaceutical Bulletin*, 41(9), 1567–1572.

Kiuchi, F., Chen, X. & Tsuda, Y. (1990) Four new phenolic constituents from licorice (root of *Glycyrrhiza* sp.). *Heterocycles*, 31(4), 629–636.

Kuang, Y., Li, B., Fan, J., Qiao, X. & Ye, M. (2018) Antitussive and expectorant activities of licorice and its major compounds. *Bioorganic & Medicinal Chemistry*, 26(1), 278–284.

Lange, D. (1998) *Europe's Medicinal and Aromatic Plants: Their Use, Trade and Conservation*. Traffic International.

Lees, R. (2012) *Sugar Confectionery and Chocolate Manufacture*. Springer Science & Business Media.

Lim, T. K. (2016) *Glycyrrhiza glabra*. In T. K. Lim (Éd.), *Edible Medicinal and Non-Medicinal Plants: Volume 10, Modified Stems, Roots, Bulbs* (pp. 354–457). Springer. https://doi.org/10.1007/978-94-017-7276-1_18

Liu, H.-M., Sugimoto, N., Akiyama, T. & Maitani, T. (2000) Constituents and their sweetness of food additive enzymatically modified licorice extract. *Journal of Agricultural and Food Chemistry*, 48(12), 6044–6047.

Liu, L. J., Pan, J. Y., Yin, X. H. & Lu, Y. S. (1985) Quantitative determination of glycyrrhizin in Glycyrrhiza and its preparations by maero-reticular resin. *Chin Trad Herb Drugs*, 16, 202–206.

Malekinejad, H., Farshid, A. A. & Mirzakhani, N. (2011) Liquorice plant extract reduces ochratoxin A-induced nephrotoxicity in rats. *Experimental and Toxicologic Pathology*, 63(1–2), 125–130.

Martins, N., Barros, L., Dueñas, M., Santos-Buelga, C. & Ferreira, I. C. (2015) Characterization of phenolic compounds and antioxidant properties of *Glycyrrhiza glabra* L. rhizomes and roots. *RSC Advances*, 5(34), 26991–26997.

Masad, A. A., Bashiti, T. A., Mosleh, F. & Abu Madi, Y. M. (2019) Antibacterial and anti-biofilm activity of selected plant extracts against some human pathogenic microorganisms. *Pakistan Journal of Nutrition*, 18(11), 1014-1020.

Minifie, B. W. & Chocolate, C. (1989) *Confectionery: Science and Technology*. Pitman.

Mitscher, L. A., Park, Y. H., Clark, D. & Beal, J. L. (1980) Antimicrobial agents from higher plants. Antimicrobial isoflavanoids and related substances from *Glycyrrhiza glabra* L. var. Typica. *Journal of Natural Products*, 43(2), 259–269.

Morteza-Semnani, K., Saeedi, M. & Shahnavaz, B. (2003) Comparison of antioxidant activity of extract from roots of licorice (*Glycyrrhiza glabra* L.) to commercial antioxidants in 2% hydroquinone cream. *Journal of Cosmetic Science*, 54(6), 551–558.

Nair, A. A. & Geetha, R. V. (2014) In vitro antibacterial activity of *Glycyrrhiza glabra* on enterococcus faecalis. *Research Journal of Pharmacy and Technology*, 7(11), 1237–1239.

Nieman, C. (1957) Licorice. In *Advances in Food Research*, 7, 339–381.

Nirmala, P. & Selvaraj, T. (2011) Anti-inflammatory and antibacterial activities of *Glycyrrhiza glabra* L. *Journal of Agricultural Technology*, 7(3), 815–823.

Nomura, T. & Fukai, T. (1998) Phenolic constituents of licorice (Glycyrrhiza species). Fortschritte der Chemie organischer Naturstoffe/Progress in the Chemistry of Organic Natural Products, Edited by De T. Fukai, T. Nomura (pp. 1–140).

Nomura, T., Fukai, T. & Akiyama, T. (2002) Chemistry of phenolic compounds of licorice (*Glycyrrhiza* species) and their estrogenic and cytotoxic activities. *Pure and Applied Chemistry*, 74(7), 1199–1206.

Numata, T., Kobayashi, Y., Ito, T., Harada, K., Tsuboi, R., & Okubo, Y. (2015) Two cases of allergic contact dermatitis due to skin-whitening cosmetics. *Allergology International*, 64(2), 194–195.

Olukoga, A. & Donaldson, D. (1998) Historical perspectives on health the history of liquorice: The plant, its extract, cultivation, commercialization and etymology. *The Journal of the Royal Society for the Promotion of Health*, 118(5), 300–304.

Onstad, D. (1996) *Whole Foods Companion*. Chelsea Green.

Orch, H., Chaachouay, N., Douiri, E. M., Faiz, N., Zidane, L. & Douira, A. (2021) Use of medicinal plants in dermato-cosmetology: An ethnobotanical study among the population of Izarène. *Jordan Journal of Pharmaceutical Sciences*, 14(3), 323–340.

Öztürk, M., Altay, V., Hakeem, K. R. & Akçiçek, E. (2018) *Liquorice: From botany to phytochemistry*. Springer.

Kovalenko, PG., Antonjuk, V. P. & Maliuta, S. S. (1989) Secondary metabolites production from transformed cells of *Glycyrrhiza glabra* and *Potentilla alba* as a producents of radioprotective compounds. *Genetic Engineering*, 1(1), 21–32.

Preedy, V. R. (2015) *Essential Oils in Food Preservation, Flavor and Safety*. Academic Press.

Quirós-Sauceda, A. E., Ovando-Martínez, M., Velderrain-Rodríguez, G. R., González-Aguilar, G. A. & Ayala-Zavala, J. F. (2016). Licorice (*Glycyrrhiza glabra* Linn.) oils. In *Essential Oils in Food Preservation, Flavor and Safety*, Edited by Victor R. Preedy (pp. 523–530).

Račková, L., Jančinová, V., Petríková, M., Drábiková, K., Nosáľ, R., Štefek, M., Košťálová, D., Prónayová, N. & Kováčová, M. (2007) Mechanism of anti-inflammatory action of liquorice extract and glycyrrhizin. *Natural Product Research*, 21(14), 1234–1241.

Raggi, M. A., Bugamelli, F., Nobile, L., Curcelli, V., Mandrioli, R. & Rossetti, A. (1995) The choleretic effects of licorice: Identification and determination of the pharmacologically active components of *Glycyrrhiza glabra*. *Bollettino Chimico Farmaceutico*, 134(11), 634–638.

Rahnama, M., Fakheri, B. A., Mashhady, M. A. & Saeidi, S. (2019) Anti-bacterial and Anti-biofilm activity of *Glycyrrhiza glabra*, *Rosmarinus officinalis* and *Saponaria officinalis* extracts on important food pathogens. *Gene, Cell and Tissue*, 6(4):e96326. doi: 10.5812/gct.96326.

Saad, B. & Said, O. (2011) *Greco-Arab and Islamic herbal medicine: Traditional system, ethics, safety, efficacy, and regulatory issues*. John Wiley & Sons.

Sahin Yaglıoglu, A., Eser, F., Tekin, S. & Onal, A. (2016) Antiproliferative activities of several plant extracts from Turkey on rat brain tumor and human cervix carcinoma cell lines. *Frontiers in Life Science*, 9(1), 69–74.

Sajid, T. M., Rashid, S. & Saeed, S. A. (1991) Inhibition of adrenaline-induced aggregation of human platelets by Pakistani medicinal plants. *Pakistan Journal of Pharmaceutical Sciences*, 4(2), 145–152.

Sayed, M. D. (1980) Traditional medicine in health care. *Journal of Ethnopharmacology*, 2(1), 19–22.

Sharma, V., Katiyar, A. & Agrawal, R. C. (2018) *Glycyrrhiza glabra*: Chemistry and pharmacological activity. *Sweeteners*, 87–100. doi.org/10.1007/978-3-319-27027-2_21

Shetty, T. K., Satav, J. G. & Nair, C. K. K. (2002) Protection of DNA and microsomal membranes in vitro by *Glycyrrhiza glabra* L. against gamma irradiation. *Phytotherapy Research: An International Journal Devoted to Pharmacological and Toxicological Evaluation of Natural Product Derivatives*, 16(6), 576–578.

Shinde, D. B., Koratkar, S. S., Sharma, N. & Shitole, A. A. (2016) Antioxidant activity and antiproliferative action of methanolic extract of liquorice (*Glycyrrhiza glabra*) in HepG2 cell line. *International Journal of Pharmacy and Pharmaceutical Sciences*, 8, 293–298.

Siracusa, L., Saija, A., Cristani, M., Cimino, F., D'Arrigo, M., Trombetta, D., Rao, F. & Ruberto, G. (2011) Phytocomplexes from liquorice (*Glycyrrhiza glabra* L.) leaves—Chemical characterization and evaluation of their antioxidant, anti-genotoxic and anti-inflammatory activity. *Fitoterapia*, 82(4), 546–556.

Somjen, D., Knoll, E., Vaya, J., Stern, N. & Tamir, S. (2004) Estrogen-like activity of licorice root constituents: Glabridin and glabrene, in vascular tissues in vitro and in vivo. *The Journal of Steroid Biochemistry and Molecular Biology*, 91(3), 147–155.

Soulef, K. & Yahia, A. (1970) Antibacterial effects of glycosides extract of *Glycyrrhiza glabra* L. from the region of Djamâa (South-East of Algeria). *Journal of Medicinal Herbs and Ethnomedicine*, (3), 22-25. doi: 10.25081/jmhe.2017.v3.3367

Sultana, S., Haque, A., Hamid, K., Urmi, K. F. & Roy, S. (2010) Antimicrobial, cytotoxic and antioxidant activity of methanolic extract of *Glycyrrhiza glabra*. *Agriculture and Biology Journal of North America*, 1(5), 957–960.

Tamir, S., Eizenberg, M., Somjen, D., Izrael, S. & Vaya, J. (2001). Estrogen-like activity of glabrene and other constituents isolated from licorice root. *The Journal of Steroid Biochemistry and Molecular Biology*, 78(3), 291–298.

Tamura, Y. (2017). The History of Licorice Applications in Maruzen Pharmaceuticals Co., Ltd. *Biological Activities and Action Mechanisms of Licorice Ingredients*, 13-35 DOI: 10.5772/65962.

Tewari, D., Mocan, A., Parvanov, E. D., Sah, A. N., Nabavi, S. M., Huminiecki, L., Ma, Z. F., Lee, Y. Y., Horbańczuk, J. O. & Atanasov, A. G. (2017) Ethnopharmacological approaches for therapy of jaundice: Part II. Highly used plant species from Acanthaceae, Euphorbiaceae, Asteraceae, Combretaceae, and Fabaceae families. *Frontiers in Pharmacology*, 8, 519.

Thakur, D., Jain, A. & Ghoshal, G. (2016) Evaluation of phytochemical, antioxidant and antimicrobial properties of glycyrrhizin extracted from roots of *Glycyrrhiza glabra*.

The Plant List. (2013) The Plant List. Version 1.1. Published on the Internet; http://www.theplantlist.org/ (accessed 1st January).

Tohma, H. S. & Gulçin, I. (2010) Antioxidant and radical scavenging activity of aerial parts and roots of Turkish liquorice (*Glycyrrhiza glabra* L.). *International Journal of Food Properties*, 13(4), 657–671.

Varsha, S., Agrawal, R. C. & Sonam, P. (2013) Phytochemical screening and determination of antibacterial and antioxidant potential of Glycyrrhiza glabra root extracts. *Journal of Environmental Research and Development*, 7(4A), 1552.

Visavadiya, N. P., Soni, B. & Dalwadi, N. (2009) Evaluation of antioxidant and anti-atherogenic properties of Glycyrrhiza glabra root using in vitro models. *International Journal of Food Sciences and Nutrition*, 60(sup2), 135–149.

Wang, Q., Qian, Y., Wang, Q., Yang, Y., Ji, S., Song, W., Qiao, X., Guo, D., Liang, H. & Ye, M. (2015) Metabolites identification of bioactive licorice compounds in rats. *Journal of Pharmaceutical and Biomedical Analysis*, 115, 515–522.

Yahara, S. & Nishioka, I. (1984). Flavonoid glucosides from licorice. *Phytochemistry*, 23(9), 2108–2109.

Yang, R., Wang, L. & Liu, Y. (2014) Antitumor activities of widely-used Chinese herb—Licorice. *Chinese Herbal Medicines*, 6(4), 274–281.

Yavuz Kocaman, A. & Güzelkokar, M. (2018) The genotoxic and antigenotoxic potential of the methanolic root extract of *Glycyrrhiza glabra* L. on human peripheral blood lymphocytes. *Drug and Chemical Toxicology*, 41(3), 368–375.

Yu, J.-Y., Ha, J. Y., Kim, K.-M., Jung, Y.-S., Jung, J.-C. & Oh, S. (2015) Anti-inflammatory activities of licorice extract and its active compounds, glycyrrhizic acid, liquiritin and liquiritigenin, in BV2 cells and mice liver. *Molecules*, 20(7), 13041–13054.

Zadeh, J. B., Kor, Z. M. & Goftar, M. K. (2013) Licorice (*Glycyrrhiza glabra* Linn) as a valuable medicinal plant. *International Journal of Advanced Biological and Biomedical Research*, 1(10), 1281–1288.

Zahedi, M., Memar Maher, B., Anarjan, N. & Hamishehkar, H. (2021). Licorice-garlic-fennel essential oils composite nanoparticles as natural food preservatives. *International Journal of Nano Dimension*, 12(3), 239–251.

Zeng, L., Li, S. H. & Lou, Z. C. (1988). Morphological and histological studies of Chinese licorice. *Acta Pharmacologica Sinica*, 23(3), 200.

Zhang, J., Cui, M., He, Y., Yu, H. & Guo, D. (2005). Chemical fingerprint and metabolic fingerprint analysis of Danshen injection by HPLC–UV and HPLC–MS methods. *Journal of Pharmaceutical and Biomedical Analysis*, 36(5), 1029–1035.

Zhao, Z.-H., Zu, Y.-G., Tang, Z.-H. & Cao, J.-G. (2005). The classify of the liquorice life cycle form. *Acta Ecologica Sinica*, 25(19), 2341–2346.

Ziegler, H. L., Hansen, H. S., Staerk, D., Christensen, S. B., Hägerstrand, H. & Jaroszewski, J. W. (2004). The antiparasitic compound licochalcone a is a potent echinocytogenic agent that modifies the erythrocyte membrane in the concentration range where antiplasmodial activity is observed. *Antimicrobial Agents and Chemotherapy*, 48(10), 4067–4071.

8 Medicinal and Commercial Uses of *Angelica* Roots

Anuj Kandwal, Rakesh Kumar Bachheti, Sonali Purohit, Archana Bachheti, and Arun Kumar Khajuria

8.1 INTRODUCTION

Biodiversity, heterogeneity, and variability of our planet have been a fountainhead of human inspiration and survival across the civilization for ages and are a product of spontaneous evolution for millions of years (Dhiman & Kumar, 2006; Bisht & Khajuria, 2014; Kumar et al., 2018). Since time immemorial, people have been using plants and plant-based products, that are not part of their usual diet, to treat various illnesses. They accumulated this knowledge from generation to generation through trial and error. Each and every community across the globe has a long history of using plants in folk medicine. Even today's pharmacopoeias and modern allopathic medicine have their roots in folk medicine and have been derived from such experiences (Gurib-Fakim, 2006; Bhutani & Gohil, 2010; Asfaw et al., 2022).

In general, about 90% of collected material from therapeutic plants is from natural sources, and, since 70% of these collections involve destructive harvesting, many of those plants are now endangered, vulnerable, or threatened in their natural habitat (Khajuria et al., 2017; Khajuria et al., 2021). One such highly important but critically endangered genus is *Angelica* because the root of the members of this genus has been in great demand to the present day (Bussmann et al., 2009; Kunwar et al., 2013; Butola & Vashistha, 2013).

Angelica L. is a complex genus that belongs to the family Apiaceae (Umbelliferae), commonly referred to as the parsley family. The genus comprises about 110–115 species, of which 60 are reported to have medicinal value (Sarker & Nahar, 2004; Butola & Vashistha, 2013; Khajuria et al., 2021). Members of this genus are perennial herbs that may grow up to 8 feet tall. The plants are extensively dispersed throughout North America, Europe, and Asia, and can be traced to altitudes from 1,500 to 4,000 masl along slow-moving streams, soggy meadows, and valleys in the forest with stony substrate, as these species need well-drained soil and partial shade for growth (Kaul, 1997; Kala et al., 2004; Kunwar et al., 2013; Butola & Vashistha, 2013).

The use of *Angelica* root can be traced to different civilizations at the time, particularly in China, Japan, and Korea. However, these countries derived roots from different species. From the records, it is clear that, in the past, the people of China used *Angelica sinensis* as a source of *Angelica* root, whereas, in Korea, people used *Angelica gigas* (Zhang et al., 2012), with *Angelica acutiloba* being a very famous root source in Japan (Hatano et al., 2004).

The most talked-about species among the 60 medicinal spp. of *Angelia*, regarding their traditional medical applications and market value in modern alternative or herbal medicine, include *A. acutiloba*, *Angelica anomala*, *Angelica archangelica*, *Angelica atropupurea*, *Angelica breweri*, *Angelica dahurica*, *Angelica decursiva*, *Angelica genuflex*, *Angelica glauca*, *A. gigas*, *Angelica grosserrata*, *Angelica laxifoliata*, *Angelica lucida*, *Angelica pubescens*, *Angelica rosaefolia*, *A. sinensis*, *Angelica sylvestris*, *Angelica tenuissima*, *Angelica uchiyamai*, and *Angelica venenosa*. Thus, the aim of this chapter is to provide in-depth information about the botany and the ethnomedicinal and other uses of the *Angelica* genus and its popular medicinal species.

DOI: 10.1201/b22924-8

8.2 BOTANY OF *ANGELICA* GENUS

The *Angelica* genus is represented by approximately 110–115 species from around the globe, mostly from the north temperate and subarctic regions of the world. Out of these species, 32 species are endemic to China. In India, *Angelica* is represented by three species, namely *A. glauca* Edgew., *A. archangelica* L. var. *himalaica* Clarke syn. *A. officinalis* Hoffm. and *A. nubigena* Clarke.

The genus, in general, can be described as:

Herbs, either perennial or biennial, with roots which are often thick, conical, or cylindric in shape. Petiolate leaves with notably inflated petiole sheaths; 1- to 4-pinnate or 1- to 3-ternate/pinnate blades. Bracts are numerous or few, seldom absent, rays are numerous, and bracteoles are numerous or few, complete. Umbels are compound, terminal, and lateral, with obsolete or ovate-triangular calyx teeth. White, irregularly shaped, ovate to obovate, and occasionally pink or dark purple petals. Strictly short conic stylopodium. Fruit is ovoid to orbicular, compressed dorsally, with filiform dorsal ribs and broad- or narrow-winged lateral ribs that split when the fruit is mature. Vittae are frequently 1–2 in each furrow and 2–4 on the commissure. Concave or slightly convex seed faces. Seed face plane or slightly concave. Carpophore 2-cleft to base (Hooker, 1875; Gaur, 1999; Wu et al., 2005).

8.3 BOTANY OF SOME IMPORTANT *ANGELICA* SPP.

Some common but highly valued members of this genus are botanically described as:

1. *Angelica glauca* Edgew. Commonly known as chandaa, chandaamshuka, kathachoraa, choru, chora, chorak, gandrayan. The plant is indigenous to Syria and is found in Central Russia, Austria, Belgium, Denmark, Germany, Hungary, Iceland, and other areas of temperate Asia. In India, the plant is reported to be distributed in parts of the western Himalayas. The plant is a perennial, glabrous herb with a thick aromatic rootstock; stem 1–3 m high, stout, fistular. Leaves are large, 1–3 pinnatipartite; segments 2.5–8 cm long, irregularly serrate or coarsely toothed; broad, inflated sheathing surrounds the leaf base. Flower white, 15–30 cm across, as compound umbels; peduncle stout to 20 cm long or more. Involucels 6–10, linear bractlets. Calyx teeth are obsolete. Five petals, ovate, emarginated, sometimes tinged purple. Stamens are five in number. Fruit oblong, flattened, lateral ridges expanded into wings, dorsal ribs conspicuous; inner seed face concave. Flowering in July to August; fruiting in September to October. The plant can be traced to an altitude from 1,500 to 4,000 masl in the stony substrate near slow-moving streams, moist meadows, and forest gullies (Kaul, 1997; Kala et al., 2004).
2. *Angelica sinensis* (Oliver) Diels. As a perennial herb, plants may grow up to 40–100 cm. A succulent, highly aromatic, cylindrical, branching root with numerous rootlets. Lower and basal petioles are 5-–20 cm long with purple-green, ovate and membranous-margined sheaths; blades are elliptical, 10–30 × 12–25 cm and are 2- to 3-ternate-pinnate. Leaflets are oblong or oval-lanceolate, 2–3.5 × 0.8–2.5 cm, 2- or 3-lobed, edge irregularly coarse-cuspidate-serrate, and sparsely papillate-hairy along nerves and margin. Pinnae are three to four pairs, with the proximal and middle pinnae being long-petiolulate. Petals white, sporadically red-purple in colour. Sub-orbicular or ellipsoidal fruit. Flowering: Jun to July; fruiting: July to September.
3. *Angelica acutiloba* (Siebold & Zuccarini) Kitagawa. Plant is a medium-sized perennial herb up to 30-–100 cm in height. Short tuberous root with a few enlarged secondary roots, yellow-brown in colour, measuring 10–25 × 1–2.5 cm, and highly aromatic. Lower and basal leaves are petiolate, with petioles 10-30 cm and sheaths that are oblong. Blades are glabrous, triangular-ovate, 10–25 cm long, and 1 or 2 ternate-pinnate. Pinnae are short-petiolulate, 2–9 × 1–3 cm, 3-lobed, segments lanceolate, edge irregularly acute-serrate, and apex acuminate to acute. Fruit narrow-oblong, flowering: Jul to Aug, fruiting: Aug to Sep.
4. *Angelica dahurica* Fischer ex Hoffmann. Plant is perennial 100–250 cm tall, stout. Brown, cylindrical, 3–5 cm thick, highly aromatic roots. Lower and basal leaves long-petiolate, sheaths oblong-inflated, glabrous; blade triangular-ovate, 30–50 × 25-40 cm, 2-3-ternate-pinnate; leaflets sessile, oblong-elliptic to oblong-lanceolate, 4-10 × 1-4 cm, base slightly

decurrent, margin white-cartilaginous and coarse-cuspidate-serrate, apex acute, pubescent along nerves adaxially. Reduced upper leaves with saccate-inflated, bladeless sheaths. Umbels 10–30 cm across; Fruit suborbicular, flowering: Jul to Aug, fruiting: Aug to Sep.

5. *Angelica sylvestris* L. Plants perennial, 80–200 cm. Conical, thick, and mildly aromatic root. Lower and basal leaves petiolate, petioles long, ovate to saccate-inflated sheaths; blade broadly triangular-ovate, 2- or 3-pinnate; leaflets sessile, lanceolate to ovate, 2.5–8 × 1–4 cm, base cuneate, margin serrulate, slightly hispidulous along nerves. White, oblong to obovate petals. Fruit broad-ovoid; flowering: Jun to Jul; fruiting: Aug to Sep.
6. *Angelica gigas* Nakai. Plants perennial, 100–200 cm, stout. Grey-brown, conical, and 2- to 5-cm thick root. Basal and lower leaves petiolate, petioles 30–45 cm, broad sheaths; blade triangular-ovate in outline, 20–40 × 20–30 cm, 2- or 3-ternate-pinnate, basal pinnae petiolulate; leaflets oblong-lanceolate, 4–15 × 1.5–5 cm, base decurrent, margin irregularly coarse-toothed, apex acute, slightly scabrous along nerves adaxially. Upper leaves are purple-green with inflated, frequently bladeless sheaths. Fruit ellipsoid, flowering: Jul to Aug; fruiting: Aug to Sep.

8.4 MEDICINAL USE OF *ANGELICA* GENUS

The genus *Angelica* is very famous and popular among different communities for its medicinal value, and a few species of the genus also hold ornamental and food values. Many species of the genus are popular in folk systems of medicine and in some codified systems, such as Ayurveda and homoeopathy.

In the Chinese system of medicine, a number of *Angelica* species are still in use, so that the genus retains its aura from the past to the present. Traditionally, *Angelica* plants in China are used to treat gynaecological diseases, regulate menstruation, apoplexy, constipation, malaria, chills, fever, and haemorrhoids (Sowndhararajan et al., 2017b).

In Korean traditional medicine, *Angelica* is used to treating anaemia, stomach and intestinal disturbances, bilious complaints, constipation, skin diseases, respiratory problems, gynaecological diseases, circulatory diseases, insomnia, headaches, analgesic, nervousness, and atrophy, and as a sedative (Pasqua et al., 2003; Chauhan et al., 2016; Wang et al., 2016; Sowndhararajan et al., 2017b).

According to European culture or folklore in the remote past, when the plague almost destroyed Europe, an angel came to a monk and offered a herb, which would effect a cure, and, since then, this herb has borne the name Angelica (Sarker & Nahar, 2004). In several North European countries, several species of the genus are used against contagion, for purifying the blood, as an anti-inflammatory, diuretic, expectorant, and diaphoretic, and as a remedy for colds, influenza, hepatitis, and arthritis.

In Asia, the genus *Angelica* is predominantly regarded as 'female ginseng' and is used to treat conditions such as dysmenorrhoea (painful menstruation), metrorrhagia (abnormal menstruation), amenorrhoea (absence of menstruation), menopausal symptoms (especially hot flushes) and assuring easy delivery (Butola & Badola, 2004; Butola & Vashistha, 2013).

In the present system of medicine, several folk preparations from different species of *Angelica* are still in use. *A. archangelica* is very famous in China, Japan, and Korea for its medicinal value. Roots are used to cure different ailments; tea made from the roots is used to cure stomach cancer, and other formulations of roots are used to cure gastrointestinal diseases, migraine, respiratory problems, insomnia, menstrual complaints, arthritis, and chronic fatigue (Pasqua et al., 2003; Sarker & Nahar, 2004; Bhat et al., 2011; Chauhan et al., 2016; Aćimović et al., 2022). The roots of *A. glauca* are used to cure a broad spectrum of diseases, from common digestive ailments to lethal diseases. Mainly, the plant is used to treat flatulence, stomach pains, atrophy, menorrhagia, rheumatism, skin disease, constipation, and urinary disorders (Gaur, 1999; Vashistha et al., 2006; Butola & Vashistha, 2013; Khajuria et al., 2021; Kumar et al., 2022). *A. sinensis* is used to cure different ailments in various systems of medicine in Japan, Korea, and China. The dried root of the plant is used to cure amenorrhoea, gynoecopathia, anaemia, dysmenorrhoea, arthritis, premenstrual and menopausal syndromes, cancer, and cardiovascular diseases, and as a uterotonic (Tsai et al., 2006; Su et al., 2013; Jian-Ping et al., 2015; Hu et al., 2022). Products of *Angelica* are available in various forms, either as crude plant parts or supplied by commercial suppliers around the globe (Table 8.1).

TABLE 8.1
Ethnomedicinal Notes on Twenty Popular *Angelica* Species from the Total of Sixty Medicinal *Angelica* Spp.

S. No	*Angelica* Species	Distribution and Ethnomedicinal Notes	References
1	*Angelica acutiloba* (Siebold & Zucc.) Kitag.	The plant is indigenous to Japan; The roots of *A. acutibola* are traditionally used in China and Japan as a tonic and used to treat phthisis and haemorrhage, as an emmenagogue, to treat anaemia, and to alleviate pain during parturition.	Sheu et al. (1987); Hwang and Yang (1997); Kim et al. (2011)
2	*Angelica anomala* Avé-Lall.	*A. anomala* is one of the most important medicinal plants and is used in various traditional Chinese medicines system as an antidote to snake bite; roots are also used in curing breast cancer, tumours, wounds, and the common cold.	Sarker and Nahar (2004); He et al. (2012)
3	*Angelica archangelica* L.	The plant is indigenous to temperate Asia; the plant is very famous in China, Japan, and Korea for its medicinal value. Roots are used to cure different ailments: tea made from the roots is used to cure stomach cancer, and other formulations of roots are used to cure gastro diseases, migraine, respiratory problems, insomnia, menstrual complaints, arthritis, and chronic fatigue.	Pasqua et al. (2003); Sarker and Nahar (2004); Bhat et al. (2011); Chauhan et al. (2016); Aćimović et al. (2022)
4	*Angelica atropurpurea* L.	The plant is used to treat fever, gout, and respiratory problems.	Nagasawa et al. (2014); USDA (2022)
5	*Angelica breweri* A. Gray	The plant is used to treat chest problems.	Sarker and Nahar (2004); USDA (2022)
6	*Angelica dahurica* (Hoffm.) Benth. & Hook. f. ex Franch. & Sav.	This species is known by different names in different countries, i.e., is called 'Zenko' in Japan. 'Jahwajeonho' in Korea and 'Zi hua qian hu' in China. The plant is used as an antidote to snake bite, to treat rheumatism, sore throat, common cold, furunculosis, headache, tumour, acne, erythema, toothache, and sinusitis.	Sarker and Nahar (2004); Pervin et al. (2014); Bai et al. (2016); Xie et al. 2016; Chen et al. (2021)
7	*Angelica decuirsiva* (Miq.) Franch. & Sav.	The plant is used to cure a broad spectrum of diseases such as fever, stomach problems, and respiratory problems, and is also used as a carminative.	O'Hara and Zhan (1994); Czerepanov and Cherepanov 1995; Sarker and Nahar (2004); USDA (2022)
8	*Angelica genuflexa* Nutt.	The plant is used to treat body pain and has purgative properties.	Sarker and Nahar (2004)
9	*Angelica gigas* Nakai	The plant is very famous in the Korean system of medicine, and the dried roots of the plants are used to cure sedative and anaemia. Roots are also used as a blood tonic agent.	Yan et al. (2004); Kim et al. (2011); Sowndhararajan and Kim (2017b)
10	*Angelica glauca* Edgew.	The plant is one of the most famous medicinal plants from the genus *Angelica* and the roots of the plant are used to cure a broad spectrum of disease from common digestive ailments to lethal diseases. Mainly, the plant is used to treat asthma, nervous system disorders, bronchitis, flatulence, stomach pains, atrophy, menorrhagia, rheumatism, skin disease, constipation, and urinary disorders. The plant also exhibits antibacterial, anti-inflammatory, anticancer, antifungal, and antioxidant activities.	Gaur, (1999); Vashistha et al. (2006); Butola and Vashistha (2013); Khajuria et al. (2021); Kumar et al. (2022)

(Continued)

TABLE 8.1 (CONTINUED)
Ethnomedicinal Notes on Twenty Popular *Angelica* Species from the Total of Sixty Medicinal *Angelica* Spp.

S. No	*Angelica* Species	Distribution and Ethnomedicinal Notes	References
11	*Angelica grosseserrata* Maxim.	The plant is used as a diuretic and diaphoretic.	Sarker and Nahar (2004)
12	*Angelica laxifoliata* Diels	The plant is used in folk medicine for the treatment of rheumatism, as a sedative and to invigorate the circulation of blood in some areas in Sichuan Province, China.	Xinyu et al. (1999); Sarker and Nahar (2004)
13	*Angelica lucida* L.	Used to cure common cold and oral problems.	Widelski et al. (2009)
14	*Angelica pubescens* Maxim.	Used to cure rheumatoid arthritis, headache, insomnia, carminative, inflammation, antiseptic, and toothache.	Tabanca et al. (2014); Sowndhararajan et al. (2017a)
15	*Angelica rosaefolia* L.	The plant is used as a diuretic.	Sarker and Nahar (2004)
16	*Angelica sinensis* (Oliv.) Diels	The plant is used to cure different ailments in different systems of medicine in Japan, Korea, and China. The dried root of the plant is used to cure amenorrhoea, gynoecopathia, anaemia, dysmenorrhoea, arthritis, premenstrual or menopausal syndromes, cancer, and cardiovascular diseases, and as a uterotonic.	Tsai et al. (2006); Su et al. (2013); Jian-Ping et al. (2015); Hu et al. (2022)
17	*Angelica sylvestris* L.	The plant is used in various Chinese systems of medicines; the plant is used to treat headache, tongue cancer, respiratory problems, and as a spasm carminative.	Sarker and Nahar (2004)
18	*Angelica tenuissima* Nakai	The plant is very famous in Korea and the roots of the plant, known by the name Gubon, are extensively used in Korean traditional medicines to cure neuralgia, stop pain, cough, headache, rheumatic, and arthralgia.	Lee et al. (2008); Islam et al. (2009); Choi et al. (2010); Choi et al. (2018)
19	*Angelica uchiyamai* A Gray	The plant is used to cure women's complaints, as an emmenagogue, and to treat gynoecopathy.	Sarker and Nahar (2004)
20	*Angelica venenosa* (J. Greenway) Fernald	The plant is used to treat orthopaedic ailments.	Sarker and Nahar (2004).

In addition to serving mankind by treating many ailments, the genus is also used in ethnoveterinary preparations and is helpful in curing winter diarrhoea in animals and improving lactation in cattle (Chauhan, 1999; Butola & Badola, 2006; Ali et al., 2019; Butola et al., 2016).

8.5 COMMERCIAL USES OF THE GENUS *ANGELICA*

The *Angelica* genus from the carrot family has a large number of species and is listed among the world's economic plants. Phytochemical compounds or secondary metabolites derived from plants are the primary sources of new drug discovery. Most *Angelica* species contain diverse phytochemicals (phenolics, flavonoids, terpenoids, steroids, tannins, saponins, glycosides), such as coumarins, valeric acid, angelic acid, and angelicine resin, imperatorin, umbelliprenin, and bergapten, a linear furanocoumarin. Further hydrodistillation of roots from different *Angelica* species confirmed the presence of essential oils containing different compounds (Figure 8.1 and Table 8.2). So this diverse range of phytochemicals makes the *Angelica* genus one of the Apiaceae's most important

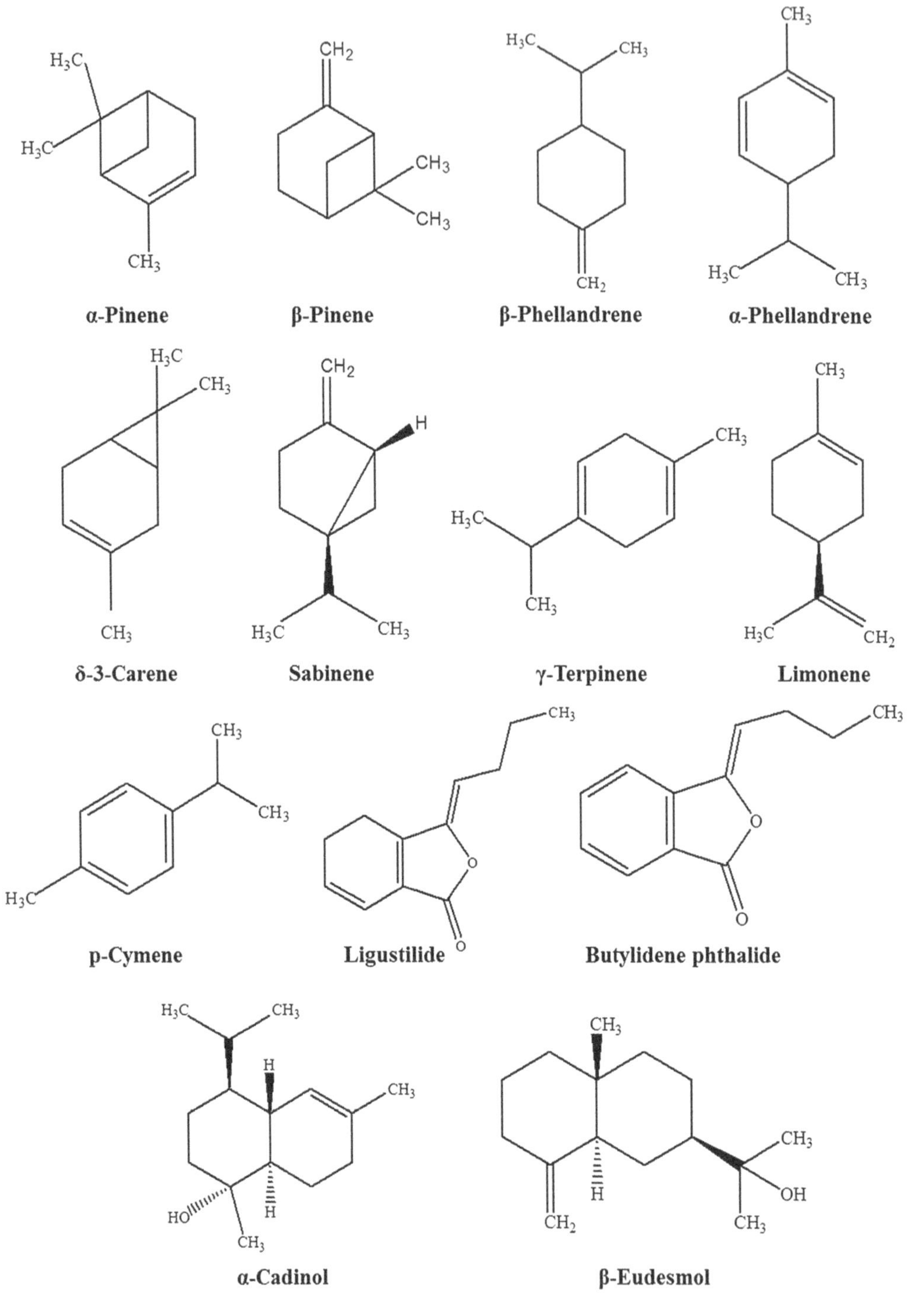

FIGURE 8.1 The chemical structure of some important essential oil components from *Angelica* species (Sowndhararajan et al. 2017b).

TABLE 8.2
Phytochemical Compounds in Essential Oil from Hydrodistillation of Root from *Angelica* spp.

S. No	*Angelica* Species	Major Components after Hydrodistillation	References
1	*Angelica acutiloba* (Siebold & Zucc.) Kitag.	Butylidene phthalide, furfural, 2-furanmethanol, 5-methyl furfural, maltol, butylidene and dihydro-phthalide.	Kim et al. (2006)
2	*Angelica archangelica* L.	α-Pinene, δ-3-carene, limonene, β-phellandrene, α-phellandrene and *p*-cymene, dillapiole and nothoapiol.	Pasqua et al. (2003); Nivinskiene et al. (2005); Fraternale et al. (2014); Chauhan et al. (2016)
3	*Angelica gigas* Nakai	α-Pinene, 2,4,6-trimethyl heptanes, α-limonene and camphene, β-eudesmol, nonanedecursin, decursinol angelate and γ-eudesmol.	Seoet al. (2007); Sowndhararajan et al. (2017a)
4	*Angelica glauca* Edgew.	(*Z*)-Ligustilide, (*Z*)-butylidene phthalide and (*E*)-butylidenephthalide.	Purohit et al. (2015)
5	*Angelica pubescens* Maxim.	α-Pinene, *p*-cymene, limonene and cryptone.	Tabanca et al. (2014)
6	*Angelica sinensis* (Oliv.) Diels	Butylidene dihydro-phthalide, butylidene phthalide, furfural, camphene, 4-pyridinol, ligustilide and di-iso-octyl phthalate.	Kim et al. (2006); Champakaew et al. (2015)
7	*Angelica dahurica* (Fisch. Ex Hoffm.) Benth. & Hook.	Dodecyl alcohol, elemene, hexadecanoic acid, ethyl ester, 1-pentadecanol, α-pinene, sabinene, myrcene, 1-dodecanol and terpinen-4-ol.	Tabanca et al. (2014); Wang et al. (2016)

commercial genera. Additional commercial importance of the genus may be attributed in different ways, i.e., pharmacology, food industry, and others. In pharmacology, the genus exhibits a range of biological activities, i.e., antibacterial (Irshad et al., 2011; Mullen et al., 2014), antifungal (Tabanca et al., 2014; Roh & Shin, 2014), antioxidant (Li et al., 2007; Wei et al., 2007; Roh & Shin, 2014), neuroprotective (Sowndhararajan and Kim 2017a), antiviral (Rajtar et al., 2017), anti-inflammatory (Zhang et al., 2015; Yao et al., 2015; Wang et al., 2016,), hepatoprotective (Yeh et al., 2003; Ji et al., 2014; Hou et al., 2021), anticancer (Wu et al., 2017), antitumour (Sigurdsson et al., 2005), hyperglycaemic (Wang et al., 2016; Bae et al., 2016), cardioprotective (Xin et al., 2007; Zhang et al., 2010), acetylcholinesterase inhibitory (Kang et al., 2001; Sigurdsson & Gudbjarnason, 2007; Kwon et al., 2017), effect on central nervous system (Kim et al., 2007; Ma et al., 2019; Kaur and Bhatti 2021), calcium blocking (Harmala et al., 1992; Wang et al., 2018), effect on female sex hormones (Usuki, 1988; Liu et al., 2001; Butola & Vashistha, 2013), and nephroprotective effect (Li et al., 1998; Cai et al., 2001).

In addition, many *Angelica* root products have local and international markets, with the root being used in many homoeopathic and Ayurvedic preparations (Figure 8.2). Furthermore, in food and other industries, several members of this genus are consumed as a vegetable or sometimes pickled, with some members being used to flavour confectionery ('flavouring honey') and liqueurs (the main flavouring ingredient of gins, vermouth, and wines to provide a particular aroma or taste) and to prepare perfumes, particularly in the Paris perfume industries (Chauhan, 1999; Butola & Badola, 2006; Butola et al., 2016).

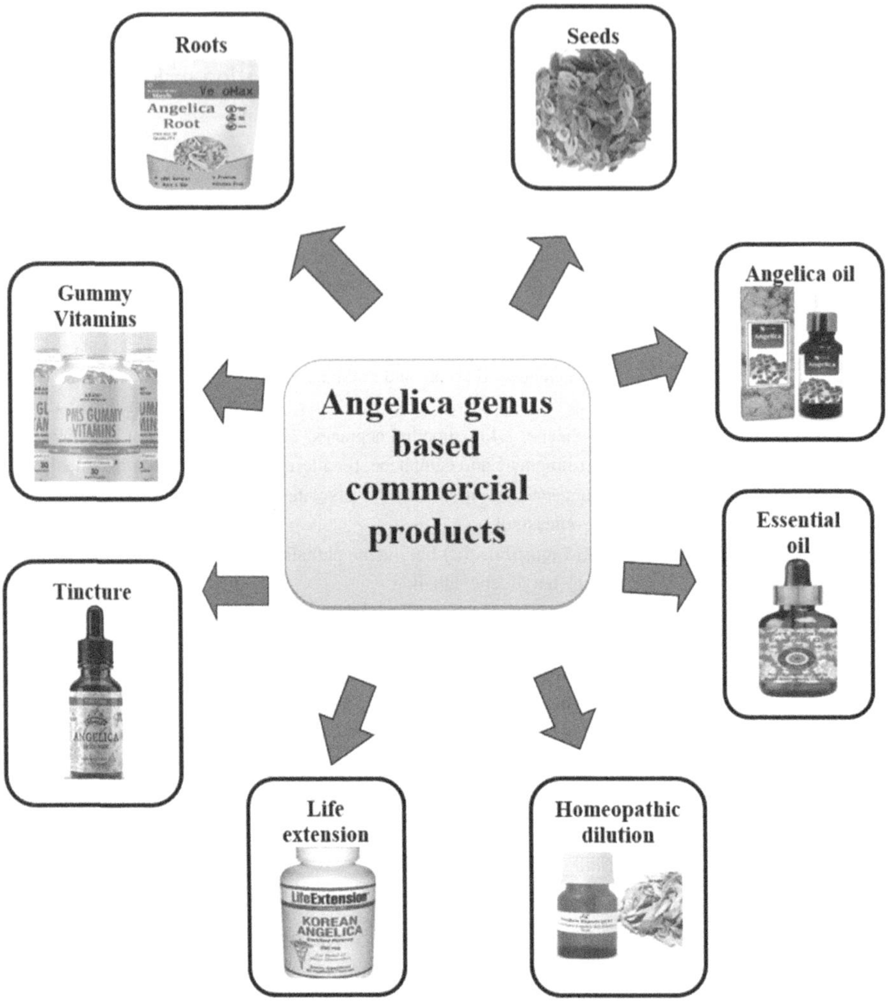

FIGURE 8.2 Various commercial products of the *Angelica* genus.

8.6 CONCLUSION

In conclusion, the genus *Angelica* is not a genus of the past, as a number of species of this genus are very popular in herbal formulations even today. The genus is equally used in codified and non-codified s of medicine to treat a broad spectrum of diseases, paralleling the number of species used for other commercial purposes. The root is the preferred part of the herb. It contains diverse phytochemicals but the illegal harvesting of plants for their roots from the wild has put several species of this genus into the endangered or threatened categories. To date, numerous studies on this genus have been published, but much work is needed to expand the already-existing knowledge for this genus in order to understand and evaluate the healing properties of this genus and to simultaneously conserve the species in their natural habitats.

REFERENCES

Aćimović, M., Rat, M., Pezo, L., Lončar, B., Pezo, M., Miljković, A. et al. (2022) Biological and chemical diversity of *Angelica archangelica* L.—Case study of essential oil and its biological activity. *Agronomy*, 12(7), 1570.

Ali, M.Y., Seong, S.H., Jannat, S., Jung, H.A. & Choi, J.S. (2019) Ethnobotany, phytochemistry, and pharmacology of Angelica decursiva Fr. et Sav. *Natural Product Sciences*, 25(3), 181–199.

Asfaw, T.B., Esho, T.B., Bachheti, A., Bachheti, R.K., Pandey, D.P. & Husen, A. (2022) Exploring important herbs, shrubs, and trees for their traditional knowledge, chemical derivatives, and potential benefits. In: Husen, A. (Eds.) *Herbs, shrubs, and trees of potential medicinal benefits*, CRC Press, Boca Raton, Florida, United State. pp. 1–26.

Bae, U.J., Choi, E.K., Oh, M.R., Jung, S.J., Park, J., Jung, T.S. et al. (2016) *Angelica gigas* ameliorates hyperglycemia and hepatic steatosis in C57BL/KsJ-db/db mice via activation of AMP-activated protein kinase signaling pathway. *The American Journal of Chinese Medicine*, 44(08), 1627–1638.

Bai, Y., Li, D., Zhou, T., Qin, N., Li, Z., Yu, Z. et al. (2016) Coumarins from the roots of *Angelica dahurica* with antioxidant and antiproliferative activities. *Journal of Functional Foods*, 20, 453–462.

Bhat, Z.A., Kumar, D. & Shah, M.Y. (2011) *Angelica archangelica* Linn. is an angel on earth for the treatment of diseases. *International Journal of Nutrition, Pharmacology, Neurological Diseases*, 1(1), 36–50.

Bhutani, K.K. & Gohil, V.M. (2010) Natural products drug discovery research in India: Status and appraisal. *Indian Journal of Experimental Biology*, 48, 197–207.

Bisht, N.S. & Khajuria, A.K. (2014) Ethno-medicinal plants of Tehsil, Kathua, Jammu & Kashmir. *Journal of Mountain Research*, 9, 1–12.

Bussmann, R.W., Sharon, D. & Garcia, M. (2009) From chamomile to aspirin? Medicinal plant use among clients at Laboratorios Beal in Trujillo, Peru. *Ethnobotany Research and Applications*, 7, 399–407.

Butola, J.S. & Badola, H.K. (2004) Effect of pre-sowing treatment on seed germination and seedling vigour in *Angelica glauca*, a threatened medicinal herb. *Current Science*, 87(6) 796–799.

Butola, J.S. & Badola, H.K. (2006) Under different growing media and environments. *Journal of Non-Timber Forest Products*, 13(2), 141–153.

Butola, J.S., Vashistha, R.K., Malik, A.R. & Rawat, M.S. (2016) Ethnomedicinal importanceof Gandrayan (Angelica glauca Edgew.) in the North-Westernpart of Indian Himalayan Region. *Medicinal Plants-International Journal of Phytomedicines and Related Industries*, 8(4), 313–318.

Butola, J.S. & Vashistha, R.K. (2013) An overview on conservation and utilization of *Angelica glauca* Edgew. in three Himalayan states of India. *Med Plants*, 5(3), 171–178.

Cai, Q., Li, X. & Wang, H. (2001) Astragali and Angelica protect the kidney against ischemia and reperfusion injury and accelerate recovery. *Chinese Medical Journal*, 114(02), 119–123.

Champakaew, D., Junkum, A., Chaithong, U., Jitpakdi, A., Riyong, D., Sanghong, R. et al. (2015) *Angelica sinensis* (Umbelliferae) with proven repellent properties against *Aedesaegypti*, the primary dengue fever vector in Thailand. *Parasitology Research*, 114(6), 2187–2198.

Chauhan, N.S. (1999) *Medicinal and aromatic plants of Himachal Pradesh*. Indus Publishing, Delhi, India.

Chauhan, R.S., Nautiyal, M.C., Cecotti, R., Mella, M. & Tava, A. (2016) Variation in the essential oil composition of *Angelica archangelica* from three different altitudes in Western Himalaya, India. *Industrial Crops and Products*, 94, 401–404.

Chen, W., Wang, G., Mei, K. & Zhu, J. (2021) Coumarins from *Angelica dahurica* and their antitumor activities in human MG-63 osteosarcoma cells. *Records of Natural Products*, 15(5), 356–362.

Choi, M., Lee, Y. & Cho, S.H. (2018) *Angelica tenuissima* Nakai ameliorates cognitive impairment and promotes neurogenesis in mouse model of Alzheimer's disease. *Chinese Journal of Integrative Medicine*, 24(5), 378–384.

Choi, U.K., Lee, O.H., Lim, S.I. & Kim, Y.C. (2010) Optimization of antibacterial activity of *Angelica tenuissima* leaf against *Pseudomonas aeruginosa* using the evolutionary operation-factorial design. *International Journal of Molecular Sciences*, 11, 3922–3932.

Czerepanov, S.K. & Cherepanov, S.K. (1995) *Vascular plants of Russia and adjacent states (the former USSR)*. Cambridge University Press Cambridge, United Kingdom.

Dhiman, A.K. & Kumar, A. (2006) *Ayurvedic drug plants*. Daya Books, Delhi, India.

Fraternale, D., Flamini, G. & Ricci, D. (2014) Essential oil composition and antimicrobial activity of *Angelica archangelica* L. (Apiaceae) roots. *Journal of Medicinal Food*, 17(9), 1043–1047.

Gaur, R.D. (1999) *Flora of the district Garhwal, North West Himalaya*. Transmedia, Garhwal- Uttarakhand, India.

Gurib-Fakim, A. (2006) Medicinal plants: Traditions of yesterday and drugs of tomorrow. *Molecular Aspects of Medicine*, 27(1), 1–93.

Harmala, P., Vuorela, H., Törnquist, K. & Hiltunen, R. (1992) Choice of solvent in the extraction of *Angelica archangelica* roots with reference to calcium blocking activity. *Planta Medica*, 58(02), 176–183.

Hatano, R., Takano, F., Fushiya, S., Michimata, M., Tanaka, T., Kazama, I. et al. (2004) Water-soluble extracts from *Angelica acutiloba* Kitagawa enhance hematopoiesis by activating immature erythroid cells in mice with 5-fluorouracil-induced anemia. *Experimental Hematology*, 32(10), 918–924.

He, Y., Hou, P., Fan, G., Song, Z., Arain, S., Shu, H. et al. (2012) Authentication of *Angelica anomala* Avé-Lall cultivars through DNA barcodes. *Mitochondrial DNA*, 23(2), 100–105.

Hooker, J.D. (1875) *The flora of British India (Vol. 1)*. L. Reeve, London.

Hou, C., Yin, M., Lan, P., Wang, H., Nie, H. & Ji, X. (2021) Recent progress in the research of *Angelica sinensis* (Oliv.) Diels polysaccharides: Extraction, purification, structure and bioactivities. *Chemical and Biological Technologies in Agriculture*, 8(1), 1–14.

Hu, Q., Wu, C., Yu, J., Luo, J. & Peng, X. (2022) *Angelica sinensis* polysaccharide improves rheumatoid arthritis by modifying the expression of intestinal Cldn5, Slit3 and Rgs18 through gut microbiota. *International Journal of Biological Macromolecules*, 209, 153–161.

Hwang, J.B. & Yang, M.O. (1997) Comparision of chemical components of *Angelica gigas* Nakai and *Angelica acutiloba* Kitagawa. *Korean Journal of Food Science and Technology*, 29(6), 1113–1118.

Irshad, M., Shahid, M., Aziz, S. & Ghous, T. (2011) Antioxidant, antimicrobial and phytotoxic activities of essential oil of *Angelica glauca*. *Asian Journal of Chemistry*, 23(5), 1947–1951.

Islam, M.N., Nam, J.W., Lee, J., Seo, E.K., Han, S.B., Kim, D.H. et al. (2009) A validated LC method for simultaneous determination of phenolic, coumarin and phthalide compounds in the ethanolic extract of *Angelica tenuissima*. *Chromatographia*, 70(7), 1079–1085.

Ji, P., Wei, Y., Sun, H., Xue, W., Hua, Y., Li, P. et al. (2014) Metabolomics research on the hepatoprotective effect of *Angelica sinensis* polysaccharides through gas chromatography–mass spectrometry. *Journal of Chromatography B*, 973, 45–54.

Jian-Ping, M.A., Zhi-Bing, G.U.O., Ling, J.I.N. & Ying-Dong, L.I. (2015) Phytochemical progress made in investigations of *Angelica* sinensis (Oliv.) Diels. *Chinese Journal of Natural Medicines*, 13(4), 241–249.

Kala, C.P., Farooquee, N.A. & Dhar, U. (2004) Prioritization of medicinal plants on the basis of available knowledge, existing practices and use value status in Uttaranchal, India. *Biodiversity & Conservation*, 13(2), 453–469.

Kang, S.Y., Lee, K.Y., Sung, S.H., Park, M.J. & Kim, Y.C. (2001) Coumarins Isolated from *Angelica gigas* inhibit Acetylcholinesterase: Structure – Activity relationships. *Journal of Natural Products*, 64(5), 683–685.

Kaul, M.K. (1997) *Medicinal plants of Kashmir and Ladakh: Temperate and cold arid Himalaya*. Indus publishing, Delhi, India.

Kaur, A. & Bhatti, R. (2021) Understanding the phytochemistry and molecular insights to the pharmacology of *Angelica archangelica*L.(garden angelica) and its bioactive components. *Phytotherapy Research*, 35(11), 5961–5979.

Khajuria, A.K., Manhas, R.K., Kumar, H. & Bisht, N.S. (2021) Ethnobotanical study of traditionally used medicinal plants of Pauri district of Uttarakhand, India. *Journal of Ethnopharmacology*, 276, 114204.

Khajuria, A.K. & Bisht, N.S. (2017) Ethnomedicinal plants used to treat Nephrolithiasis: A case study Pauri (PAURI Garhwal), Uttarakhand. *International Journal of Herbal Medicine*, 5(1), 10–13.

Kim, M.R., El-Aty, A.A., Choi, J.H., Lee, K.B. & Shim, J.H. (2006) Identification of volatile components in Angelica species using supercritical-CO2 fluid extraction and solid phase microextraction coupled to gas chromatography-mass spectrometry. *Biomedical Chromatography*, 20(11), 1267–1273.

Kim, S., Oh, H.K., Kim, J.Y., Hong, J.W. & Cho, S.I. (2011) A review of pharmacological effects of *Angelica* gigas, *Angelica* sinensis, *Angelica acutiloba* and their bioactive compounds. *The Journal of Korean Medicine*, 32(4), 1–24.

Kim, Y.K., Kim, Y.S. & Ryu, S.Y. (2007) Antiproliferative effect of furanocoumarins from the root of *Angelica dahurica* on cultured human tumor cell lines. *Phytotherapy Research*, 21(3), 288–290.

Kumar, H., Khajuria, A.K. & Bisht, N.S. (2018) Traditional Phytoremedies used to treat Urolithiasis in Pauri (Garhwal) Uttarakhand India. *Journal of Pharmacognosy and Phytochemistry*, 7(3), 2941–2944.

Kumar, P., Rana, V. & Singh, A.N. (2022) *Angelica glauca* Edgew. - A comprehensive review. *Journal of Applied Research on Medicinal and Aromatic Plants*, 31, 100397.

Kunwar, R.M., Mahat, L., Acharya, R.P. & Bussmann, R.W. (2013) Medicinal plants, traditional medicine, markets and management in far-west Nepal. *Journal of Ethnobiology and Ethnomedicine*, 9(1), 1–10.

Kwon, Y., Kim, H.P., Kim, M.J. & Chun, W. (2017) Acetylcholinesterase inhibitors from *Angelica polymorpha* stem. *Natural Product Sciences*, 23(2), 97–102.

Lee, H.W., Choi, J.H., Park, S.Y., Choo, B.K., Chun, J.M., Lee, A. et al. (2008) Constituents comparison of components in native and cultivated species of *Angelica tenuissima* Nakai. *Korean Journal of Medicinal Crop Science*, 16(3), 168–172.

Li, L., Chen, M., Tong, T. & Wang, H. (1998) The molecular basis of Chinese herbs (Astragali and Angelica) on increasing serum albumin synthesis in nephrotic rats. *Nephrology*, 4(5), 373–378.

Li, S.Y., Yu, Y. & Li, S.P. (2007) Identification of antioxidants in essential oil of radix *Angelica sinensis* using HPLC coupled with DAD-MS and ABTS-based assay. *Journal of Agricultural and Food Chemistry*, 55(9), 3358–3362.

Liu, J., Burdette, J.E., Xu, H., Gu, C., Van Breemen, R.B., Bhat, K.P. et al. (2001) Evaluation of estrogenic activity of plant extracts for the potential treatment of menopausal symptoms. *Journal of Agricultural and Food Chemistry*, 49(5), 2472–2479.

Ma, J., Huang, J., Hua, S., Zhang, Y., Zhang, Y., Li, T. et al. (2019) The ethnopharmacology, phytochemistry and pharmacology of *Angelica biserrata*–a review. *Journal of Ethnopharmacology*, 231, 152–169.

Mullen, K.A.E., Lee, A.R., Lyman, R.L., Mason, S.E., Washburn, S.P. & Anderson, K.L. (2014) An in vitro assessment of the antibacterial activity of plant-derived oils. *Journal of Dairy Science*, 97(9), 5587–5591.

Nagasawa, A., Sakasai, M., Sakaguchi, D., Moriwaki, S., Nishizawa, Y. & Kitahara, T. (2014) Angular-type Furocoumarins from the roots of *Angelica atropurpurea* and their inhibitory activity on the NFAT signal transduction pathway. *Natural Product Communications*, 9(12), 1934578X1400901218.

Nivinskienė, O., Butkienė, R. & Mockutė, D. (2005) The chemical composition of the essential oil of *Angelica archangelica* L. roots growing wild in Lithuania. *Journal of Essential Oil Research*, 17(4), 373–377.

O'Hara, E.M. & Zhan, L. (1994) Cultural and pharmacologic considerations when caring for Chinese elders: Knowledge of traditional Chinese medicine is necessary. *Journal of Gerontological Nursing*, 20(10), 11–16.

Pasqua, G., Monacelli, B. & Silvestrini, A. (2003) Accumulation of essential oils in relation to root differentiation in *Angelica archangelica* L. *European Journal of Histochemistry*, 47(1), 87–90.

Pervin, M., Hasnat, M.A., Debnath, T., Park, S.R., Kim, D.H. & Lim, B.O. (2014) Antioxidant, anti-inflammatory and antiproliferative activity of *Angelica dahurica* root extracts. *Journal of Food Biochemistry*, 38(3), 281–292.

Purohit, V.K., Andola, H.C., Haider, S.Z., Tiwari, D., Bahuguna, Y.M., Gairola, K.C. et al. (2015) Essential oil constituents of *Angelica glauca* Edgew. Roots: An endangered species from Uttarakhand Himalaya (India). *National Academy Science Letters*, 38(5), 445–447.

Rajtar, B., Skalicka-Woźniak, K., Świątek, Ł., Stec, A., Boguszewska, A. & Polz-Dacewicz, M. (2017) Antiviral effect of compounds derived from *Angelicaarchangelica* L. on Herpes simplex virus-1 and Coxsackievirus B3 infections. *Food and Chemical Toxicology*, 109, 1026–1031.

Roh, J. & Shin, S. (2014) Antifungal and antioxidant activities of the essential oil from *Angelica koreana* Nakai. *Evidence-Based Complementary and Alternative Medicine*, 1–7 https://doi.org/10.1155/2014/398503.

Sarker, S.D. & Nahar, L. (2004) Natural medicine: The genus Angelica. *Current Medicinal Chemistry*, 11(11), 1479–1500.

Seo, H.Y., Kim, J.H., Song, H.P., Kim, D.H., Byun, M.W., Kwon, J.H. et al. (2007) Effects of gamma irradiation on the yields of volatile extracts of *Angelica gigas* Nakai. *Radiation Physics and Chemistry*, 76, 1869–1874.

Sheu, S.J., Ho, Y.S., Chen, Y.P. & Hsu, H.Y. (1987) Analysis and processing of Chinese herbal drugs. VI. The study of Angelicae Radix. *Planta Medica*, 53(4), 377–378.

Sigurdsson, S. & Gudbjarnason, S. (2007) Inhibition of acetylcholinesterase by extracts and constituents from *Angelica archangelica* and *Geranium sylvaticum*. *Zeitschrift für Naturforschung C*, 62(9–10), 689–693.

Sigurdsson, S., Ögmundsdóttir, H.M. & Gudbjarnason, S. (2005) The cytotoxic effect of two chemotypes of essential oils from the fruits of *Angelicaarchangelica* L. *Anticancer Research*, 25(3), 1877–1880.

Sowndhararajan, K. & Kim, S. (2017a) Neuroprotective and cognitive enhancement potentials of *Angelica* gigas Nakai root: A review. *Scientia Pharmaceutica*, 85(2), 1–21.

Sowndhararajan, K., Deepa, P., Kim, M., Park, S.J. & Kim, S. (2017b) A review of the composition of the essential oils and biological activities of Angelica species. *Scientia Pharmaceutica*, 85(3), 1–14.

Su, Z.Y., Khor, T.O., Shu, L., Lee, J.H., Saw, C.L.L., Wu, T.Y., Huang, Y. et al. (2013) Epigenetic reactivation of Nrf2 in murine prostate cancer TRAMP C1 cells by natural phytochemicals Z-ligustilide and Radix *Angelica sinensis* via promoter CpG demethylation. *Chemical Research in Toxicology*, 26(3), 477–485.

Tabanca, N., Gao, Z., Demirci, B., Techen, N., Wedge, D.E., Ali, A. et al. (2014) Molecular and phytochemical investigation of *Angelica dahurica* and *Angelica pubescentis* essential oils and their biological activity against *Aedesaegypti*, *Stephanitispyrioides*, and Colletotrichum species. *Journal of Agricultural and Food Chemistry*, 62(35), 8848–8857.

Tsai, N.M., Chen, Y.L., Lee, C.C., Lin, P.C., Cheng, Y.L., Chang, W.L. et al. (2006) The natural compound n-butylidenephthalide derived from *Angelica sinensis* inhibits malignant brain tumor growth in vitro and in vivo3. *Journal of Neurochemistry*, 99(4), 1251–1262.

USDA, Agricultural Research Service, National Plant Germplasm System. (2022) Germplasm Resources Information Network (GRIN Taxonomy). National Germplasm Resources Laboratory, Beltsville, Maryland. Available online at http://npgsweb.ars grin.gov/gringlobal/taxon/taxonomygenus?id=674. Accessed 1 July 2022.

Usuki, S. (1988) Effects of Tokishakuyakusan and its components on rat corpus luteum function in vitro. *The American Journal of Chinese Medicine*, 16, 11–19.

Vashistha, R., Nautiyal, B.P. & Nautiyal, M.C. (2006) Conservation status and morphological variations between populations of *Angelica glauca* Edgew. And *Angelica archangelica* Linn. in Garhwal Himalaya. *Current Science*, 91(11), 1537–1542.

Wang, K., Tang, Z., Zheng, Z., Cao, P., Shui, W., Li, Q. et al. (2016) Protective effects of *Angelica sinensis* polysaccharide against hyperglycemia and liver injury in multiple low-dose streptozotocin-induced type 2 diabetic BALB/c mice. *Food & function*, 7(12), 4889–4897.

Wang, S., Li, X., Bao, J. & Chen, S. (2018) Protective potential of *Angelica sinensis* polysaccharide extract against ethylene glycol-induced calcium oxalate urolithiasis. *Renal Failure*, 40(1), 618–627.

Wei, A. & Shibamoto, T. (2007) Antioxidant activities and volatile constituents of various essential oils. *Journal of Agricultural and Food Chemistry*, 55(5), 1737–1742.

Widelski, J., Popova, M., Graikou, K., Glowniak, K. & Chinou, I. (2009) Coumarins from *Angelica lucida* L.-antibacterial activities. *Molecules*, 14(8), 2729–2734.

Wu, Z., Raven, P.H. & Hong, D. (2005) *Flora of China. Volume 14: Apiaceae Through Ericaceae*. Science Press, Beijing, China.

Wu, W., Tang, S. N., Zhang, Y., Puppala, M., Cooper, T. K., Xing, C. et al. (2017). Prostate cancer xenograft inhibitory activity and pharmacokinetics of decursinol, a metabolite of angelica gigas pyranocoumarins, in mouse models. *The American Journal of Chinese Medicine*, 45(08), 1773–1792.

Xie, Q., Li, S.X., Liao, D.F., Wang, W., Tekwani, B., Huang, H.Y. et al. (2016) Bio-pesticidal and Antimicrobial Coumarins from *Angelica dahurica* (Fisch. Ex Hoffm). *Records of Natural Products*, 10(3) 17881–17888.

Xin, Y.F., Zhou, G.L., Shen, M., Chen, Y.X., Liu, S.P., Chen, G.C. et al. (2007) *Angelica sinensis*: A novel adjunct to prevent doxorubicin-induced chronic cardiotoxicity. *Basic & Clinical Pharmacology & Toxicology*, 101(6), 421–426.

Xinyu, G., Hanqing, Z. & Nianhe, W. (1999) The chemical constituents of root of *Angelica laxifoliata* Diels. *Journal of Plant Resources and Environment*, 8(1), 1–5.

Yan, J.J., Kim, D.H., Moon, Y.S., Jung, J.S., Ahn, E.M., Baek, N.I. et al. (2004) Protection against beta-amyloid peptide-induced memory impairment with long-term administration of extract of *Angelica gigas* or decursinol in mice. *Progress in Neuro-Psychopharmacology & Biological Psychiatry*, 28, 25–30.

Yao, W., Zhang, L., Hua, Y., Ji, P., Li, P., Li, J. et al. (2015) The investigation of anti-inflammatory activity of volatile oil of *Angelica sinensis* by plasma metabolomics approach. *International immunopharmacology*, 29(2), 269–277.

Yeh, M.L., Liu, C.F., Huang, C.L. & Huang, T.C. (2003) Hepatoprotective effect of *Angelica archangelica* in chronically ethanol-treated mice. *Pharmacology*, 68(2), 70–73.

Zhang, J., Li, L., Jiang, C., Xing, C., Kim, S.H. & Lu, J. (2012) Anti-cancer and other bioactivities of Korean *Angelica* gigas Nakai (AGN) and its major pyranocoumarin compounds. *Anti-Cancer Agents in Medicinal Chemistry (Formerly Current Medicinal Chemistry-Anti-Cancer Agents)*, 12(10), 1239–1254.

Zhang, S., He, B., Ge, J., Li, H., Luo, X., Zhang, H. et al. (2010) Extraction, chemical analysis of *Angelica sinensis* polysaccharides and antioxidant activity of the polysaccharides in ischemia–reperfusion rats. *International Journal of Biological Macromolecules*, 47(4), 546–550.

Zhang, W.Q., Hua, Y.L., Zhang, M., Ji, P., Li, J.X., Zhang, L. et al. (2015) Metabonomic analysis of the anti-inflammatory effects of volatile oils of Angelica sinensis on rat model of acute inflammation. *Biomedical Chromatography*, 29(6), 902–910.

9 Traditional and Modern Medicinal Uses of *Cichorium intybus* Roots

Munir Ozturk, Volkan Altay, and
Mehdi Younessi-Hamzekhanlu

9.1 INTRODUCTION

Throughout history, plants, fungi, bee products, minerals, shells, and even some animals have been used to treat different diseases (Masoodi et al., 2020; Malik et al., 2021; Ozturk et al., 2021, 2022). Approximately 30.000 plant species are traditionally used globally by different cultures (Ozturk et al., 2014, 2017a, b, 2018b). Herbal medicines used for traditional health care are generally used as teas, tinctures, ointments, capsules or tablets, inhalants, or in herbal baths (Malik et al., 2021). In traditional systems, plants used for medicinal purposes by local people are not only a cultural heritage but also a part of a rich information system. No matter where they are in the world, plants are the main component of biodiversity. When this situation is inevitably associated with human health, there is no doubt that it will make great positive contributions. Unfortunately, despite the increasing number of conference presentations, articles, books, and book chapters on data about medicinal plants in recent years, there is still insufficiently detailed information for numerous medicinally valuable plant species (Ozturk et al., 2012b, 2017c, 2018e-g; Imanberdieva 2018a, b; Malik et al., 2021).

Cichorium intybus (chicory) (Figure 9.1) is one of the folk remedies that has been and still is commonly used worldwide. Although the plant has a long traditional use historically, it has not been reported in the European Pharmacopoeia (Street et al., 2013; Das et al., 2016; Janda et al., 2021).

This species has been reported to occur naturally in many different areas (Janda et al., 2021). The natural distribution areas given in the online database are as follows: Afghanistan, Albania, Algeria, Austria, Azores, Baleares, the Baltic states, Belarus, Belgium, Bulgaria, Cyprus, Czechoslovakia, Denmark, East Aegean Islands, Egypt, Finland, France, Germany, Greece, Hungary, Iran, Iraq, Italy, Kazakhstan, Kyrgyzstan, Lebanon-Syria, Morocco, Netherlands, North Caucasus, Norway, Pakistan, Palestine, Poland, Portugal, Romania, Russia, Spain, Sweden, Switzerland, Tadzhikistan, Transcaucasus, Tunisia, Turkiye, Turkmenistan, Ukraine, Uzbekistan, West Himalayas, and the former Yugoslavia (www.powo.science.kew.org).

The places where this plant has been introduced are: Argentina, Arizona, Australia (especially western and southern areas), Bangladesh, Bolivia, Borneo, Brazil, California, Cambodia, Cape Provinces, Chile, China, Costa Rica, Cuba, Ethiopia, Guatemala, India, Ireland, Kuwait, Libya, Madagascar, Mauritania, Mexico Mongolia, Myanmar, New Zealand, Oman, Paraguay, Philippines, Saudi Arabia, South Australia, Taiwan, Tasmania, Thailand, Tibet, Uruguay, the USA., Venezuela, Vietnam, West Siberia, Yemen, Zaïre, and Zimbabwe (www.powo.science.kew.org).

Chicory commonly grows as a wild plant but is also cultivated in the temperate climate zone. Ecologically, the habitats of this plant are plains, low mountain areas, meadows, grasslands, cultivated fields, and waste places, from sea level to 3,050 m above sea level (Davis, 1975; Ozturk, 1980; Janda et al., 2021). It is one of the characteristic ruderal plants in many floristic and ecological

DOI: 10.1201/b22924-9

FIGURE 9.1 Some morphological features of the plant (Photos: Mustafa Keskin from Turkiye) 1: General view of the plant, 2: Cauline leaves of the plant, 3: Flower bud of the plant, 4: Leaves and stem of the plant, 5: Flower of the plant, and 6: Albino flowers of the plant.

studies. It is reported to show typical ecotype formation in response to trampling pressure (Ozturk, 1980; Yarcı et al., 2007; Osma et al., 2010; Altay et al., 2010, 2012, 2015a, b, 2020; Severoğlu et al., 2011; Ozturk et al., 2012; Tarakçı et al., 2012; Eskin et al., 2012; Yarcı & Altay, 2016).

Chicory thrives well under mild climatic conditions but requires well-distributed rainfall. It can also be cultivated under controlled irrigation systems. If the plant is raised from roots or seed a hot and humid climate is required for its cultivation. A minimum temperature of 21°C is required for successful seed germination. The seeds need 90 days of chilling to break dormancy, and the same applies to the growth stage transition from the vegetative to the reproductive phase. Chicory can tolerate extreme temperatures during its vegetative and reproductive phases of growth. However, moderate and uniform temperatures with an optimum of 18–24°C are suitable for plant growth. During the vegetative phases, these plants tolerate wide ranges of precipitation. The distribution of rain and snowfall controls good seed production. Vegetative growth is perfect under moderate rainfall and high sunshine levels. The seeds develop partially in areas with heavy rains from July to August. Light soils are suitable for the growth and development of chicory crops. Deep fertile loams with a slight excess of clay are better for higher yields and high-quality roots. Alluvial soils are ideal for growing in hilly areas to obtain thick and fleshy roots, whereas somewhat sandy and moderately rich soil is preferred in the plains (Baert, 1997; Bais & Ravishankar, 2001; Poli et al., 2002; Street et al., 2013).

This chapter will provide a detailed account, summarizing the historical reports up to the recent research literature, along with traditional and modern medicinal uses of chicory roots and data on its phytochemistry in detail.

9.2 THE TRADITIONAL USES OF *CICHORIUM INTYBUS* ROOTS

Chicory has been reported to be one of the plants used in ancient Rome, Greece, and Egypt almost 2,500 years ago (Aisa et al., 2020; Janda et al., 2021). To date, different parts of this species have been used for different medicinal purposes, such as anaemia, diabetes, haemorrhoids, gastrointestinal disorders, gallstones, jaundice, and malaria (Aisa et al., 2020). In addition, the young leaves of this species are used in salads. It has been reported that the fresh roots of the plant are also used in milk soup (Bussmann et al., 2017; Aisa et al., 2020). The hot beverage prepared from the roasted roots of the plant is consumed as coffee (Figure 9.2) (Aisa et al., 2020) and it has been evaluated in different cultures as a potential alternative source of coffee.

Information on traditional medicinal uses of *C. intybus* roots in different countries and cultures are summarized in the following section (Table 9.1).

9.3 ROLE OF *CICHORIUM INTYBUS* ROOTS IN PRESENT-DAY MEDICINE AND ITS PHARMACOLOGICAL ACTIVITIES

The roots of chicory are used to treat loss of appetite and mild dyspeptic disorders. It has also been reported that the herbal root tea is used to treat digestive disorders and act to achieve appetite suppression (Madius, 1938; Scholz, 2006). In addition, the herbal medicines prepared from the roots are used as a tonic and blood purifier and to treat internal and external bleeding, jaundice, and stomach flu (Aftab & Hakeem, 2021). The decoctions prepared from the roots have also been mentioned as a mild laxative and bitter tonic for stomach remedies and supporting liver ailments (Poletti et al., 1989).

Madius (1938) reports that, on a daily basis, 6 g should be taken in two glasses of water as tea infusion during the day. Poletti et al. (1989) mention that a handful of comminuted chicory roots should be cooked for 30 min in 1 litre of water, and one cup of this should be taken before each meal. Scholz (2006) proposes that the root should be taken in the form of 3 g of comminuted roots as an infusion or 2–4 g as a decoction in 250 mL of water once daily.

The use of the roots to treat indigestion and loss of appetite is registered in ‘A German Commission E monograph’ and it is suggested that the daily dose can be 3 g of powder (Blumenthal, 1998). The medicinal uses of chicory roots are also reported and recommended in the Chinese Pharmacopoeia (Aisa et al., 2020).

Many studies have been undertaken to define and determine the pharmacological activities of *C. intybus*. The groupings, based on activities, are presented in detail below in light of the current literature:

FIGURE 9.2 Coffee-making stages of the plant (Photos: Mustafa Keskin from Turkiye) 1: Collecting the roots of the plant; 2: Finely chop the cleaned roots; 3: Roasting of roots; 4: Turning the roots into powder with the help of a grinder; 5: The brewing process in a filter coffee machine; and 6: Making it ready to drink.

9.3.1 Antimicrobial and Antiparasitic Activities

Disruption of the cytoplasmic membrane, inhibitory effect against biofilm formation, inhibition of cell wall formation, and inhibition of microbial DNA replication are the main mechanisms for inducing bactericidal effects of chicory roots; it is thus used as a bactericidal agent for treating many infectious diseases (Hamid et al., 2015; Upadhayay et al., 2021).

Antimicrobial activities against Gram-positive, and Gram-negative bacteria and yeast have been reported for chicory root extract (Shaikh et al., 2016; Rahimullah, 2019; Jasim, 2019; Choudhary et

TABLE 9.1
Traditional Medicinal Uses of *Cichorium intybus* Roots in Different Countries

Country	Preparation	Disease Treated	References
Afghanistan	Extract	Malaria	Bischoff et al. (2004)
Azerbaijan	Decoction	Anaemia, antirheumatic, cholagogue, cold, diabetes, dysentery, hepatic disorders, jaundice, kidney disorders, and malaria	Ozturk et al. (2018c, d)
Bosnia and Herzegovina	Decoction	Antiseptic, cancer, cholesterol, diarrhoea, liver disorders, pulmonary hangover, purification of biliary tract, spasmolytic, and strengthening the prostate	Hanlidou et al. (2004); Šarić-Kundalić et al. (2011)
Bulgaria	Decoction	Cholagogue stimulant for gastric secretion, and hypoglycaemic	Leporatti and Ivancheva (2003)
India	Decoction	Cancer, cough relief, diabetes, gout, jaundice, liver enlargement, and rheumatism	Ahmed et al. (2003); Pushparaj et al. (2007); Masoodi et al. (2020)
Iran	Decoction, and distillation	Antipyretic, blood purifier, choleretic, constipation, depurative, diabetes, eupeptic, facilitating delivery, heart pain, hypotension, jaundice, laxative, lipid disorders, liver disorders, nervous disorders, skin diseases, skin itching, stomach pain, and tonic	Miraldi et al. (2001); Ozturk et al. (2018c, d); Younessi-Hamzekhanlu et al. (2020)
Italy	Decoction	Antiarthritis, antispasmodic, arteriosclerosis, blood purification, choleretic, digestive, hepatoprotective against jaundice, high blood pressure, hypoglycaemic, and mild laxative	Guarrera et al. (2005); Loi et al. (2005); Leporatti and Ivancheva (2003)
Morocco	Decoction	Diabetes and kidney disorders	Jouad et al. (2001); El-Hilaly et al. (2003)
Pakistan	Decoction	Diabetes	Ahmad et al. (2009); Ozturk et al. (2018a)
Poland	Infusion	Digestive complaints, and lack of appetite	European Medicines Agency (2013)
Serbia	Decoction, and infusion	Anti-inflammatory, cholagogue, diarrhoea, digestive, diuretic, hypoglycaemic, laxative, liver complaints, and high blood sugar	Kokoska et al. (2002); Jarić et al. (2007); Šavikin et al. (2013)
South Africa	Decoction	Jaundice, and tonic	van Wyk et al. (1997)
Turkiye	Boiled, cooked, decoction, infusion, and oinment	Abdominal pain, analgesic, asthma, balding, bronchitis, cancer, cardiac diseases, cough, dermatitis, diabetes, eczema, haemorrhoids, hypertension, kidney stones, liver diseases, prostate treatment, sedative, sunburn, warts, ulcer, urinary disorders, vulnerary, and wounds	Yeşilada et al. (1999); Sezik et al. (2001); Tuzlacı et al. (2010); Süntar et al. (2012); Tetik et al. (2013); Altay et al. (2015a); Altay and Karahan (2017); Ozturk et al. (2017a, 2018a, b, c, d)
In the high mountains of the East Mediterranean countries (Cyprus, Jordan, Lebanon, Palestine, Syria, and Turkiye)	Cooked, decoction, and infusion	Anaemia, antiseptic, asthma, biligenic, constipation, high cholesterol, diabetes, diaphoretic, diarrhoea, diuretic, eczema, gout, internal haemorrhage, jaundice, laxative, liver cirrhosis, liver enlargement, orexigenic, rheumatism, sedative, skin diseases, stomach pain, tonic, ulcer, and vomiting	Ozturk et al. (2017b)

al., 2020). Alcoholic extracts of the plant root show antimicrobial activity against specific microorganisms that cause gingivitis (Doshi et al., 1981).

It also has antifungal activity (Rub & Sasikumar, 2016).

Petrovic et al. (2004) studied the antibacterial activity of chicory roots using water, ethanol, or ethyl acetate extracts of the roots and reported positive results. However, ethyl acetate extract was reported to be the most active (Petrovic et al., 2004).

C. intybus has shown great potential against parasites in all living beings. It synthesizes many bioactive compounds with antiparasitic activity. Much work has been reported on the role of sesquiterpene lactones in antiparasitic activity (Peña-Espinoza et al., 2018; Janda et al., 2021). The efficiency of chicory root extracts against *Cryptosporidium parvum* has also been reported (Woolsey et al., 2019). A study has also been carried out on lambs that had acquired parasites by natural transmission (Marley et al., 2003). Their findings have revealed that the effects of chicory on total helminth parasites have shown that the use of chicory as a forage reduces the number of adult abomasal helminth parasites in lambs (Marley et al., 2003).

9.3.2 Anti-diabetes Effects

Diabetes is the most common chronic disease today (Durmuskahya & Ozturk, 2013; Ozturk et al., 2018a). *C. intybus* is reported to show anti-diabetic properties, owing to the presence of bioactive compounds such as alkaloids, phenolic compounds, and inulin (40%), acting through the increased insulin secretion or limiting glucose uptake into the gut, either of which has a beneficial effect for people who have type 2 diabetes (Ning et al., 2017; Salleh et al., 2021). The inulin in the root of this plant is considered to be a prebiotic fibre and is trapped inside the root cell walls as a storage carbohydrate (Puhlmann & de Vos, 2021). This prebiotic fibre affects satiety and insulin response to obesity (Rao et al., 2019).

Pushparaj et al. (2007) have studied the effects of whole chicory plant alcoholic extracts on rats with streptozotocin-induced diabetes (STZ, 50 mg/kg). The cholesterol and triglycerides are reduced following an application of the whole-plant ethanolic extract at a dose of 125 mg/kg body mass (Street et al., 2013; Janda et al., 2021). The serum glucose, triglycerides, and total cholesterol showed a reduction of 20, 91, and 16%, respectively, after feeding the rats with chicory extract (125 mg/kg body weight) once a day for 2 weeks (Pushparaj et al., 2007). The chicory plants are rich in caffeine and chlorogenic acids, both potent inductors of insulin secretion; these show ability as antidiabetics, affecting glucose reuptake in muscles. Cichoric acid is another compound with similar properties, which displays insulin-sensitizing activity and induction of dose-dependent glucose tolerance (Azay-Milhau et al., 2013; Janda et al., 2021).

Rendón-Huerta et al. (2012) have studied modulating body weight, hyperglycaemia, and hypercholesterolaemia in different situations (*in vivo*, in normal, obese, diabetic, and non-diabetic subjects) by using chicory root fructans. Metabolic disorders (such as diabetes) could be controlled by inulin and its fructans, which exist in *C. intybus* root extract (Lepczyński et al., 2017). Moreover, numerous polysaccharides in chicory root have a protective effect on the liver in NASH (non-alcoholic steatohepatitis) (Wu et al., 2018), and they can control blood sugar (Rendón-Huerta et al., 2012).

9.3.3 Hepatoprotective Activity

The high levels of SGOT (serum glutamic oxaloacetic transaminase), SGPT (serum glutamic-pyruvic transaminase), SALP (serum alkaline phosphatase), and total bilirubin levels are related to liver disorders, mainly necrosis of liver cells (Mossa et al., 1991). In this regard, research by Nallamilli et al. (2013) has demonstrated a reduction in SGOT, SGPT, SALP, and bilirubin levels in response to chicory root; owing to its phenolic compounds, which act as a liver protective, it prevents oxidative damage to DNA in rat liver tissue, whereas it can protect against hepatotoxicity in albino Wistar rats induced by carbon tetrachloride (Mahboob & Noori, 2009). Rasmussen et al. (2011) fed pigs orally with dried root of chicory for two weeks. They studied the actions followed *in vivo* on cytochrome P450 enzymes. Their results revealed an increase in the mRNA abundance of the genes encoding protective liver enzymes. Many researchers have recently confirmed the hepatoprotective properties of root extracts of the chicory plant (Zafar & Mujahid, 1998; Street et al., 2013; Bukhari et al., 2015; Elgengaihi et al., 2016; Mohammadifar et al., 2020). The beneficial effect of the extracts is generally related to their antioxidant activity (Asadi et al., 2018; Janda et al., 2021). Ahmed et al. (2003) have reported that chicory root can be a good choice to improve liver function and eliminate the toxic effects of chemicals on it due to its high content of elements and vitamins (Ahmed et al., 2003).

9.3.4 Gastroprotective Effects and Digestive Support

A gastric or peptic ulcer is a common gastrointestinal disease that occurs in the gastric mucosa due to the function of gastric juice and also due to the effect of the bacterium *Helicobacter pylori*. The presence of secondary metabolites (such as inulin, lactones, and vitamins.) in the chicory root is a factor in the anti-ulcer activity of this plant (Saxena et al., 2011). The experiments by these authors on rats have revealed that ethanol-induced gastric ulcers (direct effect on the epithelium and formation of necrotic lesions) or pyloric ligation (automatic digestion of the stomach and rupture of the mucosal barrier) can be reduced by the hydroalcoholic extract of chicory root (HACL). The protective activity of this extract has been related to the inhibition of free radicals (Saxena et al., 2011). Mahboob and Noori (2009) have proposed the decoction of dried root of *C. intybus* as a treatment for stomach acidity.

An evaluation of the gastric ulcer in a rat model using methanol extract of chicory roots has revealed its gastroprotective effect when used orally (Gürbüz et al., 2002).

Inulin increases the growth of beneficial bacteria in the gut, thus improving digestion and preventing constipation. The presence of the prebiotic fibre inulin in chicory root prevents digestive problems such as acid reflux and indigestion. Also, the root extract of this plant reduces digestive disorders caused by intestinal gastric problems (Thorat & Raut, 2018). Research has shown that the root of *C. intybus* leads to fat breakdown by increasing bile flow and thus has a positive effect on digestion (Bais & Ravishankar, 2001). Chicory root, due to its essential oils, also helps to eliminate intestinal worms (Choudhary et al., 2020).

9.3.5 Anti-inflammatory Activity

A proprietary bioactive extract of *C. intybus* root has been found to play a potential role in the management of osteoarthritis (Olsen et al., 2010). These effects have been investigated by Rizvi et al. (2014) using chicory root extracts in rats. A significant dose-dependent decrease in paw oedema has been recorded, and a significant attenuation of malondialdehyde levels together with an increase in the activities of the antioxidant enzymes catalase and glutathione peroxidase. A placebo-controlled, double-blind, dose-escalating trial has been carried out on patients with osteoarthritis, to determine the safety and tolerance of a proprietary bioactive extract of *C. intybus* roots. The roots show anti-inflammatory effects and reduce the pain caused by atherosclerosis. In this regard, chicory extract is helpful in treating joint problems such as arthritis and rheumatism (Thorat & Raut, 2018).

9.3.6 Cytotoxicity Activity

According to Khan et al. (2020) and Janda et al. (2021), extracts of the chicory plant have a cytotoxic effect on cells of prostate cancer, breast cancer, renal adenocarcinoma, amelanotic melanoma, and leukaemia cells. Imam et al. (2019) and Khan et al. (2020) studied cytotoxicity using *C. intybus* extracts and revealed its anti-tumour potential. The guaianolide 8-deoxylactucin found in chicory root extract is reported to inhibit cyclooxygenase-2 (COX-2), an enzyme associated with inflammation and pain, in human colon carcinoma cells (HT29 cells) (Cavin et al., 2005). It has also shown a tumour-inhibitory effect in Ehrlich ascites carcinoma in mice (Street et al., 2013).

9.3.7 Other Pharmacological Activities

The chicory plant root extract is used to treat pathological conditions because it is a diuretic, cleanses the kidneys and bladder, and crushes and dissolves the stones (Das et al., 2016). Glucosinolates in chicory roots play an essential role in improving kidney and liver function. Also, an excretion of stone-forming crystals has been reported via stimulation of the kidneys and increased urination. This action of chicory root extract eventually reduces potassium concentration (Khodadadi et al., 2016). The presence of active ingredients in the roots prevents calcium deposition in the form of calcium oxalate crystals and calcium phosphate, thereby helping to excrete urine and thus reducing blood calcium (Hassan and Ahmed 2021). Hassan and Ahmed (2021) report that the root extract of chicory also acts on the fragmentation of kidney stones. The antioxidants in chicory root extract improve kidney tissue. In this regard, it has a positive effect on eliminating infections caused by kidney stone formation and related bacterial infections (Tranchimand et al., 2010). The kidney and liver are the main detoxification channels in the body. The most important health benefits of chicory root are increased detoxification by stimulating urine and bile secretion (Thorat & Raut, 2018).

Inulin and fructo-oligosaccharides are used in the diet for weight loss. This prebiotic fibre suppresses appetite without raising blood sugar or calorie levels and provides a feeling of satiety. The presence of oligosaccharides also helps maintain the balance of the gut biome (beneficial effect on growth). It thus plays a key role in weight loss due to the proper functioning of the digestive system (Thorat & Raut, 2018).

Inulin lowers bad cholesterol (low-density lipoprotein, LDL), thus blocking atherosclerosis. LDL may increase the risk of heart attack and stroke (Thorat & Raut, 2018).

Karimi et al. (2014) have studied the effect of ethanolic extract of chicory root on the immune system by targeting mouse dendritic cells. The extract has inhibited the proliferation of allogenic T-cells at higher concentrations. It has changed the cytokine level so that the concentration of interleukin IL-4 decreased and that of interferon (IFN-γ) increased at lower concentrations. Several beneficial effects on pyridoxine-induced peripheral neuropathy in male rats have been reported by Hasannejad et al. (2019). Lactucopicrin and lactucin, the sesquiterpene lactones found in chicory roots, show antimalarial activity *in vitro* against an HB3 clone of a Hondurus-1 strain of *Plasmodium falciparum* (Bischoff et al., 2004).

9.4 THE PHYTOCHEMICAL COMPONENTS OF CHICORY ROOTS

In studies on the identification of the phytochemicals in this plant, it has been emphasized that the main component is cichoric acid, together with aliphatic compounds and their derivatives; terpenoids are present at slightly lower concentrations in the plant (Aisa et al., 2020; Janda et al., 2021). Nwafor et al. (2017) have reported that chicory leaves contain inulin, vitamins A, B_1, B_2 and C, elements Ca, K, Mg, Na, Fe, Cu, Mn, and Zn, and phenolic compounds. Although essential oils vary according to the part of the plant, with the aerial parts and roots being particularly rich in these chemicals (Janda et al., 2021).

The chemical constituents in the roots are bitter intybin glycosides (0.01–0.02%), reducing sugars (9–15%), and inulinb (40–60%); however, no starch is found (Nwafor et al., 2017; Puhlmann and de Vos 2021; Janda et al., 2021). The plant's roots contain sesquiterpene lactones and guajanolides, which give a bitter taste. The taraxane-type triterpenes, such as taraxasterol and phenolic acids, are also found in the chicory plant.

The phytochemical compounds detected in the roots of *C. intybus* are presented in detail in Table 9.2.

TABLE 9.2
Somc Phytochemical Constituents in *Cichorium intybus* Roots (modified from Aisa et al. 2020; and Janda et al. 2021)

Phytochemical Groups	Phytochemical Compounds
Flavonoids	quercetin-7-*O*-glucoside
	quercetin-7-*O*-glucuronide
	pelargonidin-3-*O*-monoglucuronide
	kaempferol-7-*O*-glucuronide
Triterpenoids and steroids	taraxerone
	ar-amyrin
	taraxasterol
	Q-taraxasterol
	β-sitosterol
	daucosterol
	bauerenyl acetate
	intybusoloid
	up-12,20(29)-dien-3b-ol-3b-l-arabino furanosyl-20-hexadecanoate
	lup-12,20(29)-dien-3b-olyl hexadecanoate
Sesquiterpenoids	11b, 13-dihydrolactucin
	cichoriolide A
	crepidiaside A
	crepidiaside B
	sonchuside A
	picriside B
	cichotyboside
	sonchuside C
	cichoriosides A
	cichoriosides B
	cichoriosides C
	cichorioside D
	8-deoxylactucin
	iIxerisoside D
	Ierisoside D
	lactucopicrin
	11-epiartesin
	santamarine
	intybulide
	crepidiaside B
	jacquinelin
	10a-hydroxycichopumilide
	8a-a ngeloyloxycichoralexin
	artesin
	magnolialide
Nitrogen-containingcompounds	caffeine
	theophylline
	8-hydroxyquinoline
	1-phenyl-3-pyrazolidone
	4-methyl-3-nitrobenzoic acid
	2,3,4,9-tetrahydro-1H-pyrido-(3,4-b)indole-3-carboxylic acid
	zeatin riboside

(*Continued*)

TABLE 9.2 (CONTINUED)
Some Phytochemical Constituents in *Cichorium intybus* Roots (modified from Aisa et al. 2020; and Janda et al. 2021)

Phytochemical Groups	Phytochemical Compounds
Essential oils	kaempferol
	octane
	octen-3-ol-1
	2-pentylfuran
	n-nonanal
	n-tridecane
	(2*E*,4*E*)-decadienal
	(2*E*,4*Z*)-decadienal
	(2*E*,4*E*)-heptadienal
	β-elemene
	(*E*)-caryophyllene
	β-ylangene
	(*E*)-β-farnesene
	geranyl acetone
	allo-aromadendrene
	dehydro-aromadendrene
	β-ionone
	pentadecane
	trans-β-guaiene
	(2*E*)-undecenol acetate
	sesquicineole
	(2*E*)-tridecanol
	pentenyl salicilate
	n-hexadecane
	tetradecanal
	2-pentadecanone
	n-nonadecane
	n-eicosane
	n-heicosane
Minerals	Ca
	Mg
	Na
	Cu
	Zn
	Mn
	Se
	N
	P
	K
	S
	B
	Fe
	Mo
	Cd
Polysaccharides	inulin

(*Continued*)

TABLE 9.2 (CONTINUED)
Some Phytochemical Constituents in *Cichorium intybus* Roots (modified from Aisa et al. 2020; and Janda et al. 2021)

Phytochemical Groups	Phytochemical Compounds
Other compounds	syringaldehyde coniferyl aldehyde coniferyl alcohol coniferin syringin cichorin A cichorin B cichorin C saponins tannins

9.5 CONCLUSION

The drugs obtained from the root of *C. intybus* are used in treating different diseases in traditional medicine. Local communities in different parts of the world also use these for different purposes. All Indigenous knowledge of this plant and its various medicinal uses has been supported by phytochemical isolation and investigations into their biological activity. The species shows great potential for future medicinal evaluation, due to their richness in cichoric acid and aliphatic compounds and their derivatives, especially flavonoids, triterpenoids, steroids, sesquiterpenoids, nitrogen-contained compounds, and essential oils. The antidiabetic, antioxidant, anti-inflammatory, antimicrobial, antiparasitic, and hepatoprotective effects of this plant, its digestive support, and cytotoxic activities are well reported. However, the plant root extracts and their different bioactivities need to be evaluated at length in future investigations.

Further research is required to better understand the phytochemicals against various diseases (Nandi & Lyndem, 2016). As Nandi and Lyndem (2016) pointed out, many of its constituents have yet to be explored for their pharmacological potential. There is a need for detailed investigations by ethnopharmacologists on the secondary metabolites in chicory and authentication of their future uses (Ozturk et al., 2022). As Nandi and Lyndem (2016) pointed out, we conclude that more work must be carried out on extracting, identifying, and isolating active components from *C. intybus*.

REFERENCES

Aftab, T. & Hakeem, K.R. (Eds.) (2021) *Medicinal and aromatic plants: Healthcare and industrial applications. Springer nature.*

Ahmad, M., Qureshi, R., Arshad, M., Khan, M.A. & Zafar, M. (2009) Traditional herbal remedies used for the treatment of diabetes from district Attock (Pakistan). *Pakistan Journal of Botany*, 41(6), 2777–2782.

Ahmed, B., Al-Howiriny, T.A. & Siddiqui, A.B. (2003) Antihepatotoxic activity of seeds of *Cichorium intybus*. *Journal of Ethnopharmacology*, 87(2–3), 237–240.

Aisa, H.A., Xin, X.L. & Tang, D. (2020) Chemical constituents and their pharmacological activities of plants from *Cichorium* genus. *Chinese Herbal Medicines*, 12(3), 224–236.

Altay, V. & Karahan, F. (2017) Anadolu geleneksel tıbbında güneş çarpması ve güneş yanığı tedavisinde kullanılan tıbbi bitkiler. *Erzincan University Journal of Science and Technology*, 10(1), 124–137.

Altay, V., Keskin, M. & Karahan, F. (2015a) An assessment of the plant biodiversity of Mustafa Kemal University Tayfur Sokmen Campus (Hatay-Turkey) for the view of human health. *International Journal of Scientific and Technological Research*, 1(2), 83–103.

Altay, V., Ozyigit, I.I., Osma, E., Bakir, Y., Demir, G., Severoglu, Z. et al. (2015b) Environmental relationships of the vascular flora alongside the railway tracks between Haydarpasa and Gebze (Istanbul-Kocaeli/Turkey). *Journal of Enviromental Biology*, 36, 153–162.

Altay, V., Ozyıgıt, I.I. & Yarcı, C. (2010) Urban flora and ecological characteristics of the Kartal District (Istanbul): A contribution to urban ecology in Turkey. *Scientific Research and Essays*, 5(2), 183–200.

Altay, V., Ozyıgıt, I.I. & Yarcı, C. (2012) Plant communities in urban habitats of Istanbul-Turkey. *Pakistan Journal of Botany*, 44, 177–186.

Altay, V., Šilc, U., Yarcı, C., Kavgacı, A., Čarni, A. & Ozturk, M. (2020) Urban vegetation of the Anatolian side of Istanbul. *Phytocoenologia*, 50(2), 101–121.

Asadi, M., Mohammadi, M., Mohammadian, B., Shahriari, A. & Foruozandeh, H. (2018) The protective effect of *Cichorium intybus* L. hydroalcoholic extract against methotrexate-induced oxidative stress in rats. *Jundishapur Journal of Natural Pharmaceutical Products*, 13(4), e59556.

Azay-Milhau, J., Ferrare, K., Leroy, J., Aubaterre, J., Tournier, M., Lajoix, A.D. et al. (2013) Antihyperglycemic effect of a natural chicoric acid extract of chicory (*Cichorium intybus* L.): A comparative in vitro study with the effects of caffeic and ferulic acids. *Journal of Ethnopharmacology*, 150(2), 755–760.

Baert, J.R. (1997) The effect of sowing and harvest date and cultivar on inulin yield and composition of chicory (*Cichorium intybus* L.) roots. *Industrial Crops and Products*, 6(3–4), 195–199.

Bais, H.P. & Ravishankar, G.A. (2001) *Cichorium intybus* L. - Cultivation, processing, utility, value addition and biotechnology, with an emphasis on current status and future prospects. *Journal of the Science of Food and Agriculture*, 81(5), 467–484.

Bischoff, T.A., Kelley, C.J., Karchesy, Y., Laurantos, M., Nguyen-Dinh, P. & Arefi, A.G. (2004) Antimalarial activity of lactucin and lactucopicrin: Sesquiterpene lactones isolated from *Cichorium intybus* L. *Journal of Ethnopharmacology*, 95(2–3), 455–457.

Blumenthal, M. (1998) *The complete German Commission E monographs*, p. 109. Austin, TX: American Botanical Council.

Bukhari, S., Ali, M., Anwar, H., Farooq, M., Ercisli, S., Dima, L. et al. (2015) Antioxidant potential of *Cichorium intybus* and Lentinus edodes ameloriates carbontetrachloride-induced liver toxicity. *Oxidation Communications*, 38, 2006–2015.

Bussmann, R.W., Paniagua-Zambrana, N.Y., Sikharulidze, S., Kikvidze, Z., Kikodze, D., Tchelidze, D. et al. (2017) Plant and fungal use in Tusheti, Khevsureti, and Pshavi, SakArtVelo (Republic of Georgia) Caucasus. *Acta Societatis Botanicorum Poloniae*, 86(2), 3517.

Cavin, C., Delannoy, M., Malnoe, A., Debefve, E., Touché, A., Courtois, D. & Schilter, B. (2005) Inhibition of the expression and activity of cyclooxygenase-2 by chicory extract. *Biochemical and Biophysical Research Communications*, 327(3), 742–749.

Choudhary, M., Rajput, P. & Sharma, R.A. (2020) Screening of antioxidant, antimicrobial and phytochemicals composition of various plant parts of *Cichorium intybus*. Retrieved from www.researchgate.net/publication/342080388.

Das, S., Vasudeva, N. & Sharma, S. (2016) *Cichorium intybus*: A concise report on its ethnomedicinal, botanical, and phytopharmacological aspects. *Drug Development and Therapeutics*, 7(1), 1.

Davis, P.H. (1975) *Flora of Turkey and the East Aegean islands*, 5. Edinburg: Edinburgh University Press.

Doshi, J.S., Patel, V.K. & Venkatakrishna Bhatt, H. (1981) Dental illness in handicapped blinds and the role of a dentist. *Journal of Indian Dental Association*, 53, 179–180.

Durmuskahya, C. & Ozturk, M. (2013) Ethnobotanical survey of medicinal plants used for the treatment of diabetes in Manisa, Turkey. *Sains Malaysiana*, 42(10), 1441–1448.

Elgengaihi, S., Mossa, A.-T.H., Refaie, A.A. & Aboubaker, D. (2016) Hepatoprotective efficacy of *Cichorium intybus* L. extract against carbon tetrachloride-ınduced liver damage in rats. *Journal of Dietary Supplements*, 13(5), 570–584.

El-Hilaly, J., Hmammouchi, M. & Lyoussi, B. (2003) Ethnobotanical studies and economic evaluation of medicinal plants in Taounate province (Northern Morocco). *Journal of Ethnopharmacology*, 86(2–3), 149–158.

Eskin, B., Altay, V., Ozyıgıt, I.I. & Serın, M. (2012) Urban vascular flora and ecological characteristics of the Pendik District (Istanbul-Turkey). *African Journal of Agricultural Research*, 7(4), 629–646.

European Medicines Agency (2013) *Assessment report on Cichorium intybus L., radix. EMA/HMPC/113041/2010*. EMA.

Guarrera, P.M., Forti, G. & Marignoli, S. (2005) Ethnobotanical and ethnomedicinal uses of plants in the district of Acquapendente (Latium, Central Italy). *Journal of Ethnopharmacology*, 96(3), 429–444.

Gürbüz, I., Ustün, O., Yesilada, E., Sezik, E. & Akyürek, N. (2002) *In vivo* gastroprotective effects of five Turkish folk remedies against ethanol-induced lesions. *Journal of Ethnopharmacology*, 83(3), 241–244.

Hamid, R., Kamili, A.N., Mahmooduzzafar, S., Gücel, M., Öztürk, M. & Ahmad, P. (2015) Analysis of physiobiochemical attributes, some key antioxidants and esculin content through HPLC in in vitro grown *Cichorium intybus* L. treated with ethylmethane sulfonate. *Plant Growth Regulation*, 76(3), 233–241.

Hanlidou, E., Karousou, R., Kleftoyanni, V. & Kokkini, S. (2004) The herbal market of Thessaloniki (N Greece) and its relation to the ethnobotanical tradition. *Journal of Ethnopharmacology*, 91(2–3), 281–299.

Hasannejad, F., Ansar, M.M., Rostampour, M., Mahdavi Fikijivar, E. & Khakpour Taleghani, B. (2019) Improvement of pyridoxine-induced peripheral neuropathy by *Cichorium intybus* hydroalcoholic extract through GABAergic system. *Journal of Physiological Sciences*, 69(3), 465–476.

Hassan, H.S.K. & Ahmed, M.S. (2021) The protective role of aqueous extract of plant roots *Cichorium intybus* L. in a number of biochemical variables in the blood serum and tissues of rabbits infected with experimental kidney stones. *Plant Archives*, 21 Suppl.-1, 1128–1136.

Imam, K.M.S.U., Xie, Y., Liu, Y., Wang, F. & Xin, F. (2019) Cytotoxicity of *Cichorium intybus* L. metabolites. *Oncology Reports*, 42(6), 2196–2212.

Imanberdieva, N., Imankul, B., Severoğlu, Z., Altay, V. & Ozturk, M. (2018b) Potential impacts of climate change on plant diversity of Sary-Chelek Biosphere Reserve in Kyrgyzstan. In: Egamberdieva, D. & Ozturk, M. (Eds.), *Vegetation of Central Asia and environs*. Switzerland: Springer Nature Switzerland AG, pp. 349–364.

Imanberdieva, N., Severoğlu, Z., Kurmanbekova, G., Altay, V. & Ozturk, M. (2018a) Plant Diversity of Ala-Archa Nature Park in Kyrgyzstan with emphasis on its economic potential. In: Egamberdieva, D. & Öztürk, M. (Eds.), *Vegetation of Central Asia and environs*. Switzerland: Springer Nature Switzerland AG, pp. 365–381.

Janda, K., Gutowska, I., Geszke-Moritz, M. & Jakubczyk, K. (2021) The common cichory (*Cichorium intybus* L.) as a source of extracts with health-promoting properties - A review. *Molecules*, 26(6), 1814.

Jarić, S., Popović, Z., Mačukanović-Jocić, M., Djurdjević, L., Mijatović, M., Karadžić, B. et al. (2007) An ethnobotanical study on the usage of wild medicinal herbs from Kopaonik Mountain (Central Serbia). *Journal of Ethnopharmacology*, 111(1), 160–175.

Jasim, R.S. (2019) Antioxidant, antimicrobial activities and phytochemical constituents of *Cichorium intybus* L. aerial parts. *International Journal of Botany*, 14(1), 24–29.

Jouad, H., Haloui, M., Rhiouani, H., El Hilaly, J. & Eddouks, M. (2001) Ethnobotanical survey of medicinal plants used for the treatment of diabetes, cardiac and renal diseases in the North centre region of Morocco (Fez-Boulemane). *Journal of Ethnopharmacology*, 77(2–3), 175–182.

Karimi, M.H., Ebrahimnezhad, S., Namayandeh, M. & Amirghofran, Z. (2014) The effects of *Cichorium intybus* extract on the maturation and activity of dendritic cells. *Daru*, 22(1), 28.

Khan, M.F., Nasr, F.A., Noman, O.M., Alyhya, N.A., Ali, I., Saoud, M. et al. (2020) Cichorins D-F: Three new compounds from *Cichorium intybus* and their biological effects. *Molecules*, 25(18), 4160.

Khodadadi, M., Mousavinasab, S., Khamesipour, F. & Katsande, S. (2016) The effect of *Cichorium intybus* L. ethanol extraction on the pathological and biomedical indexes of the liver and kidney of broilers reared under heat stress. *Brazilian Journal of Poultry Science*, 18(3), 407–412.

Kokoska, L., Polesny, Z., Rada, V., Nepovim, A. & Vanek, T. (2002) Screening of some Siberian medicinal plants for antimicrobial activity. *Journal of Ethnopharmacology*, 82(1), 51–53.

Lepczyński, A., Herosimczyk, A., Ożgo, M., Marynowska, M., Pawlikowska, M., Barszcz, M. et al. (2017) Dietary chicory root and chicory inulin trigger changes in energetic metabolism, stress prevention and cytoskeletal proteins in the liver of growing pigs–a proteomic study. *Journal of Animal Physiology and Animal Nutrition*, 101(5), e225–e236.

Leporatti, M.L. & Ivancheva, S. (2003) Preliminary comparative analysis of medicinal plants used in the traditional medicine of Bulgaria and Italy. *Journal of Ethnopharmacology*, 87(2–3), 123–142.

Loi, M.C., Maxia, L. & Maxia, A. (2005) Ethnobotanical comparison between the villages of Escolca and Lotzorai (Sardinia, Italy). *Journal of Herbs, Spices, and Medicinal Plants*, 11(3), 67–84.

Madius, G. (1938) *Lehrbuch der Biologischen Heilmittel. Abt. 1. Heilpflanzen*, I. Leipzig: Thieme.

Mahboob, T. & Noori, S. (2009) Ameliorative effect of ethanolic extract of *Cichorium intybus* on cisplatin-induced nephrotoxicity in rats. *Biological Sciences-PJSIR*, 52(4), 208–216.

Malik, K., Ahmad, M., Ozturk, M., Altay, V., Zafar, M. & Sultana, S. (2021) *Herbals of Asia - Prevalent diseases and their treatments*. Springer International Publishing.

Marley, C.L., Cook, R., Keatinge, R., Barrett, J. & Lampkin, N.H. (2003) The effect of birdsfoot trefoil (*Lotus corniculatus*) and chicory (*Cichorium intybus*) on parasite intensities and performance of lambs naturally infected with helminth parasites. *Veterinary Parasitology*, 112(1–2), 147–155.

Masoodi, K.Z., Amin, I., Mansoor, S., Ahmed, N., Altay, V. & Ozturk, M. (2020) Botanicals from the Himalayas with anticancer potential – An emphasis on Kashmir Himalayas. In: Ozturk, M., Egamberdieva, D. & Pešić, M. (Eds.) *Biodiversity and biomedicine – Our future*. Cambridge, MA: Academic Press, pp. 189–234.

Miraldi, E., Ferri, S. & Mostaghimi, V. (2001) Botanical drugs and preparations in the traditional medicine of West Azerbaijan (Iran). *Journal of Ethnopharmacology*, 75(2–3), 77–87.

Mohammadifar, M., Taghizadeh, M., Abed, A., Soltani, A., Tamtaji, O., Khamechian, T. et al.(2020) Effect of *Ziziphus jujuba* Mill., *Cichorium intybus* L. and *Silybum marianum* (L.) Gaertn. combination extract on non-alcoholic fatty liver disease in rats. *Journal of Medicinal Plants*, 18(72), 133–142.

Mossa, J.S., Tariq, M., Mohsin, A., Ageel, A.M., Al-Yahya, M.A., Al-Said, M.S. et al. (1991) Pharmacological studies on aerial parts of *Calotropis procera*. *American Journal of Chinese Medicine*, 19(3–4), 223–231.

Nallamilli, B.R., Kumar, C.S.P., Reddy, K.V., Prasanna, M.L., Maruthi, V. & Sucharita, P. (2013) Hepatoprotective activity of *Cichorium intybus* (Linn.) root extract against carbon tetrachloride induced hepatotoxicity in albino Wistar rats. *Drug Invention Today*, 5(4), 311–314.

Nandi, S. & Lyndem, L.M. (2016) *Clerodendrum viscosum*: Traditional uses, pharmacological activities and phytochemical constituents. *Natural Product Research*, 30(5), 497–506.

Ning, C., Wang, X., Gao, S., Mu, J., Wang, Y., Liu, S. et al. (2017) Chicory inulin ameliorates type 2 diabetes mellitus and suppresses JNK and MAPK pathways in vivo and in vitro. *Molecular Nutrition and Food Research*, 61(8), 1600673.

Nwafor, I.C., Shale, K. & Achilonu, M.C. (2017) Chemical composition and nutritive benefits of chicory (*Cichorium intybus*) as an ideal complementary and/or alternative livestock feed supplement. *Scientific World Journal*, 2017, article ID 7343928.

Olsen, N.J., Branch, V.K., Jonnala, G., Seskar, M. & Cooper, M. (2010) Phase 1, placebo-controlled, dose escalation trial of chicory root extract in patients with osteoarthritis of the hip or knee. *BMC Musculoskeletal Disorders*, 11, 56.

Osma, E., Ozyıgıt, I.I., Altay, V. & Serın, M. (2010) Urban vascular flora and ecological characteristics of Kadikoy district, Istanbul, Turkey. *Maejo International Journal of Science and Technology*, 4(1), 64–87.

Ozturk, M., Altay, V., Altundağ, E., Ibadullayeva, S.J., Aslanipour, B. & Gönenç, T.M. (2018d) Herbals in Iğdır (Turkey), Nakhchivan (Azerbaijan), and Tabriz (Iran). In: Ozturk, M. & Hakeem, K.R. (Eds.) *Plant and human health*, Vol. 1. Cham: Springer, pp. 197–266.

Ozturk, M., Altay, V. & Efe, R. (Eds.) (2021) Biodiversity, conservation and sustainability in Asia, Volume I: Prospects and challenges in West Asia and Caucasus. Springer nature Switzerland AG. https://doi.org/10.1007/978-3-030-59928-7.

Ozturk, M., Altay, V. & Gonenç, T.M. (2017b) Herbal from high mountains in the East Mediterranean. In: Bhojraj, S., Talas-Ogros, T. Adam, S. & Madhunapantula, S. R. V. (Eds.) *Drug discovery from Herbs-Approaches and Applications*. New Delhi-India: NAM S & T Centre, DAYA Publishing House, pp. 327–367.

Ozturk, M., Altay, V., Gucel, S. & Aksoy, A. (2012) Aegean Grasslands as endangered ecosystems in Turkey. *Pakistan Journal of Botany*, 44, 7–18.

Ozturk, M., Altay, V., Gucel, S. & Altundağ, E. (2017a) Plant diversity of the drylands in southeast Anatolia-Turkey: Role in human health and food security. In: Ansari, A.A. & Gill, S.S. (Eds.), *Plant biodiversity*. UKCABI, pp. 83–124.

Ozturk, M., Altay, V., Gucel, S. & Guvensen, A. (2014) Halophytes in the East Mediterranean – Their medicinal and other economical values. In: Khan, M. A., Böer, B. Ozturk, M. Abdessalaam, T. Z. Clüsener-Godt, M. & Gul, B. (Eds.) *Sabkha ecosystems: Volume IV: Cash crop halophyte and biodiversity conservation*. Tasks for Vegetation Science Vol. 47. Springer Netherlands. Springer Science+Business Media Dordrecht, 247–272.

Ozturk, M., Altay, V., Hakeem, K.R. & Akçiçek, E. (2017c) Liquorice from botany to phytochemistry. In *SpringerBriefs in plant science*. https://doi.org/10.1007/978-3-319-74240-3.

Ozturk, M., Altay, V., Latiff, A., Asad Ziaee, M., Choudhry, M.I., Shaheen, F. et al. (2018a) A comparative analysis of the medicinal plants used for diabetes mellitus in the traditional medicine in Turkey, Pakistan, and Malaysia. In Ozturk, M. & Hakeem, K.R. (Eds.), *Plant and human health*, Vol. 1. Springer International Publishing AG, part of Springer Nature, pp. 409–461.

Ozturk, M., Altay, V., Latiff, A., Salman, T. & Choudhry, I. (2018e) A comparative analysis of the medicinal pteridophytes in Turkey, Pakistan, and Malaysia. In: M. Ozturk & K.R. Hakeem (Eds.), *Plant and human health*, Vol. 1. Springer International Publishing AG, part of Springer Nature, pp. 349–390.

Ozturk, M., Altay, V., Latiff, A., Shareef, S., Shaheen, F. & Choudhry, M.I. (2018b) Potential medicinal plants used in the hypertension in Turkey, Pakistan, and Malaysia. In: Ozturk, M. & Hakeem, K.R. (Eds.), *Plant and human health*, Vol. 1. Springer International Publishing AG, part of Springer Nature, pp. 595–618.

Ozturk, M., Altay, V., Orçen, N., Yaprak, A.E., Tuğ, G.N. & Güvensen, A. (2018g) A little known and a little consumed natural resource: *Salicornia*. In: Ozturk, M., Hakeem, K.R., Ashraf, M. & Ahmad, M.S.A. (Eds.), *Global perspectives on underutilized crops*, Springer International Publishing AG, part of Springer Nature, pp. 83–108.

Ozturk, M., Altundağ, E., Ibadullayeva, S.J., Altay, V. & Aslanıpour, B. (2018c) A comparative analysis of medicinal and aromatic plants used in the traditional medicine of Iğdır (Turkey), Nakhchivan (Azerbaijan), and Tabriz (Iran). *Pakistan Journal of Botany*, 50(1), 337–343.

Ozturk, M., Gökler, I. & Altay, V. (2018f) Medicinal bryophytes distributed in Turkey. In: Ozturk, M. & Hakeem, K.R. (Eds.), *Plant and human health*, Vol. 1. Springer International Publishing AG, part of Springer Nature, pp. 323–348.

Ozturk, M., Gucel, S., Altay, V. & Altundağ, E. (2012b) Alliums, an underutilized genetic resource in the East Mediterranean. *Proceedings of the 6th IS on edible Alliaceae. Acta Horticulturae*, 969(969)(ISHS 2012), 303–309.

Ozturk, M., Khan, S.M., Altay, V., Efe, R., Egamberdieva, D. & Khassanov, F. (Eds.) (2022) Biodiversity, conservation and sustainability in Asia, II. *Prospects and challenges in south and middle Asia. Springer nature Switzerland AG.* https://doi.org/10.1007/978-3-030-73943-0.

Ozturk, M. (1980) Population differentiation in *Cichorium intybus*. I – Ecad formation. *Tubitak*, VII. Sci. Congress, Kusadası (pp. 647–656). Izmir.

Peña-Espinoza, M., Valente, A.H., Thamsborg, S.M., Simonsen, H.T., Boas, U., Enemark, H.L. et al. (2018) Antiparasitic activity of chicory (*Cichorium intybus*) and its natural bioactive compounds in livestock: A review. *Parasites and Vectors*, 11(1), 1–14.

Petrovic, J., Stanojkovic, A., Comic, L. & Curcic, S. (2004) Antibacterial activity of *Cichorium intybus. Fitoterapia*, 75(7–8), 737–739.

Poletti, A., Schilcher, H. & Müller, A. (1989) *Heilkraftige Pflanzen*. Germany: Walter Hadecke Verlag.

Poli, F., Sacchetti, G., Tosi, B., Fogagnolo, M., Chillemi, G., Lazzarin, R. et al. (2002) Variation in the content of the main guaianolides and sugars in *Cichorium intybus* var. 'Rosso di Chioggia' selections during cultivation. *Food Chemistry*, 76(2), 139–147.

Puhlmann, M.L. & de Vos, W.M. (2021) Corrigendum: Back to the roots: Revisiting the use of the fiber-rich *Cichorium intybus* L. taproots. *Advances in Nutrition*, 12(4), 1598.

Pushparaj, P.N., Low, H.K., Manikandan, J., Tan, B.K.H. & Tan, C.H. (2007) Anti-diabetic effects of *Cichorium intybus* in streptozotocin-induced diabetic rats. *Journal of Ethnopharmacology*, 111(2), 430–434.

Rahimullah, T.G. (2019) Phytochemical and antibacterial screening of *Cichorium intybus* seeds use in traditional medicine systems in Pakistan. *International Journal of Basic Medical Sciences and Pharmacy (IJBMSP)* 8, 4963, 2049.

Rao, M., Gao, C., Xu, L., Jiang, L., Zhu, J., Chen, G. et al. (2019) Effect of inulin-type carbohydrates on insulin resistance in patients with type 2 diabetes and obesity: A systematic review and meta-analysis. *Journal of Diabetes Research*, 2019(1), Article ID 5101423.

Rasmussen, M.K., Zamaratskaia, G. & Ekstrand, B. (2011) In vivo effect of dried chicory root (*Cichorium intybus* L.) on xenobiotica metabolising cytochrome P450 enzymes in porcine liver. *Toxicology Letter*, 200(1–2), 88–91.

Rendón-Huerta, J.A., Juárez-Flores, B., Pinos-Rodríguez, J.M., Aguirre-Rivera, J.R. & Delgado-Portales, R.E. (2012) Effects of different sources of fructans on body weight, blood metabolites and fecal bacteria in normal and obese non-diabetic and diabetic rats. *Plant Foods for Human Nutrition*, 67(1), 64–70.

Rizvi, W., Fayazuddin, M., Shariq, S., Singh, O., Moin, S., Akhtar, K. et al. (2014) Anti-inflammatory activity of roots of *Cichorium intybus* due to its inhibitory effect on various cytokines and antioxidant activity. *Ancient Science of Life*, 34(1), 44–49.

Rub, R.A. & Sasikumar, S. (2016) Antimicrobial screening of *Cichorium intybus* seed extracts. *Arabian Journal of Chemistry*, 9, S1569–S1573.

Salleh, N.H., Zulkipli, I.N., Yasin, H.M., Ja'afar, F., Ahmad, N., Ahmad, W.A.N., Ahmad, W. et al. (2021) Systematic review of medicinal plants used for treatment of diabetes in human clinical trials: An ASEAN perspective. *Evidence-Based Complementary and Alternative Medicine: eCAM*, 2021 article ID 5570939..

Šarić-Kundalić, B., Dobeš, C., Klatte-Asselmeyer, V. & Saukel, J. (2011) Ethnobotanical survey of traditionally used plants in human therapy of east, north and north-east Bosnia and Herzegovina. *Journal of Ethnopharmacology*, 133(3), 1051–1076.

Šavikin, K., Zdunić, G., Menković, N., Živković, J., Ćujić, N., Tereščenko, M. et al. (2013) Ethnobotanical study on traditional use of medicinal plants in South-Western Serbia, Zlatibor District. *Journal of Ethnopharmacology*, 146(3), 803–810.

Saxena, R., Belemkar, S., Apte, K.G. & Saxena, R.C. (2011) Antiulcerogenic and antioxidant activity of hydro alcoholic extract of the root of *Cichorium intybus* L. in experimentally induced ulcer in rats. *Inventi*, 2011(2), 3–6.

Scholz, E. (2006) Monographien: Cichorium, *Cichorium intybus* L., Cichorii folia et radix, Cichorii radix. In der (Eds. Drogen, H. H., and R. O. M. Hager). Heidelberg: Springer Medizin Verlag.

Severoğlu, Z., Altay, V., Keskin, M., Serin, M. & Yarcı, C. (2011) Some ecological characteristic and the flora of Gölcük District and its environs (Kocaeli-Turkey). *Scientific Research and Essays*, 6(4), 847–875.

Sezik, E., Yeşilada, E., Honda, G., Takaishi, Y., Takeda, Y. & Tanaka, T. (2001) Traditional medicine in Turkey X. Folk medicine in central Anatolia. *Journal of Ethnopharmacology*, 75(2–3), 95–115.

Shaikh, T., Rub, R.A. & Sasikumar, S. (2016) Antimicrobial screening of *Cichorium intybus* seed extracts. *Arabian Journal of Chemistry*, 9, S1569–S1573.

Street, R.A., Sidana, J. & Prinsloo, G. (2013) *Cichorium intybus*: Traditional uses, phytochemistry, pharmacology, and toxicology. *Evidence-Based Complementary and Alternative Medicine: eCAM*, 2013. article ID 579319.

Süntar, I., Akkol, E.K., Keles, H., Yesilada, E., Sarker, S.D. & Baykal, T. (2012) Comparative evaluation of traditional prescriptions from *Cichorium intybus* L. for wound healing: Stepwise isolation of an active component by in vivo bioassay and its mode of activity. *Journal of Ethnopharmacology*, 143(1), 299–309.

Tarakçı, S., Altay, V., Keskin, M. & Sümer, S. (2012) Beykoz ve çevresi (İstanbul)'nin kent florası. *Karadeniz Fen Bilimleri Dergisi*, 2(7), 47–66.

Tetik, F., Civelek, S. & Cakilcioglu, U. (2013) Traditional uses of some medicinal plants in Malatya (Turkey). *Journal of Ethnopharmacology*, 146(1), 331–346.

Thorat, B.S. & Raut, S.M. (2018) Chicory the supplementary medicinal herb for human diet. *Journal of Medicinal Plants Studies*, 6(2), 49–52.

Tranchimand, S., Brouant, P. & Iacazio, G. (2010) The rutin catabolic pathway with special emphasis on quercetinase. *Biodegradation* 21(6), 833–859.

Tuzlacı, E., Is-bilen, D.P.A. & Bulut, G. (2010) Turkish folk medicinal plants. VIII: Lalapasa (Edirne). *Marmara Pharmaceutical Journal* 14, 47–52.

Upadhayay, V.K., Pandey, D., Taj, G., Samantaray, S.D. & Singh, S.P. (2021) Contemporary insights on antibacterial and pharmacological importance of *Cichorium intybus*-A compendious review. *Biomedical Journal of Scientific & Technical Research*, 35(2), 27409–27418.

van Wyk, B.E., van Oudtshoorn, B. & Gericke, N. (1997) *Medicinal plants of South Africa*. Pretoria, South Africa: Briza Publications.

Woolsey, I.D., Valente, A.H., Williams, A.R., Thamsborg, S.M., Simonsen, H.T. & Enemark, H.L. (2019) Anti-protozoal activity of extracts from chicory (*Cichorium intybus*) against *Cryptosporidium parvum* in cell culture. *Scientific Reports*, 9(1), 20414.

Wu, Y., Zhou, F., Jiang, H., Wang, Z., Hua, C. & Zhang, Y. (2018) Chicory (*Cichorium intybus* L.) polysaccharides attenuate high-fat diet induced non-alcoholic fatty liver disease via AMPK activation. *International Journal of Biological Macromolecules*, 118(A), 886–895.

www.powo.science.kew.org. (as accessed on 05.05.2022).

Yarcı, C. & Altay, V. (2016) Kocaeli ve çevresindeki tarım alanlarının yabani ot florası. *Erzincan University Journal of Science and Technology*, 9(2), 148–171.

Yarcı, C., Serin, M. & Altay, V. (2007) The segetal vegetation of Kocaeli Province (Turkey). *Ekoloji*, 16(63), 23–33.

Yeşilada, E., Sezik, E., Honda, G., Takaishi, Y., Takeda, Y. & Tanaka, T. (1999) Traditional medicine in Turkey IX. Folk medicine in north-west Anatolia. *Journal of Ethnopharmacology*, 64(3), 195–210.

Younessi-Hamzekhanlu, M., Ozturk, M., Altay, V., Nojadeh, M.S. & Alakbarli, F. (2020) Ethnopharmacological study of medicinal plants from Khoy city of West Azerbaijan-Iran. *Indian Journal of Traditional Knowledge*, 19(2), 251–267.

Zafar, R. & Mujahid Ali, S. (1998) Anti-hepatotoxic effects of root and root callus extracts of *Cichorium intybus* L. *Journal of Ethnopharmacology*, 63(3), 227–231.

10 Therapeutic Importance of *Withania somnifera* (Ashwagandha) in Medicine

Sonali Purohit, MC Purohit, Arun Kumar Khajuria, Rakesh Kumar Bachheti, and Anuj Kandwal

10.1 INTRODUCTION

Plants are the chief source of all basic needs, i.e., food, shelter, medicine, fibres, etc. The vital role of plants in human life has given a special place throughout the ages, i.e., they are worshipped and have sacred values in almost every culture (even in very early civilizations) and society. From very early civilization, owing to their extravagant qualities, medicinal plants were used in society for different rituals, ceremonies, and magical powers. The recognition of medically beneficial plants is usually because of certain bioactive compounds found in various parts of medicinal plants (Samant et al., 1998; Dhar, 2002; Asfaw et al., 2022).

Traditionally, several practices in medical systems, viz. Ayurveda, Naturopathy, Homoeopathy, Yoga, Siddha, and Unani, have been practised for total health care. The Ayurveda system of medicine not only cures diseases but also teaches how to maintain and protect life and health. It is mentioned that a person in an area/zone can be cured by the local flora present in the surrounding area (Dhiman, 2005). There are numerous drugs have been made by plants all over the world since ancient times. Thus, pharmaceutically important plants are the valuable sources of significant and potential medicines, and their therapeutical importance is mainly due to certain chemical compounds in the various parts of plants. Medicinal plants are used either solely or as a combination of one or more plant species or a combination of various parts of the same plant. Plants with medicinal properties provide alternative therapies along with enormous opportunities. They offer accessible and economical drugs to needy people and create income sources for economically developing nations. They can treat various diseases, from common flu to critically life-threatening illnesses, including tuberculosis, malaria, dengue, fascioliasis, schistosomiasis, psoriasis, asthma, pneumonia, swine flu, etc. (Fakim, 2006; Bisht, 2009; Abate et al., 2021, 2022; Bachheti et al., 2022). The preference for plant-based formulations over synthetic chemicals is due to their rare side-effects.

Withania somnifera, commonly called Indian ginseng, Indian winter cherry, or Ashwagandha, is a member of the Solanaceae family and is widely used in both modern and traditional systems of medical sciences because of the existence of various important phytochemical constituents. Leaves of *W. somnifera* are extensively utilized in Ayurvedic and Unani systems for treating tubercular glands, sexual debility, asthma, anxiety, skin ulcers, and tumours. Due to its pronounced stress-relieving properties, the species name of the plant has been given a Latin word, '*somnifera*' meaning 'sleep inducer'. Since the pharmacological activities and traditional folkloric utilization of Ashwagandha are similar and analogous to Korean ginseng tea, Ashwagandha is also called Indian ginseng. The modern medical system has also begun growing in connection with the traditional system of medical science that appeared in the examination of various medicinal plants to understand and examine the phytochemical components

DOI: 10.1201/b22924-10

responsible for their specific pharmaceutical properties. Plant biological properties are linked with several parts, such as flowers, leaves, fruits, seeds, or the whole plant itself (Ale et al., 2021; Khanal et al., 2022; Tewari et al., 2022). Several withanolides, steroids specific to the Solanaceae, have been isolated from Ashwagandha, which may be the reason for its wide range of pharmacological activities, such as antibacterial, antifungal, antitumour, anti-Parkinson's, analgesic, anti-anxiety, antidepressant, anti-ulcer, anti-Alzheimer's, anticancer, cardioprotective, antidiabetic and anti-inflammatory activities. Roots of Ashwagandha plants have been used globally by Ayurvedic practitioners for centuries as a tonic to enhance longevity and to treat diverse health conditions like arthritis and tumours (Dar et al., 2015; Dar & Ahmad, 2020; Verma et al., 2021). Recently the plant has been used to produce pharmaceutical drugs, energy drinks, and energy boosters (Wankhede et al., 2015). Thus, this chapter intends to give detailed knowledge about the therapeutic importance of *W. somnifera* in traditional and modern systems of treatment.

10.2 TAXONOMY OF *WITHANIA SOMNIFERA*

Kingdom: Plantae
- Sub-kingdom: Tracheobionta (vascular plants)
- Super division: Spermatophyta (seed plants)
- Division: Angiosperm (flowering plants)
- Class: Dicotyledons
- Order: Tubiflorae
- Family: Solanaceae (potato, nightshade family)
- Genus: *Withania*
- Species, authority: *somnifera* Dunal (RajeshwarRao et al., 2012; Gupta & Rana, 2007).
- **Common names:** Ashwagandha; Indian Ginseng, Indian winter cherry.

10.3 BOTANY AND CYTOLOGY OF *WITHANIA SOMNIFERA*

10.3.1 Botany of *Withania somnifera*

Plant is perennial, branched undershrub or shrub, may reach up to 150 cm in height; stem stellate-tomentose throughout. All parts are covered with whitish, stellate trichomes. Leaves ovate or elliptic-oblong, 4–12 × 2.5–7 cm, acute or decurrent base, entire or rarely crenate, acute or obtuse; petioles 1.5–3 cm long. Flower sessile, light yellow, in 1- to few-flowered axillary fascicles, pedicels short. Calyx about 4.5 mm long, hairy; lobes triangular, acute. Corolla greenish-yellow, 5–8 mm across; lobes ovate, obtuse. Ovary glabrous. Berries 8–12 mm across, yellow-orange, globose, enclosed within the inflated, 5- to 10-ribbed calyx; seed discoid, muricate. Distributed throughout the warmer parts of India, Sri Lanka, Pakistan, and the Mediterranean region (Hooker, 1875: Gaur, 1999; Shishkin, 1936; Hepper, 1991).

10.3.2 Cytology

Cytological studies of *W. somnifera* at the global level by karyotype analysis suggested that plant's chromosome number is 2n= 48 ($4\ A_{sm} + 4\ B_m + 14\ C_m + 4\ D_{sm} + 2\ E_{st} + 18\ F_m + 2\ G_{sm}$) chromosomes (Iqbal & Datta, 2007; Gulati et al., 2017; Puri, 2002).

10.3.3 Phytochemistry

Various parts of medically beneficial plants have been used from ancient times in several systems of medical science like Ayurveda, Naturopathy, Homoeopathy, Unani, Siddha, and Yoga for total health care. New drug discoveries from medicinal plants, based on traditional knowledge, involves the isolation of secondary metabolites from the various parts of the plant and identification of their biological and medicinal properties.

Ashwagandha has a diverse range of secondary metabolites isolated from several parts of plants like the leaves, roots, fruits, seeds, etc., as listed in Table 10.1. More than ten alkaloids, nearly 40 withanolides, and various known and unknown sitoindosides have been isolated from Ashwagandha.

10.3.4 Therapeutic Importance of Ashwagandha in Ayurvedic Literature

The word Ashwagandha is coined from two different words: *Ashwa*, meaning horse, and *Gandha*, meaning smell, which collectively means the roots smell similar to a horse. In other references, the meaning is also taken as one that gives aphrodisiac properties and strength like a horse. Common synonyms mentioned in the Ayurvedic literature itself signify the morphology and actions of the plants, such as Gandhapatri (leaves also smell like a horse), Varahkarni (resemblance of the leaves to a pig's ear), Ashwarohak (its intake gives sexual strength like that of the horse), Putrada (increases the potency, maintains reproductive health, thereby increasing the chances of progeny), Kaamrupini (provides radiant and lustrous looks), Balda (provides strength) and Vrisha (increases sexual potency). This shows that, even during that ancient time, the functions of Ashwagandha were well known (Hegde & Harini, 2019; Dar et al., 2015; Bhandari, 1970; Changhadi, 1938).

In the Ayurvedic literature, plants are described as producing action (Karma) in a different form depending on their properties (Guna), taste (Rasa), potency (Veerya), post-digestive Rasa (Vipaka), and specific potency (Prabhava). The properties (Guna) of Ashwagandha are Oily (Snigdha) and Light (Laghu), taste (Rasa) is somewhat Bitter (Tikta), Sweet (Madhur), and Astringent (Kashaya), potency (Veerya) is Hot (Ushna) and post-digestive rasa (Vipaka) is Sweet (Madhur) (Sharma, 2009; Mishra, 2004).

Charak Samhita, an important source of Ayurvedic literature, includes Ashwagandha in different categories. These categories include drugs that impart strength (Balya), nourish the whole body (Brihaniya), and a group of sweet drugs (Madhur skandha), depending upon the properties (Guna) and action (Karma) it imparts (Sastri & Chaturvedi, 2016).

Ayurvedic literature describes the role of Ashwagandha in different systems of the human body. Ashwagandha's fruitful and valuable parts are roots, ashes, salts, and leaves. The external uses of Ashwagandha include the local application of a paste made from its leaves and roots that helps reduce oedema and pain, and massage from its oil is done to treat weakness. It is also helpful in skin conditions such as blisters, itching, wounds, ulcers, and vitiligo. Ayurvedic texts also mention juice obtained from Ashwagandha leaves to be used as eardrops to treat ear discharge (Chunekar, 2020; Gogte, 2020).

As for internal uses, different actions of Ashwagandha include anti-inflammatory (Shothahar), and analgesics (Vednasthapana) (Sharma, 2009). It is helpful in treating respiratory issues, such as cough and bronchial asthma, and acts as a good expectorant. In the nervous system, it acts as a nervine, sedative, hypnotic, antistress, and cognitive agent, thus playing a good role in treating dementia also.

Ashwagandha is a prominent rejuvenator (Rasayan) that helps promote general health, provides nourishment to undernourished or emaciated children, and acts as an immunomodulator.

For the urinary system, it acts as a diuretic in cases of oliguria and anuria. It is also effective to treat people with blood-related and hypertensive disorders. It has excellent action in female joint pain, backaches, and leucorrhoea. It also acts against abdominal pain, constipation, and abdominal worms. The antitumour effects of Ashwagandha are also known from the Ayurvedic literature (Chunekar, 2020; Gogte, 2020; Nadkarni & Nadkarni, 1982).

TABLE 10.1
Biologically Active Phytochemicals in the Various Parts of Ashwagandha (*Withania somniferum*)

S.No.	Part of Ashwagandha	Biologically Active Phytochemicals Present	Biological Activities	References
1.	Leaf	Withaferin-A; withanolide-E; withanolide-B; withanone; withanolide-D; 7-hydroxywithanolide; withanolide-F; withanolide-Z; withanoside-IV; 2, 3-dihydrowithaferin-A; 4-(1-hydroxy-2, 2-dimethylcyclopropanone) -2,3-dihydrowithaferin A; Sitoindoside IX; 24, 25-dihydro-27-desoxywithaferin-A; physagulin-D; physagulin-D (1–>6)-β-D-glucopyranosyl-(1–>4)-β-D-glucopyranoside; 27-*O*-β-D-glucopyranosyl viscosalactone-B; viscosalactone-B; 4,16-dihydroxy-5β,6β-epoxyphysagulin-D; 2, 3-dihydrowithaferin-A; 27-*O*-β-D-glucopyranosyl physagulin-D; 5, 20α R-dihydroxy-6α and 7α-epoxy-1-oxo-5α-2,24-withadienolide.	Antibacterial, antifungal, anti-diabetic, neuro-protective, anti-Parkinson's, cardioprotective and anticancer activities.	(Anjaneyulu et al., 1997; Kirson et al., 1970; Sharada et al., 2007; Jayaprakasam et al., 2003; Glotter et al., 1977; Parmanick et al., 2008; Menssen & Stapel,1973; Dar et al., 2015; Girish et al., 2006; Widodo et al., 2007; Tewari et al., 2022; Udaykumar et al., 2009; RajaSankar et al., 2009a; Konar et al., 2011; Singh & Kumar, 2011; Bisht & Rawat, 2014).
2.	Root	Sitoindosides VII; Sitoindosides VIII (two new acylsteryl-glucosides); stigmasterol; stigmasterol glucoside; withanolide-A; viscosa lactone-B; β-sitosterol; withananine; withanine; sitoindosides IX; sitoindosides X; ashwagandhanolide.	Anti-inflammatory, antioxidant, antistress, anticancer, cardio-protective, anti-diabetic, anti-arthritic, anti-Parkinson's, neuroprotective, and anti-Alzheimer's activities.	(Bhattacharya et al., 1987; Ghosal et al., 1988; Praveen & Murthi, 2010; Majumdar, 1955; Misra et al., 2008 Mirjalili et al., 2009; Subbaraju et al., 2006; Debnath et al., 2011; Gorelick et al., 2015; Dar et al., 2015; Pingali et al., 2014; Dey et al., 2014; Kuboyama et al., 2014 and RajaSankar et al., 2009b)
3.	Fruit	6α,7α-epoxy-5α,14α,17α,23β-tetrahydroxy-1-oxo-22 (*R*)-2,24-withadienolide and 5β, 6α, 14α, 17β, 20β-pentahydroxy-1-oxo-20(*S*), 22(*R*)-witha-2,24-dienolide.	Anti-oxidant activities.	(Abou-Douh, 2002; Jayaprakasam et al., 2004; Alam et al., 2012).

10.3.5 Pharmacological Activities of Ashwagandha

10.3.5.1 Antimicrobial Activity

W. somnifera extract has shown a broad-spectrum antibacterial activity against both Gram-positive and Gram-negative bacteria (Singh & Kumar 2011; Alam et al., 2012; Bisht & Rawat 2014). Ashwagandha suppresses acid tolerance, acid generation, and biofilm production of *Streptococcus sobrinus* and *Streptococcus mutans* (oral bacteria), even at minimum inhibitory concentration (MIC) levels (Pandit et al., 2013). *W. somnifera* also exhibits fungistatic effects on phytopathogenic fungi when tested against *Fusarium verticilloides*, *Aspergillus flavus*, and F. *oxysporum* by suppressing hyphal growth and spore germination (Dar et al., 2015; Girish et al., 2006; Kalra et al., 2017).

10.3.5.2 Anti-inflammatory Activity

W. somnifera exhibits anti-inflammatory activities against several diseases. The root extract of Ashwagandha shows many restorative and anti-inflammatory effects by settling oedema, neutrophil infiltration, and necrosis in TNBS (trinitro benzene sulphonic acid) -induced inflammatory bowel disease, thus helping in the treatment of Crohn's disease and ulcerative colitis (Debnath et al., 2013; Pawar et al., 2011). Root powder of the plant showed a significant inhibitory effect on inflammatory markers like nephritis, proteinuria, and cytokines including interleukin-6, and TNF-αtumour necrosis factor-α (TNF-α), reactive oxygen species (ROS) and the reactive nitrogen species, nitric oxide (NO) (Dar et al., 2015; Minhas et al., 2011; Minhas et al., 2012). Withaferin-A isolated from Ashwagandha was found to suppress phorbol-12-myristate-13-acetate in human umbilical vein endothelial cells (HUVEC) by suppressing TNF-α and interleukin-1β. Withaferin-A also inhibits NFκ-β (nuclear factor kappa-β) and IL-8 in cellular models of cystic fibrosis (Ku et al., 2014; Dar et al., 2015; Maitra et al., 2009).

10.3.5.3 Antidiabetic Activity

In various studies, W. *somnifera* showed strong antidiabetic activity (Gauttam & Kalia 2013). Root powder of W. *somnifera* was given to patients and it stabilized blood sugar to a level comparable to daonil (an oral hypoglycaemic medicine) when administered for one month. Moreover, treatment with W. *somnifera* effectively improved the insulin sensitivity index in adult-onset diabetes mellitus (type-2 diabetes) in mice (Andallu & Radhika, 2000; Dar et al., 2015; Anwer et al., 2008). Ashwagandha root and leaf extracts increased glucose assimilation in a dose-dependent manner in adipocytes and skeletal myotubes. Additionally, leaf and root extracts of Ashwagandha markedly regulate blood glucose, urine sugar, and tissue glycogen levels in the case of insulin-dependent diabetes mellitus (alloxan-induced) in rats (Udayakumar et al., 2009; Gorelick et al., 2015; Udayakumar et al., 2010).

10.3.5.4 Antistress Activity

Ashwagandha induced stress tolerance in human and animal subjects (Bhattacharya et al., 1995). In a clinical investigation, the root extract of Ashwagandha resulted in a reduction in serum cortisol levels (a positive indicator of stress) in the human subject without creating any marked side-effects (Chandrasekhar et al., 2012). Glycowithanolides isolated from Ashwagamdha were found to cause antidepressant and antistress effects against pentylenetetrazole-induced anxiousness and diminished the brain levels of tribulin in rats (Bhattacharya et al., 2000). Treatment with a *W. somnifera* aqueous root extract caused upregulation of Th1 cytokines and relieved persistent anxiety-induced reduction of T-cells in mice (Khan et al., 2006).

10.3.5.5 Anticancer Activity

Ashwagandha and several of its phytochemical components have been utilized to attenuate several types of carcinoma or cancer-related changes (Vaishnavi et al., 2012; Tewari et al., 2022). Ashwagandha leaf extract eliminated carcinoma cells by various routes like G2-M DNA damage regulation, apoptosis signalling, GM-CFS (granulocyte-macrophage colony-stimulating factor) signalling, p53 signalling, and death receptor signalling pathways. One of the chemical components, withaferin-A, showed anticancer activity by affecting Bcl-2/Bim and Bcl-2/Bax proportions via generating ROS-induced apoptosis in melanotic carcinoma cells (Widodo et al., 2008). Withaferin-A treatment caused the overexpression of TNF receptor-1 and eliminated the expression of Bid. Moreover, this phytochemical obstructed the binding of NFk-β to DNA and thus, through activated caspase-3, it increased nuclear cleavage of p65/Rel (Malik et al., 2007). Apoptosis generated by radiation and withaferin-A critically involves increased formation of ROS, cleavage of poly(ADP-ribose) polymerase (PARP) proteins, downregulation of Bcl-2 and a cell line-specific act of MAPK (mitogen-activated protein kinase) signalling in human lymphoma U 937 cells (Hahm et al., 2014; Dar et al., 2015; Yang et al., 2011a; Yang et al., 2011b). Withaferin-A is also used in treating human renal carcinoma cells through excessive formation of ROS and via dephosphorylation of AKT-suppressing BCl-2. Withaferin-A also inhibits mammary cancer cells in genetically modified rats by suppressing the population of breast carcinoma cells and the magnitude of the tumour (Kim & Singh, 2014). Withaferin-A experimentally also inhibits human breast cancer growth by reducing vimentin protein expression via interrupting β-tubulin of the cytoskeletal architecture (Hahm et al., 2011; Hahm et al., 2013b). Widodo et al. (2007) reported that withaferin A from the Ashwagandha leaf extract particularly activated p53 in tumour cells, causing apoptosis and the inhibition of the growth of tumour cells. In another instance, Hahm et al. (2011) reported apoptosis due to the generation of ROS through withaferin A because of the suppression of mitochondrial respiration. Ghosh et al. (2016) revealed that, apart from apoptosis, paraptosis (a non-apoptotic method) is produced through the production of ROS via the use of withaferin A. Following treatment with withaferin A, these results showed the reverse upregulation of the endogenic paraptosis suppressor Alix/AIP-1 (Actin Interacting Protein-1) (Figure 10.1). Moreover, even the levels of survivin proteins, cIAP-2 (cellular inhibitor of apoptosis protein-2), and XIAP (X-linked inhibitor of apoptosis protein) were diminished significantly in response to treatment with withaferin A on MCF-7 and MDA-MB-231 cells (Hahm & Singh, 2013a). Withaferin-A mediated inhibition was associated only with survivin protein suppression, thus highlighting the importance of survivin suppression in Withaferin-A induced apoptosis. These results suggest that withaferin A-mediated suppression of survivin protein led to apoptosis. Withaferin-A treatment induced apoptosis in a kidney cancer cell line by downregulation of the STAT-3 pathway. *W. somnifera* eliminated laryngeal carcinoma Hep-2 cells

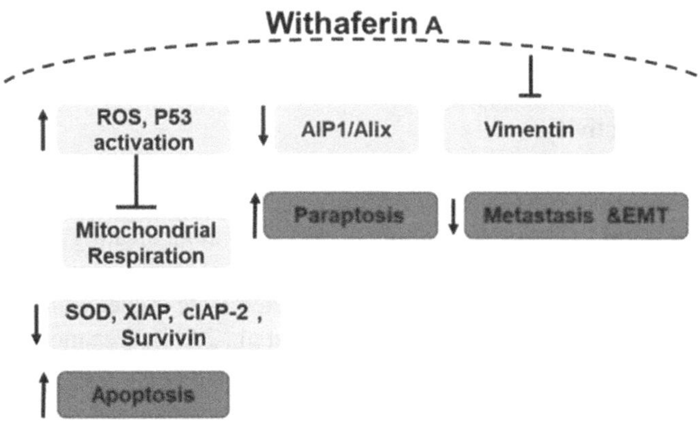

FIGURE 10.1 Role of withaferin A in breast cancer where '⊥' indicates inhibition, '↑' indicates an increase, '↓' indicates decrease (Adapted from Dutta et al., 2019).

in humans, which caused capture of the cell cycle and blockage of angiogenesis (Choi et al., 2011; Um et al., 2012; Dar et al., 2015; Mathur et al., 2006).

10.3.5.6 Anti-Parkinson's Activity:

Several research studies suggest that *W. somnifera* can play a major role in controlling Parkinson's disease. In rats, *W. somnifera* extract showed the anti-Parkinsonian effects in the 6-hydroxydopamine (OHDA)-induced Parkinson's disease rat model and restoration of striatal dopamine by attenuation of reduced glutathione content (GSH), lipid peroxidation (LPO), and effects of the antioxidant enzymes superoxide dismutase (SOD), glutathione *S*-transferase (GST), and glutathione reductase (GR) (Ahmad et al., 2005). Moreover, plant extract has also resulted in the rise of surviving dopaminergic neurons. The root extract of Ashwagandha regulated the catecholamine level in the mid-brain region of MPTP (methyl phenyl tetrahydro pyridine)-induced Parkinsonism in the mouse by restoring antioxidant status and reducing oxidative stress (RajaSankar et al., 2009a; Dar et al., 2015; Sankar et al., 2007; RajaSankar et al., 2009b). Root powder of Ashwagandha was found to decrease rotenone toxicity in the brain's striatum and cerebellum of mouse via its anti-inflammatory and antioxidant effects and by rectifying mitochondrial dysfunctions (Manjunath & Murlidhara, 2013, 2015).

10.3.5.7 Anti-Alzheimer's Activity

The literature has reported marked activity against Alzheimer's disease in various parts of Ashwagandha. Withanolide A achieved suppression of acetyl cholinesterase (AChE) which could contribute to symptom improvement in Alzheimer's disease (Yadav et al., 2010). The root extract of Ashwagandha upregulates the lipoprotein receptor-linked protein in the liver and thus reverses the physiological inadequacy and pathological cues associated with elimination of the amyloid-beta (Aβ) protein in Alzheimer's disease (Pingali et al., 2014; Sehgal et al., 2012).

10.3.5.8 Neuroprotective Activity

Many research studies have demonstrated that the Ashwagandha leaf extract and one of its phytochemicals possess neuroprotective effects (Durg et al., 2015; Wollen, 2010; Singh et al., 2008; Kuboyama et al., 2014). Withanone prevents scopolamine-generated toxic change in glial cells by attenuating GFAP (glial fibrillary acidic protein) and attenuating neuronal cell markers like PSD-95, NF-H, GAP-43, and MAP-2. Leaf extract of Ashwagandha also reduces oxidative stress (Konar et al., 2011). Glycowithanolides isolated from *W. somnifera* showed appreciable antioxidant activity in the cortex and striatum of mouse brain via normalizing the increased levels of superoxide (SOD) and lipid peroxidation (LPO) activities and improved activities of glutathione peroxidase (GPx) and catalase (CAT). Withanolide IV isolated from *W. somnifera* prevented loss of both axonal and dendritic regeneration and synaptic restoration in cultured rat neurons damaged by amyloid-β (Aβ) (Bhattacharya et al., 1997; Kuboyama et al., 2002; Dar et al., 2015; Kuboyama et al., 2006).

10.3.5.9 Cardioprotective activity

Many studies have demonstrated that W. *somnifera* showed significant cardioprotection in mice models by stimulating phase-II detoxification enzymes and activating the nuclear factor erthyroid-2-related transcript factor (Nrf-2) (Das et al., 1964; Reuland et al., 2013; Mohan et al., 2006; Ojha & Arya, 2009; Thirunavukkarasu et al., 2006). Root extracts of Ashwagandha were found to improve haematopoiesis (Aphale et al., 1998), whereas root and leaf extract of Ashwagandha blocked doxorubicin-induced cardiotoxicity in rats and restored the biochemical change (Dar et al., 2015; Hamza et al., 2008).

10.3.5.10 Antiarthritic activity

There is ample evidence in the literature to indicate a significant effect of Ashwagandha extract on arthritis. Crude Ashwagandha ethanolic extract appreciably inhibited lipopolysaccharide (LPS) -induced

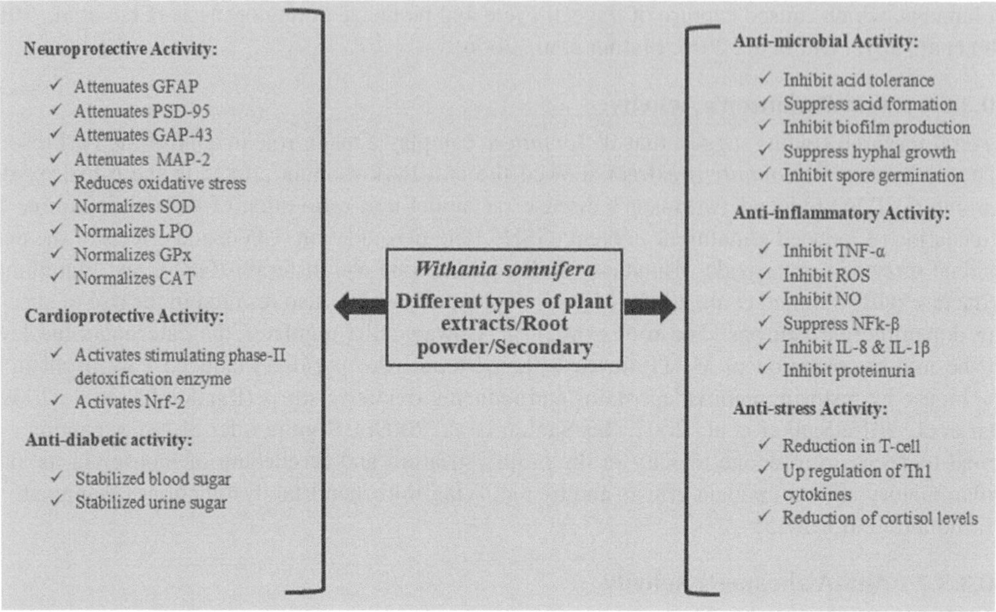

FIGURE 10.2 Potential therapeutic uses of *Withania somnifera* with the probable mode of action.

generation of pro-inflammatory cytokines such as IL-1β, TNF-α, and IL-12p40 in synovial fluid mononuclear cells by suppressing nuclear translocation of the transcription factor (NFk-β), activator protein-1, and phosphorylation of Ik-β (Singh et al., 2007). A polyherbal formulation containing *W. somnifera* suppressed NO generation and TNF-α without showing any cytotoxic effect in rheumatoid arthritis in mouse macrophage cell lines (Figure 10.2) (Ganesan et al., 2011; Dey et al., 2014).

10.4 CONCLUSION

The role of Ashwagandha is well-established in the traditional system of medicine. In Ayurveda, Ashwagandha is described as showing local as well as systemic properties, thus benefiting external and internal body conditions, respectively. Systemically, a prominent rejuvenator (Rasayana), an immunomodulator, helps to promote general health conditions and to act against emaciation. Externally, it acts on skin conditions such as ulcers, blisters, wounds, itching, and vitiligo. The traditional system of medicine has also widely used preparations of this plant to act against disorders related to nervous, respiratory, urinary, and hypertension-related disorders. It also protects against abdominal pains, backaches, leucorrhoea, and intestinal worms. It also has antitumour effects.

Many scientific studies have also proved that the chemical constituents and extracts derived from various parts of Ashwagandha possess a wide range of pharmacological activities and also help in modern systems of medicine with properties such as antibacterial, antifungal, antitumour, antioxidant, anti-Alzheimer, anticancer, anti-inflammatory, antiarthritic, and antidiabetic activities. It also acts as a cardioprotector and neuroprotector. Thus, the diverse role of Ashwagandha from the traditional system of medicine up to the modern system of medicine, with knowledge of chemical constituents obtained from today's scientific research, has confirmed the benefits of Ashwagandha. Therefore, Ashwagandha is benefiting humanity with its various therapeutic actions in traditional and modern systems of medicine.

REFERENCES

Abate, L., Tadesse, M.G., Bachheti, A. & Bachheti, R.K. (2022) Traditional and phytochemical bases of herbs, shrubs, climbers, and trees from Ethiopia for their anticancer response. *BioMed Research International*. https://doi.org/10.1155/2022/1589877

Abate, L., Bachheti, A., Bachheti, R.K. & Husen, A. (2021) Antibacterial properties of medicinal plants: recent trends, progress, and challenges. In: Husen, A. (Eds.) *Traditional Herbal Therapy for the Human Immune System*, CRC Press, pp. 13–54.

Asfaw, T.B., Esho, T.B., Bachheti, A., Bachheti, R.K., Pandey, D.P. & Husen, A. (2022) exploring important herbs, shrubs, and trees for their traditional knowledge, chemical derivatives, and potential benefits. In: Husen, A. (Eds.) *Herbs, Shrubs, and Trees of Potential Medicinal Benefits*, CRC Press, pp. 1–26.

Abou-Douh, A.M. (2002) New withanolides and other constituents from the fruit of *Withania somnifera. Archiv der Pharmazie: An International Journal Pharmaceutical and Medicinal Chemistry*, 335(6), 267–276.

Ahmad, M., Saleem, S., Ahmad, A.S., Ansari, M.A., Yousuf, S., Hoda, M.N. & Islam, F. (2005) Neuroprotective effects of *Withania somnifera* on 6-hydroxydopamine induced Parkinsonism in rats. *Human & Experimental Toxicology*, 24(3), 137–147.

Alam, N., Hossain, M., Mottalib, M.A., Sulaiman, S.A., Gan, S.H. & Khalil, M.I. (2012) Methanolic extracts of *Withania somnifera* leaves, fruits and roots possess antioxidant properties and antibacterial activities. *BMC Complementary and Alternative Medicine*, 12(1), 1–8.

Ale, Y., Sharma, S., Chaudhary, A. & Singh, A. (2021) A review on therapeutic use of *Withania somnifera* (Ashwagandha). *Annals of the Romanian Society for Cell Biology*, 25(7), 577–585.

Andallu, B. & Radhika, B. (2000) Hypoglycemic, diuretic and hypocholesterolemic effect of winter cherry (*Withania somnifera*, Dunal) root. *NISCAIR-CSIR, India*, 38(6), 607–609.

Anjaneyulu, A.S.R., Rao, D.S. & Lequesne, P.W. (1997) Withanolides, biologically active natural steroidal lactones: A review. *Studies in Natural Products Chemistry*, 20, 135–261.

Anwer, T., Sharma, M., Pillai, K.K. & Iqbal, M. (2008) Effect of *Withania somnifera* on insulin sensitivity in non-insulin-dependent diabetes mellitus rats. *Basic & Clinical Pharmacology & Toxicology*, 102(6), 498–503.

Aphale, A.A., Chibba, A.D., Kumbhakarna, N.R., Mateenuddin, M.O.H.O. & Dahat, S.H. (1998) Subacute toxicity study of the combination of ginseng (Panax ginseng) and ashwagandha (*Withania somnifera*) in rats: A safety assessment. *Indian Journal of Physiology and Pharmacology*, 42, 299–302.

Bachheti, R.K., Worku, L.A., Gonfa, Y.H., Zebeaman, M., Pandey, D.P. & Bachheti, A. (2022) Prevention and treatment of cardiovascular diseases with plant phytochemicals: A review. *Evidence-Based Complementary and Alternative Medicine*. https://doi.org/10.1155/2022/5741198

Bhandari, C.R. (1970) Ashwagandha (*Withania somnifera*) "Vanaushadhi Chandroday" (An Encyclopedia of Indian Herbs). *Publisher: CS Series of Varanasi Vidyavilas Press, Varanasi, India*, 1, 96–97.

Bhattacharya, S.K. & Satyan, K.S. (1997) Experimental methods for evaluation of psychotropic agents in rodents: I--Anti-anxiety agents. *Indian Journal of Experimental Biology*, 35(6), 565–575.

Bhattacharya, S.K., Bhattacharya, A., Sairam, K. & Ghosal, S. (2000) Anxiolytic-antidepressant activity of *Withania somnifera* glycowithanolides: An experimental study. *Phytomedicine*, 7(6), 463–469.

Bhattacharya, S.K., Goel, R.K., Kaur, R. & Ghosal, S. (1987) Antistress activity of sitoindosides VII and VIII, new acylsterylglucosides from *Withania somnifera. Phytotherapy Research*, 1(1), 32–37.

Bhattacharya, S.K., Kumar, A. & Ghosal, S. (1995) Effects of glycowithanolides from *Withania somnifera* on an animal model of Alzheimer's disease and perturbed central cholinergic markers of cognition in rats. *Phytotherapy Research*, 9(2), 110–113.

Bisht, P. (2009). *Ecophysiological and biochemical studies of two medicinally important Polygonatum species from Garhwal Himalaya*, PhD Thesis Submitted HNB Garhwal University.

Bisht, P. & Rawat, V. (2014). Antibacterial activity of *Withania somnifera* against Gram-positive isolates from pus samples. *Ayu*, 35(3), 330–332.

Chandrasekhar, K., Kapoor, J. & Anishetty, S. (2012) A prospective, randomized double-blind, placebo-controlled study of safety and efficacy of a high-concentration full-spectrum extract of ashwagandha root in reducing stress and anxiety in adults. *Indian Journal of Psychological Medicine*, 34(3), 255–262.

Changhadi, G.S. (1938) Ashwagandharishta—Rastantra Sar Evam Sidhyaprayog Sangrah. *Krishna-Gopal Ayurveda Bhawan (Dharmarth Trust), Nagpur*, 743–774.

Choi, M.J., Park, E.J., Min, K.J., Park, J.W. & Kwon, T.K. (2011) Endoplasmic reticulum stress mediates withaferin A-induced apoptosis in human renal carcinoma cells. *Toxicology in Vitro*, 25(3), 692–698.

Chunekar K.C. (2020) *Bhavprakash Nigantu (Indian Materia Medica)*. Chaukhamba Bharati Academy, Varanasi.

Dar, N.J. & Ahmad, M. (2020). Neurodegenerative diseases and *Withania somnifera* (L.): An update. *Journal of Ethnopharmacology*, 256, 112769.

Dar, N.J., Hamid, A. & Ahmad, M. (2015) Pharmacologic overview of *Withania somnifera*, the Indian Ginseng. *Cellular and Molecular Life Sciences*, 72(23), 4445–4460.

Das, P.K., Malhotra, C.L. & Prasad, K. (1964) Cardiotonic activity of Ashwagandhine and Ashwagandhinine, two alkaloids from withania ashwagandha, Kaul. *Archives Internationales de Pharmacodynamie et de Therapie*, 150, 356–362.

Debnath, T., Kim, D.H. & Lim, B.O. (2013) Natural products as a source of anti-inflammatory agents associated with inflammatory bowel disease. *Molecules*, 18(6), 7253–7270.

Dey, D., Chaskar, S., Athavale, N. & Chitre, D. (2014) Inhibition of LPS-induced TNF- α and NO production in mouse macrophage and inflammatory response in rat animal models by a novel ayurvedic formulation, BV-9238. *Phytotherapy Research*, 28(10), 1479–1485.

Dhar, U. (2002) Conservation implications of plant endemism in high-altitude Himalaya. *Current Science*, 82(2), 141–148.

Dhiman, A.K. (2005) *Wild medicinal plants of India (With ethnomedicinal use).* Bishen Singh Mahendra Pal Singh, Dehradun (India).

Durg, S., Dhadde, S.B., Vandal, R., Shivakumar, B.S. & Charan, C.S. (2015) *Withania somnifera* (Ashwagandha) in neurobehavioural disorders induced by brain oxidative stress in rodents: A systematic review and meta-analysis. *Journal of Pharmacy and Pharmacology*, 67(7), 879–899.

Dutta, R., Khalil, R., Green, R., Mohapatra, S.S. & Mohapatra, S. (2019) *Withania somnifera (Ashwagandha) and withaferin A: Potential in integrative oncology. International Journal of Molecular Sciences*, 20(21), 5310.

Fakim, A.G. (2006) Medicinal plants: Tradition of yesterday and drug of tomorrow. *Molecular Aspect of Medicine*, 27(1), 1–93.

Ganesan, K., Sehgal, P.K., Mandal, A.B. & Sayeed, S. (2011) Protective effect of *Withania somnifera* and Cardiospermum halicacabum extracts against collagenolytic degradation of collagen. *Applied Biochemistry and Biotechnology*, 165(3), 1075–1091.

Gaur, R.D. (1999) *Flora of the district Garhwal, North West Himalaya*. Srinagar, Garhwal, India: Transmedia.

Gauttam, V.K. & Kalia, A.N. (2013) Development of polyherbal antidiabetic formulation encapsulated in the phospholipids vesicle system. *Journal of Advanced Pharmaceutical Technology & Research*, 4(2), 108–117.

Ghosal, S., Kaur, R. & Srivastava, R.S. (1988) Sitoindosides IX and X, new glycowithanolides from *Withania somnifera. Indian Journal of Natural Products(Sagar)*, 4, 12–13.

Ghosh, K., De, S., Das, S., Mukherjee, S. & Sengupta Bandyopadhyay, S. (2016). Withaferin A induces ROS-mediated paraptosis in human breast cancer cell lines MCF-7 and MDA-MB-231. *PLoS One*, 11(12), 1–15.

Girish, K.S., Machiah, K.D., Ushanandini, S., Harish Kumar, K., Nagaraju, S., Govindappa, M., Vedavathi, M. & Kemparaju, K. (2006) Antimicrobial properties of a non-toxic glycoprotein (WSG) from *Withania somnifera* (Ashwagandha). *Journal of Basic Microbiology*, 46(5), 365–374.

Glotter, E., Abraham, A., Günzberg, G. & Kirson, I. (1977) Naturally occurring steroidal lactones with a 17α-oriented side chain. Structure of withanolide E and related compounds. *Journal of the Chemical Society, Perkin Transactions*, 1(4), 341–346.

Gogte, V.V.M. (2020) *Ayurvedic pharmacology and therapeutic uses of medicinal plants.* Chaukhamba Publication, New Delhi.

Gorelick, J., Rosenberg, R., Smotrich, A., Hanuš, L. & Bernstein, N. (2015) Hypoglycemic activity of withanolides and elicitated *Withania somnifera. Phytochemistry*, 116, 283–289.

Gulati, S., Madan, V.K., Singh, S., Singh, I. & Dushyant (2017) Chemical and phytochemical composition of Ashwagandha (*Withania somnifera* L.) roots. *Asian Journal of Chemistry*, 29(8), 1683–1686.

Gupta, G.L. & Rana, A.C. (2007) *Withania somnifera* (Ashwagandha): A review. *Pharmacognosy Reviews*, 1(1). 129–136.

Hahm, E.R. & Singh, S.V. (2013a) Withaferin A-induced apoptosis in human breast cancer cells is associated with suppression of inhibitor of apoptosis family protein expression. *Cancer Letters*, 334(1), 101–108.

Hahm, E.R., Lee, J. & Singh, S.V. (2014) Role of mitogen-activated protein kinases and Mcl-1 in apoptosis induction by withaferin A in human breast cancer cells. *Molecular Carcinogenesis*, 53(11), 907–916.

Hahm, E.R., Lee, J., Kim, S.H., Sehrawat, A., Arlotti, J.A., Shiva, S.S., Bhargava, R. & Singh, S.V. (2013b) Metabolic alterations in mammary cancer prevention by withaferin A in a clinically relevant mouse model. *Journal of the National Cancer Institute*, 105(15), 1111–1122.

Hahm, E.R., Moura, M.B., Kelley, E.E., Van Houten, B., Shiva, S. & Singh, S.V. (2011) Withaferin A-induced apoptosis in human breast cancer cells is mediated by reactive oxygen species. *PloS One*, 6(8), 1–12.

Hamza, A., Amin, A. & Daoud, S. (2008) The protective effect of a purified extract of *Withania somnifera* against doxorubicin-induced cardiac toxicity in rats. *Cell Biology and Toxicology*, 24(1), 63–73.

Hegde, P.K. & Harini, A. (2019) *A textbook of Dravyaguna Vijnana, Vol. II.* Chaukhamba Publication, New Delhi.

Hepper, F.N. (1991). *Old World Withania (Solanaceae): A taxonomic review and key to the species. Solanaceae III: Taxonomy, chemistry, evolution.* Royal Botanic Gardens Kew and Linnean Society of London, London.

Hooker, J.D. (1875) *The flora of British India (Vol. 1).* London: L. Reeve.

Iqbal, M. & Datta, A.K. (2007) Cytogenetic studies in *Withania somnifera* Dun. (Solanaceae). *Cytologia*, 72(1), 43–47.

Jayaprakasam, B. & Nair, M.G. (2003) Cyclooxygenase-2 enzyme inhibitory withanolides from *Withania somnifera* leaves. *Tetrahedron*, 59(6), 841–849.

Jayaprakasam, B., Strasburg, G.A. & Nair, M.G. (2004) Potent lipid peroxidation inhibitors from *Withania somnifera* fruits. *Tetrahedron*, 60(13), 3109–3121.

Kalra, R. & Kaushik, N. (2017) Withania somnifera (Linn.) Dunal: A review of chemical and pharmacological diversity. *Phytochemistry Reviews*, 16(5), 953–987.

Khan, B., Ahmad, S.F., Bani, S., Kaul, A., Suri, K.A., Satti, N.K., Athar, M. & Qazi, G.N. (2006) Augmentation and proliferation of T lymphocytes and Th-1 cytokines by *Withania somnifera* in stressed mice. *International Immunopharmacology*, 6(9), 1394–1403.

Khanal, P., Chikhale, R., Dey, Y.N., Pasha, I., Chand, S., Gurav, N., Patil, B.M. & Gurav, S. (2022) Withanolides from *Withania somnifera* as an immunity booster and their therapeutic options against COVID-19. *Journal of Biomolecular Structure and Dynamics*, 40(12), 5295–5308.

Kim, S.H. & Singh, S.V. (2014) Mammary cancer chemoprevention by withaferin a is accompanied by in vivo suppression of self-renewal of cancer stem cells. *Cancer Prevention Research*, 7(7), 738–747.

Kirson, I., Glotter, E., Abraham, A. & Lavie, D. (1970) Constituents of *Withania somnifera* dun—XI: The structure of three new withanolides. *Tetrahedron*, 26(9), 2209–2219.

Konar, A., Shah, N., Singh, R., Saxena, N., Kaul, S.C., Wadhwa, R. & Thakur, M.K. (2011) Protective role of Ashwagandha leaf extract and its component withanone on scopolamine-induced changes in the brain and brain-derived cells. *PloS One*, 6(11), 1–12.

Ku, S.K., Han, M.S. & Bae, J.S. (2014) Withaferin A is an inhibitor of endothelial protein C receptor shedding in vitro and in vivo. *Food and Chemical Toxicology*, 68, 23–29.

Kuboyama, T., Tohda, C. & Komatsu, K. (2006) Withanoside IV and its active metabolite, sominone, attenuate Aβ (25–35)-induced neurodegeneration. *European Journal of Neuroscience*, 23(6), 1417–1426.

Kuboyama, T., Tohda, C. & Komatsu, K. (2014) Effects of Ashwagandha (roots of *Withania somnifera*) on neurodegenerative diseases. *Biological and Pharmaceutical Bulletin*, 37(6), 892–897.

Kuboyama, T., Tohda, C., Zhao, J., Nakamura, N., Hattori, M. & Komatsu, K. (2002) Axon-or dendrite-predominant outgrowth induced by constituents from Ashwagandha. *Neuroreport*, 13(14), 1715–1720.

Maitra, R., Porter, M.A., Huang, S. & Gilmour, B.P. (2009) Inhibition of NFκB by the natural product Withaferin A in cellular models of Cystic Fibrosis inflammation. *Journal of Inflammation*, 6(1), 1–5.

Majumdar, D.N. (1955) *Withania somnifera* Dunal. Part II: alkaloidal constituents and their chemical characterization. *Indian Journal of Pharmaceutical*, 17, 158–161.

Malik, F., Kumar, A., Bhushan, S., Khan, S., Bhatia, A., Suri, K.A., Qazi, G.N. & Singh, J. (2007) Reactive oxygen species generation and mitochondrial dysfunction in the apoptotic cell death of human myeloid leukemia HL-60 cells by a dietary compound withaferin A with concomitant protection by N-acetyl cysteine. *Apoptosis*, 12(11), 2115–2133.

Manjunath, M.J. & Muralidhara. (2013) Effect of *Withania somnifera* supplementation on rotenone-induced oxidative damage in cerebellum and striatum of the male mice brain. *Central Nervous System Agents in Medicinal Chemistry (Formerly Current Medicinal Chemistry-Central Nervous System Agents)*, 13(1), 43–56.

Manjunath, M.J. & Muralidhara. (2015) Standardized extract of *Withania somnifera* (Ashwagandha) markedly offsets rotenone-induced locomotor deficits, oxidative impairments and neurotoxicity in Drosophila melanogaster. *Journal of Food Science and Technology*, 52(4), 1971–1981.

Mathur, R., Gupta, S.K., Singh, N., Mathur, S., Kochupillai, V. & Velpandian, T. (2006) Evaluation of the effect of *Withania somnifera* root extracts on cell cycle and angiogenesis. *Journal of Ethnopharmacology*, 105(3), 336–341.

Menssen, H.G. & Stapel, G. (1973) Uber ein C28-Steroidlacton aus der Wurzel von *Withania somnifera*. *Planta Medica*, 24(05), 8–12.

Minhas, U., Minz, R. & Bhatnagar, A. (2011) Prophylactic effect of *Withania somnifera* on inflammation in a non-autoimmune prone murine model of lupus. *Drug Discoveries & Therapeutics*, 5(4), 195–201.

Minhas, U., Minz, R., Das, P. & Bhatnagar, A. (2012) Therapeutic effect of *Withania somnifera* on pristane-induced model of SLE. *Inflammopharmacology*, 20(4), 195–205.

Mirjalili, M.H., Moyano, E., Bonfill, M., Cusido, R.M. & Palazón, J. (2009) Steroidal lactones from *Withania somnifera*, an ancient plant for novel medicine. *Molecules*, 14(7), 2373–2393.

Mishra, B. (2004) *Ashwagandha–Bhavprakash Nigantu (Indian Materia Medica)*. Chaukhambha Bharti Academy, Varanasi, 393–394.

Misra, L., Mishra, P., Pandey, A., Sangwan, R.S., Sangwan, N.S. & Tuli, R. (2008) Withanolides from *Withania somnifera* roots. *Phytochemistry*, 69(4), 1000–1004.

Mohan, I.K., Kumar, K.V., Naidu, M.U.R., Khan, M. & Sundaram, C. (2006) Protective effect of CardiPro against doxorubicin-induced cardiotoxicity in mice. *Phytomedicine*, 13(4), 222–229.

Nadkarni, A.K. & Nadkarni, A.K. (1982) *Indian material medica, popular prakashan pvt ltd*. India: Bombay.

Ojha, S.K. & Arya, D.S. (2009) *Withania somnifera* Dunal (Ashwagandha): A promising remedy for cardiovascular diseases. *World Journal of Medical Sciences*, 4(2), 156–158.

Pandit, S., Chang, K.W. & Jeon, J.G. (2013) Effects of *Withania somnifera* on the growth and virulence properties of Streptococcus mutans and Streptococcus sobrinus at sub-MIC levels. *Anaerobe*, 19, 1–8.

Pawar, P., Gilda, S., Sharma, S., Jagtap, S., Paradkar, A., Mahadik, K., Ranjekar, K. & Harsulkar, A. (2011) Rectal gel application of *Withania somnifera* root extract expounds anti-inflammatory and muco-restorative activity in TNBS-induced inflammatory bowel disease. *BMC Complementary and Alternative Medicine*, 11(1), 1–9.

Pingali, U., Pilli, R. & Fatima, N. (2014) Effect of standardized aqueous extract of *Withania somnifera* on tests of cognitive and psychomotor performance in healthy human participants. *Pharmacognosy Research*, 6(1), 12.

Pramanick, S., Roy, A., Ghosh, S., Majumder, H.K. & Mukhopadhyay, S. (2008) Withanolide Z, a new chlorinated withanolide from *Withania somnifera*. *Planta Medica*, 74(14), 1745–1748.

Praveen, N. & Murthy, H.N. (2010) Production of withanolide-A from adventitious root cultures of *Withania somnifera*. *Acta Physiologiae Plantarum*, 32(5), 1017–1022.

Puri, H.S. (2002). *Rasayana: Ayurvedic herbs for longevity and rejuvenation*. London: CRC Press.

RajaSankar, S., Manivasagam, T. & Surendran, S. (2009a) Ashwagandha leaf extract: A potential agent in treating oxidative damage and physiological abnormalities seen in a mouse model of Parkinson's disease. *Neuroscience Letters*, 454(1), 11–15.

RajaSankar, S., Manivasagam, T., Sankar, V., Prakash, S., Muthusamy, R., Krishnamurti, A. & Surendran, S. (2009b) *Withania somnifera* root extract improves catecholamines and physiological abnormalities seen in a Parkinson's disease model mouse. *Journal of Ethnopharmacology*, 125(3), 369–373.

RajeswaraRao, B.R., Rajput, D.K., Nagaraju, G. & Adinarayana, G. (2012) Opportunities and challenges in the cultivation of Ashwagandha {*Withania somnifera* (L.) Dunal}. *Journal of Pharmacognosy*, 3(2), 88–91.

Reuland, D.J., Khademi, S., Castle, C.J., Irwin, D.C., McCord, J.M., Miller, B.F. & Hamilton, K.L. (2013) Upregulation of phase II enzymes through phytochemical activation of Nrf2 protects cardiomyocytes against oxidant stress. *Free Radical Biology and Medicine*, 56, 102–111.

Samant, S.S., Dhar, U. & Palni, L.M.S. (1998) *Medicinal plants of Indian Himalayas: Diversity Distribution Potential values*, Gyanodaya Prakashan, Nainital.

Sankar, S.R., Manivasagam, T., Krishnamurti, A. & Ramanathan, M. (2007) The neuroprotective effect of *Withania somnifera* root extract in MPTP-intoxicated mice: An analysis of behavioral and biochemical variables. *Cellular & Molecular Biology Letters*, 12(4), 473–481.

Sastri, K.N. & Chaturvedi, G.N. (2016) *The Charaka Samhita*. Chaukhamba Bharati Academy, Varanasi.

Sehgal, N., Gupta, A., Valli, R.K., Joshi, S.D., Mills, J.T., Hamel, E., Khana, P., Jain, S.C., Thakur, S.S. & Ravindranath, V. (2012) *Withania somnifera* reverses Alzheimer's disease pathology by enhancing low-density lipoprotein receptor-related protein in liver. *Proceedings of the National Academy of Sciences*, 109(9), 3510–3515.

Sharada, M., Ahuja, A., Suri, K.A., Vij, S.P., Khajuria, R.K., Verma, V. & Kumar, A. (2007) Withanolide production by in vitro cultures of *Withania somnifera* and its association with differentiation. *Biologia Plantarum*, 51(1), 161–164.

Sharma, P.V. (2009). *Dravyaguna Vijnana. Vol. II*. Chaukhambha Bharti Academy, Varanasi.

Shishkin, B.K. (1936) *Flora of the USSR Volume 6, Centrospermae*. Moscow: Isdatel'stvo Akademii Nauk SSSR [translated from Russian. 1970. Washington, DC: Smithsonian Institution and National Science Foundation].

Singh, D., Aggarwal, A., Maurya, R. & Naik, S. (2007) *Withania somnifera* inhibits NF- κB and AP-1 transcription factors in human peripheral blood and synovial fluid mononuclear cells. *Phytotherapy Research*, 21(10), 905–913.

Singh, G. & Kumar, P. (2011) Evaluation of antimicrobial efficacy of flavonoids of *Withania somnifera* L. *Indian Journal of Pharmaceutical Sciences*, 73(4), 473–478.

Singh, R.H., Narsimhamurthy, K. & Singh, G. (2008) Neuronutrient impact of Ayurvedic Rasayana therapy in brain aging. *Biogerontology*, 9(6), 369–374.

Subbaraju, G.V., Vanisree, M., Rao, C.V., Sivaramakrishna, C., Sridhar, P., Jayaprakasam, B. & Nair, M.G. (2006) Ashwagandhanolide, a bioactive dimeric thiowithanolide isolated from the roots of *Withania somnifera*. *Journal of Natural Products*, 69(12), 1790–1792.

Tewari, D., Chander, V., Dhyani, A., Sahu, S., Gupta, P., Patni, P., Kalick, L.S. & Bishayee, A. (2022) Withania somnifera (L.) dunal: Phytochemistry, structure-activity relationship, and anticancer potential. *Phytomedicine*, 98 153949. https://doi.org/10.1016/j.phymed.2022.153949

Thirunavukkarasu, M., Penumathsa, S., Juhasz, B., Zhan, L., Bagchi, M., Yasmin, T., Shara, M.A., Thatte, H.S., Debasis, B. & Maulik, N. (2006) Enhanced cardiovascular function and energy level by a novel chromium (III)-supplement. *Biofactors*, 27(1–4), 53–67.

Udayakumar, R., Kasthurirengan, S., Mariashibu, T.S., Rajesh, M., Anbazhagan, V.R., Kim, S.C., Ganapathi, A. & Choi, C.W. (2009) Hypoglycaemic and hypolipidaemic effects of *Withania somnifera* root and leaf extracts on alloxan-induced diabetic rats. *International Journal of Molecular Sciences*, 10(5), 2367–2382.

Udayakumar, R., Kasthurirengan, S., Vasudevan, A., Mariashibu, T.S., Rayan, J.J.S., Choi, C.W., Ganpathi, A. & Kim, S.C. (2010). Antioxidant effect of dietary supplement *Withania somnifera* L. reduces blood glucose levels in alloxan-induced diabetic rats. *Plant Foods for Human Nutrition*, 65(2), 91–98.

Um, H.J., Min, K.J., Kim, D.E. & Kwon, T.K. (2012) Withaferin A inhibits JAK/STAT3 signaling and induces apoptosis of human renal carcinoma Caki cells. *Biochemical and Biophysical Research Communications*, 427(1), 24–29.

Vaishnavi, K., Saxena, N., Shah, N., Singh, R., Manjunath, K., Uthayakumar, M., Kanaujia, S.P., Kaul, S.C., Sekar, K. & Wadhwa, R. (2012) Differential activities of the two closely related withanolides, Withaferin A and Withanone: Bioinformatics and experimental evidence. *Plos One*, 7(9), 1–13.

Verma, N., Gupta, S.K., Tiwari, S. & Mishra, A.K. (2021) Safety of ashwagandha root extract: A randomized, placebo-controlled, study in healthy volunteers. *Complementary Therapies in Medicine*, 57, 1–6.

Wankhede, S., Langade, D., Joshi, K., Sinha, S.R. & Bhattacharyya, S. (2015) Examining the effect of *Withania somnifera* supplementation on muscle strength and recovery: A randomized controlled trial. *Journal of the International Society of Sports Nutrition*, 12(1), 1–11.

Widodo, N., Kaur, K., Shrestha, B.G., Takagi, Y., Ishii, T., Wadhwa, R. & Kaul, S.C. (2007) Selective killing of cancer cells by leaf extract of Ashwagandha: Identification of a tumor-inhibitory factor and the first molecular insights to its effect. *Clinical Cancer Research*, 13(7), 2298–2306.

Widodo, N., Takagi, Y., Shrestha, B.G., Ishii, T., Kaul, S.C. & Wadhwa, R. (2008) Selective killing of cancer cells by leaf extract of Ashwagandha: Components, activity and pathway analyses. *Cancer Letters*, 262(1), 37–47.

Wollen, K.A. (2010) Alzheimer's disease: The pros and cons of pharmaceutical, nutritional, botanical, and stimulatory therapies, with a discussion of treatment strategies from the perspective of patients and practitioners. *Alternative Medicine Review*, 15(3), 223–44.

Yadav, C.S., Kumar, V., Suke, S.G., Ahmed, R.S., Mediratta, P.K. & Banerjee, B.D. (2010) Propoxur-induced acetylcholine esterase inhibition and impairment of cognitive function: Attenuation by *Withania somnifera*. *Indian Journal of Biochemistry and Biophysics*, 47, 117–120.

Yang, E.S., Choi, M.J., Kim, J.H., Choi, K.S. & Kwon, T.K. (2011a) Combination of withaferin A and X-ray irradiation enhances apoptosis in U937 cells. *Toxicology in Vitro*, 25(8), 1803–1810.

Yang, E.S., Choi, M.J., Kim, J.H., Choi, K.S. & Kwon, T.K. (2011b) Withaferin A enhances radiation-induced apoptosis in Caki cells through induction of reactive oxygen species, Bcl-2 downregulation and Akt inhibition. *Chemico-Biological Interactions*, 190(1), 9–15.

11 Phytoconstituents and Medicinal Importance of *Chlorophytum borivilianum* Tuber

Sisay Awoke, Melaku Assefa, and Mesfin Getachew

ABBREVIATIONS

IC_{50}: Half-maximal inhibitory concentration
DPPH: 2,2-diphenyl-1-picrylhydrazyl
BHV-1: Bovine herpesvirus-1
FMDV: Foot-and-mouth disease viruses
LC_{50}: Lethal Concentration, 50%
EC_{50}: 50% Effective Concentration
INF: Interferon
NMR: Nuclear Magnetic Resonance
MS: Mass Spectrometry
Th1: T helper 1
Th2: T helper 2
HCT-116: Human colorectal carcinoma

11.1 INTRODUCTION

Natural products and traditional medicines have been of great importance since ancient times. Chinese medicine, Ayurveda, Kampo, traditional Korean medicine, and Unani are some forms of traditional medicine that have been practised in particular areas of the world (Veeresham, 2012; Beshah et al., 2020; Asfaw et al., 2022). The era of modern drugs began in the nineteenth century (Beutler, 2019; Chen, 2020; Paine, 2020; Truax, 2020). In 1805, the first pharmacologically active compound, morphine, was isolated from the opium plant and blossomed into regulated systems of medicine (Joo, 2014). Natural products possess a unique chemical diversity that leads to diverse biological activities and medicinal properties (Veeresham, 2012; Abate et al., 2021, 2022).

The genus *Chlorophytum*, belonging to the family Asparagaceae, has primarily secured its place as a commercial plant with a wide range of uses, from medicinal to ornamental, with promising economic benefits (Chauhan, 2016). *Chlorophytum* includes many economically important species, well-known for medicinal, ornamental, and horticultural values (Tripathi, 2012). *Chlorophytum borivilianum* is the most eminent medicinal plant, and is commonly known as safed musli (Khanam et al., 2013). The plant is well known for its tuberous roots containing bioactive constituents like flavonoids, alkaloids, saponins, phenols, steroids, triterpenoids, vitamins, and tannins (Nakasha, 2017). Consumption of this plant improves physical

DOI: 10.1201/b22924-11

stamina and physical and mental health, improves sexual desire, increases semen volume, and keeps the body healthy and energetic (Thakur et al., 2008a). The main aim of this chapter was to review the main aspects of this valuable plant, regarding its chemical constituents and medicinal importance.

11.2 GEOGRAPHICAL DISTRIBUTION AND BOTANICAL ASPECTS OF *C. BORIVILIANUM*

The species of the genus *Chlorophytum* are well known for their use in traditional medicine and their therapeutic efficacy (Chauhan, 2016). *Chlorophytum* is a large genus comprising over 300 species distributed in the tropical and subtropical regions of the world, mainly in Africa, India, and Australia (Tripathi, 2012). Of these species, 13 are available in India (Katoch, 2010). *C. borivilianum* is an endangered geophyte with a comprehensive traditional history and medicinal importance (Khanam et al., 2013; Patil, 2016). It is commonly known as 'safed musli' in Hindi (meaning 'white tubers'). *C. borivilianum* is a widely grown species and an integral part of Ayurvedic, Unani, Homoeopathic, and Allopathic systems of medicine, where the tuber of the plant holds the principal place (Veeresham, 2012). Its scientific classification is presented as follows (Khanam et al., 2013):

Kingdom: Plantae
Clade: Tracheophytes
Clade: Angiosperms
Clade: Monocots
Order: Asparagales
Family: Asparagaceae
Subfamily: Agavoideae
Genus: Chlorophytum
Species: *C. borivilianum*
Binomial name: *Chlorophytum borivilianum*

11.3 MEDICINAL IMPORTANCE OF *C. BORIVILIANUM*

Its tubers contribute mostly to its pharmacological activities and are known to have hepatoprotective, antioxidant, immunomodulatory, and anti-inflammatory functions (Tripathi, 2012). *C. borivilianum* is one of the most important medicinal plants and is used for treating different health problems, such as diabetes, arthritis, and physical weakness (Kumar et al., 2012). Reports have also shown the impact of safed musli on rheumatism and joint pain (Ashraf et al., 2014). Other pharmacological studies on safed musli tubers have revealed their antiviral, anticancer, antioxidant, antistress, antimicrobial, hypolipidaemic, and anti-inflammatory properties (Acharya, 2009).

C. borivilianum showed antioxidant (Chakraborthy Guno, 2008; Deore & Khadabadi, 2008; Sharma and Kumar, 2012), cytotoxic, antiulcer (Panda, 2011), hypolipidaemic (Deore & Khadabadi, 2009b), anticancer (Deore & Khadabadi, 2010), hepatoprotective (Sharma & Kumar, 2011), immunostimulant (Thakur et al., 2007), anti-inflammatory (Deore & Khadabadi, 2008), antitumour, antimutagenic, and chemomodulatory activities (Kumar et al., 2010). Saponins from the tubers of *C. borivilianum* were found to possess antiinflammatory and antiarthritic properties (Lande, 2015). The plant also showed larvicidal (Deore & Khadabadi, 2009a), antibacterial, and antifungal activities (Sharma et al., 2020). *C. borivilianum* tuber extracts are used to treat arsenic-induced toxicity (Sharma et al., 2012) and oxidative stress-induced disorders (Kenjale et al., 2007). Figure 11.1 shows the pharmacological activities of crude extracts

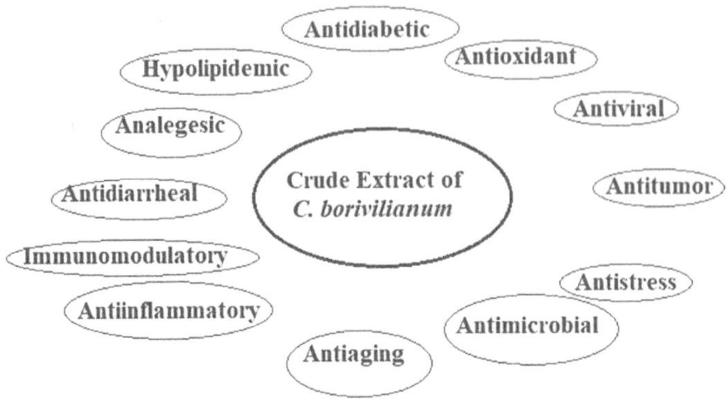

FIGURE 11.1 Pharmacological activities of crude extract of *Chlorophytum borivilianum*.

of *C. borivilianum*. Its tuber extract had no toxic effect up to a dose of 800 mg/ kg bodyweight/ day (Kumar et al., 2010). The tubers of *C. borivilianum* can also increase aphrodisiac potential (Kenjale et al., 2008; Das et al., 2016). A formulation containing *C. borivilianum* and velvet bean improved sleep quality (McCarthy et al., 2012).

11.3.1 Antioxidant

Extracts of *C. borivilianum* leaves were screened for *in-vitro* antioxidant properties using standard procedures and have been reported by Chakraborthy et al. (2008b). The successive extracts, such as petroleum ether, ethyl acetate, methanol, and water, and 50% crude methanol extracts exhibited IC_{50} values in DPPH and nitric oxide radical inhibition assays and the values were comparable with antioxidant standards such as ascorbic acid and quercetin (Chakraborthy, 2008b). The radical scavenging activity of the aqueous *C. borivilianum* tuber extract was measured using DPPH (Sharma & Kumar, 2012; Giribabu, 2014). *C. borivilianum* tuber extract prevents impairment in sperm characteristics and morphology by preventing elevation of oxidative stress, apoptosis, and the concentration of free radicals in the sperm in diabetes (Deore, 2008; Tripathi, 2012; Giribabu, 2014).

11.3.2 Anti-inflammatory

The aqueous, hydroalcoholic, and methanolic extracts of the tubers were investigated for anti-inflammatory activity using the carrageenan-induced rat paw oedema model (Deore, 2008) that shows significantly more activity than the alcoholic extract (Deore, 2008; Singh et al., 2012; Lande, 2015; Goyal, 2017).

11.3.3 Anti-ulcer

The anti-ulcer activity of the tubers of *C. borivilianum* has been reported. The methanolic extract showed a significant dose-dependent antiulcer activity at concentrations of 200 and 400 mg/kg bodyweight, using pyloric ligation and aspirin-induced ulceration models. The results indicated that the tubers exhibit effective anti-ulcer activity (Panda, 2011).

11.3.4 Anticancer and Antitumour

The potential anticancer and antitumour activity of this plant has been assessed (Deore & Khadabadi, 2010). In developed and developing countries, cancer is a deadly and increasingly common disease.

Numerous modern restorative agents and treatments are accessible in the market but have some serious side-effects on human organs. These therapeutic agents are costly and not readily available in developing nations. Different scientific reports have appeared that chemoprevention through naturally derived herbal phytochemicals are a pioneering alternative therapeutic tool against diverse cancer types. The chemopreventive property of many of them, including *C. borivilianum*, has been successfully demonstrated by inducing apoptosis equivalent to that achieved by certain drugs without causing any side-effects (Singh et al., 2018). The anticancer activity of crude extracts of the tuber of *C. borivilianum* was evaluated using the antiproliferative assay and DNA fragmentation assay (Deore & Khadabadi, 2010).

Antitumour properties of the tuber of *C. borivilianum* have been demonstrated using several antitumour assays to estimate the total number of papillomas, tumour incidence, tumour burden, the average weight of the tumour, and estimate of tumour based on the protocol proposed by Prashar (Prashar, 1995). In one study, oral administration of tuber extract at pre-, peri-, and post-initiation stages of papillomagenesis showed a significant reduction in the tumour incidence (56.7±3.3), cumulative number of papillomas (22±1.15), tumour burden (3.88±0.11) and the average weight of the tumour (127.33±5.17) compared with the control group, 55.8, 20.5, 3.9, and 124.2, respectively (Kumar et al., 2010). There was a less significant increase recorded in the average latent period (11.59±0.25 weeks) as compared with the control group (10.81±0.10 weeks) (Kumar et al., 2010).

11.3.5 Antiviral

The antiviral properties of aqueous extract of *C. borivilianum* tubers and its possible mechanism against BHV-1 and FMDV viruses as model viral agents were evaluated (Kaur, 2020). Mechanistic studies were carried out on its inductive effect on IFN-c and other cytokines in Wistar albino rats (Singh, 2013).

11.3.6 Aphrodisiac

C. borivilianum (Figure 11.2) is primarily used as ethnic medicine by local healers of Indigenous communities of India. In Ayurveda, this plant belongs to the group of *Vajikaran Rasayana*, corroborated by its rejuvenating, aphrodisiac, natural sex tonic properties and effectiveness in alleviating sexual disorders (Khanam et al.,, 2013). *Safed musli* is used for its aphrodisiac property, to improve impotence and cure sexual disorders. This herb has been scientifically studied *in vitro* to evaluate its aphrodisiac property in a rat model (Grover, 2021).

11.3.7 Other Activities

The larvicidal properties of *C. borivilianum* extracts against the mosquito species *Anopheles stephensi*, *Culex quinquefasciatus*, and *Aedes aegypti* have been reported. Methanolic tuber extract, crude saponins extract, and purified saponin fractions were used as test solutions. The concentration to kill 50% larvae and the concentration to inhibit the emergence of 50% adults, that is, LC_{50} and EC_{50} values, respectively, were calculated, and showed the methanol extract and the saponin-containing crude extract to be larvicidal. However, purified saponins were found to be the most effective (Deore & Khadabadi, 2009).

The plant extract or secondary metabolites, which can ameliorate pathological and immunological conditions, can be considered attractive candidates for developing new treatments against leishmaniasis (Kaur, 2020). Leishmaniasis is a disease that progresses with suppression of the protective Th1 type of immune responses and upregulation of the non-protective Th2 type of immune responses (Samaranayake, 2020). Ethanolic extract of the tubers of *C. borivilianum* and its sapogenin were evaluated for their immunomodulatory activity (Thakur et al., 2008a, 2008b). *C. borivilianum* has been reported to possess antihyperglycaemic, hypolipidaemic, and antidiabetic properties (Thakur et al., 2008a; Ashraf et al., 2014).

11.4 PHYTOCONSTITUENTS OF *C. BORIVILIANUM*

The tuber dry weight of this species consists of carbohydrates (42%), proteins (5–10%), fibre (3–4%), and saponins (17–20%) (Kumar et al., 2012). The plant's medicinal properties are attributed to the presence of saponins (Chakraborthy, 2008a). Saponins present in *C. borivilianum* are synthesized by the mevalonate and non-mevalonate pathways as shown in Figure 11.2 (Kumar et al., 2012).

Biologically active compounds have been reported from this plant, following isolation from column fractions and characterization. From an ethanol extract of the tubers, four new spirostane-type saponins, named borivilianosides E–H (**1–4**), and two known steroid saponins (**5** and **6**) were isolated and characterized (Figure 11.3). Borivilianosides F (**2**), G (**3**), and H (**4**), as well as the two already-known chemicals, were further examined for cytotoxicity using the human colon cancer cell lines HT-29 and HCT 116 (Acharya, 2009).

Further investigation of the butanol-soluble fraction from the ethanol extract of the tubers of *C. borivilianum* led to the isolation of four spirostane-type saponins, borivilianosides E–H (**7–10**), as well as two steroid saponins (**11** and **12**), as shown in Figure 11.4 (Acharya, 2009).

As Figure 11.5 shows, three compounds (**13–15**), together with stigmasterol (**16**) and one saponin, chlorophytoside-I (**17**), were also isolated from the tubers of *C. borivilianum*

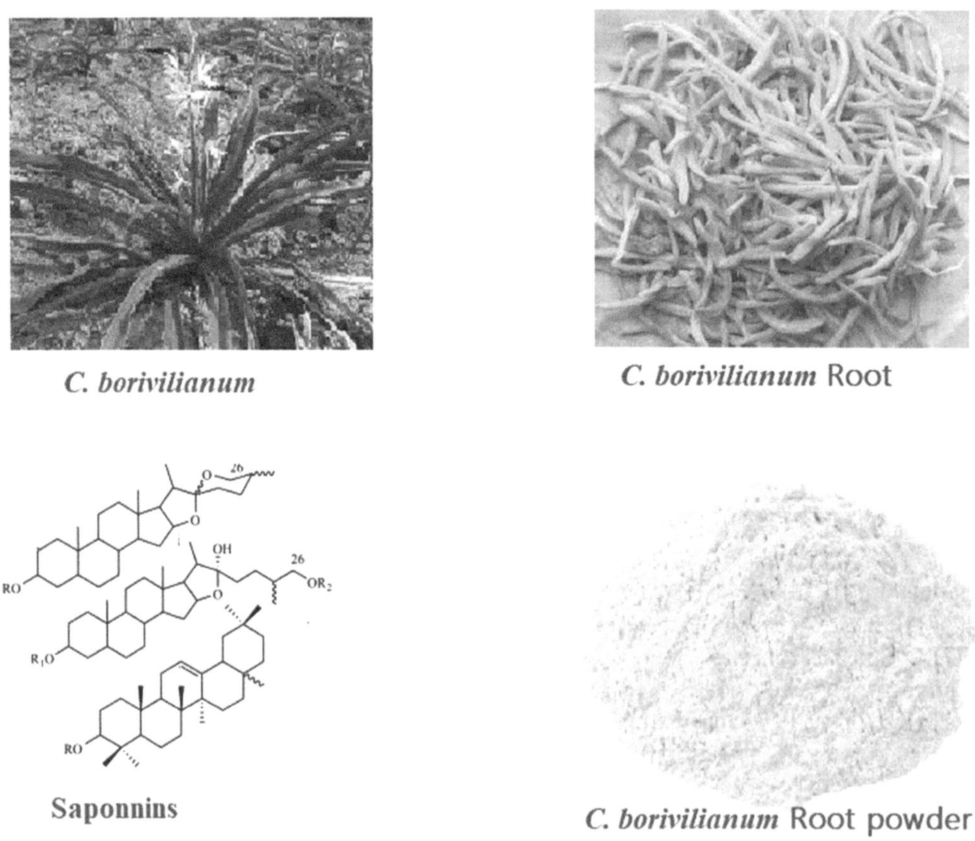

FIGURE 11.2 Chemical constituents of the tuber of *Chlorophytum borivilianum*.

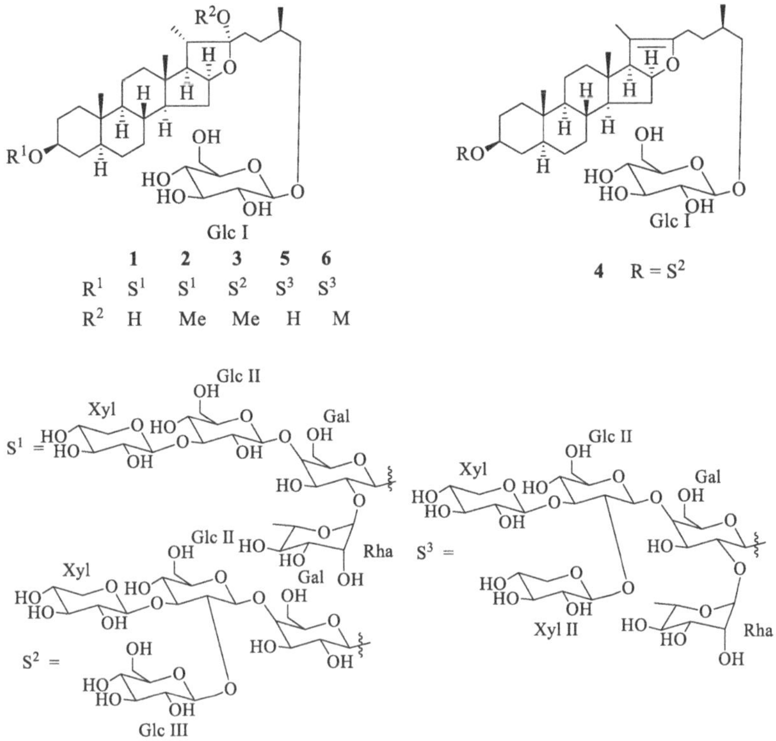

FIGURE 11.3 Borivilianosides saponnins from the tuber of *Chlorophytum borivilianum*.

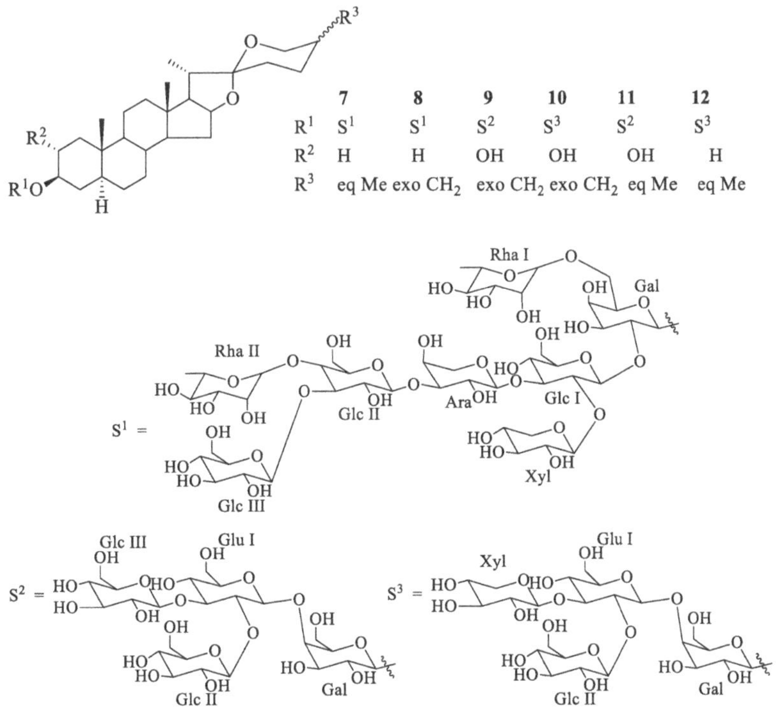

FIGURE 11.4 Spirostane and steroid saponins from the tuber of *Chlorophytum borivilianum*.

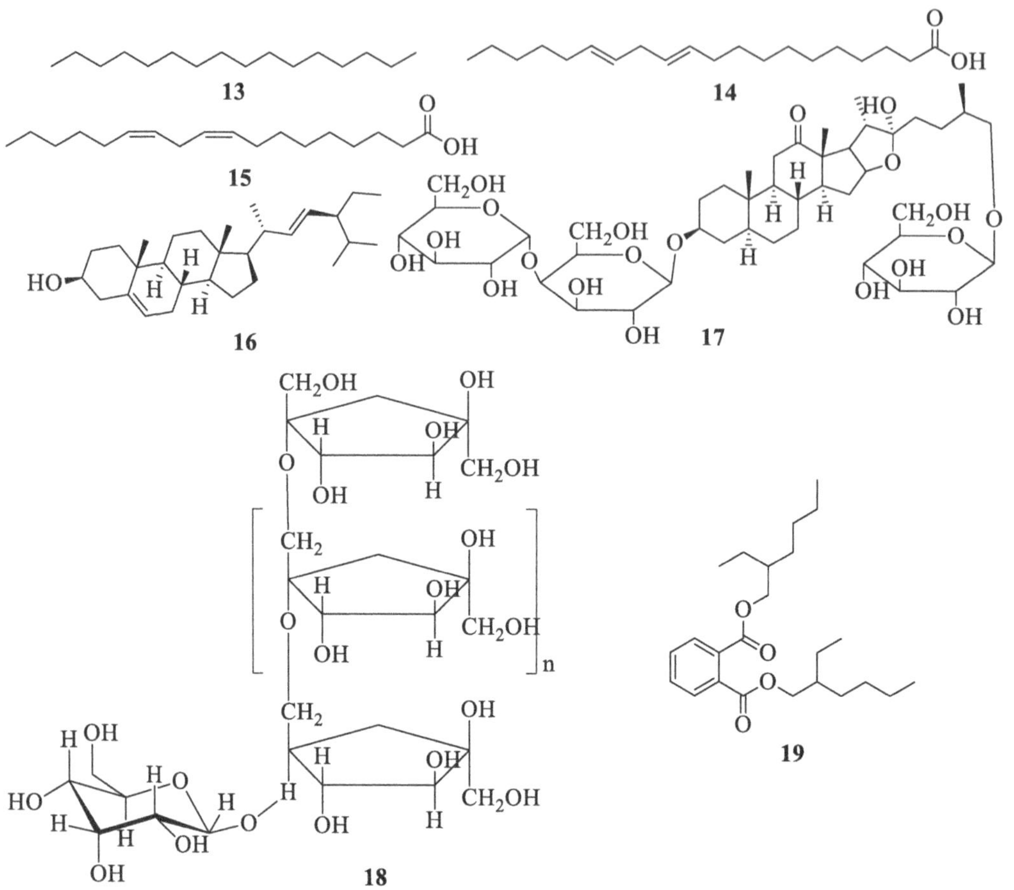

FIGURE 11.5 Fatty acids, stigmasterol, chlorophytoside, and oligosaccharides from the tuber of *Chlorophytum borivilianum*.

(Deore & Khadabadi, 2010). A fructo-oligosaccharide (**18**, (n=5–30) (Narasimhan et al., 2006) and a bis(2-ethylhexyl) benzene-1,2-dicarboxylate (**19**) were isolated and characterized from the tubers of *C. borivilianum*.

11.5 COMMERCIAL ASPECTS AND SOCIO-ECONOMIC IMPORTANCE OF *C. BORIVILIANUM*

Traditionally, *C. borivilianum* is well known for treating male impotence in India. This plant's multi-therapeutic and nutritional importance is attributed to the rich source of phytochemicals, particularly saponins. Recently, the plant has gained a well-established market for being the herbal alternative to 'Viagra' without any side-effects. Under the trade name 'Nai Chetna', the state government of Gujarat, India, has launched a novel potency drug from *C. borivilianum* (Khanam et al., 2013). The impact of *C. borivilianum* on human well-being for its aphrodisiac, immunomodulator, adaptogen, and antidiabetic properties is widely appreciated; however, the main threat to the way of commercialization of effective drugs from this plant is the non-availability of high-quality raw materials in sufficient quantity of the tubers (Thakur et al., 2009). The increased commercial exploitation of *C. borivilianum* and the low productivity of this endangered plant has raised concern over its conservation (Khanam et al., 2013). The economic importance of species of *Chlorophytum* has led to their uncontrolled harvesting and habitat destruction (Patil, 2016). The largest global markets for Safed musli are China, France, Germany, Italy, Japan, Spain, the UK, and the USA. Japan has the world's highest per capita consumption of botanical medicines (Giribabu, 2014). On average, 4 to 6 tonnes fresh weight of musli can be harvested from *C. borivilianum* per hectare and, after drying to 20 % (0.8-1.2 tonnes), dry musli is finally obtained. The market value of dry musli is around 37 USD per kg in the Indigenous Indian market and more than 68 USD per kg in the international market.

11.6 CONCLUSION

C. borivilianum is a multi-therapeutic plant which is well known for its tubers. The saponin components within the plant are related to various medicinal properties. This chapter presented an overview of the antioxidant, anti-inflammatory, anticancer, antiviral, larvicidal, and many other activities of this valuable plant. This plant is considered to be a precious gift of nature within the Ayurvedic therapeutic system and is used in numerous Ayurvedic formulations. It is used primarily for its aphrodisiac and immunomodulatory property. Moreover, the plant needs to be researched more to identify its pharmacological activities. In addition, the plant holds significant economic value and is used as a health-promoting drug. However, overexploitation, deforestation, low productivity, and insufficient knowledge of its production are factors responsible for the species being included in the Red Data Book as a critically endangered species. Therefore, conservation strategies and innovative plans should be aligned to protect the species from becoming extinct.

REFERENCES

Acharya, D., Mitaine-Offer, A., Kaushik, N., Miyamoto, T., Paululat, T., Mirjolet, J.F., Duchamp, O. & Lacaille-Dubois, M.A. (2009) Cytotoxic spirostane-type saponins from the roots of chlorophytum borivilianum. *Journal of Natural Products*, 72(1), 177–181.

Abate, L., Tadesse, M.G., Bachheti, A. & Bachheti, R.K. (2022) Traditional and phytochemical bases of herbs, shrubs, climbers, and trees from Ethiopia for their anticancer response. *BioMed Research International*. https://doi.org/10.1155/2022/1589877.

Abate, L., Bachheti, A., Bachheti, R.K. & Husen, A. (2021) Antibacterial properties of medicinal plants: Recent trends, progress, and challenges. In: *Traditional herbal therapy for the human immune system*, CRC Press, Taylor & Francis, Florida, pp. 13–54.

Asfaw, T.B., Esho, T.B. Bachheti, A., Bachheti, R.K. & Pandey, D.P. & Husen, A. (2022) Exploring important herbs, shrubs, and trees for their traditional knowledge, chemical derivatives, and potential benefits. In: *Herbs, shrubs, and trees of potential medicinal benefits*, CRC Press, Taylor & Francis, Florida, pp. 1–26.

Ashraf, M, Farshad, A., Maheran, A., Kemat, N. & Ismail, I. (2014) Effect of cytokinin types, concentrations and their interactions on in vitro shoot regeneration of Chlorophytum borivilianum Sant. & Fernandez. *Electronic Journal of Biotechnology*, 17(6), 275–279.

Beutler, J.A. (2019) Natural products as a foundation for drug discovery. *Current Protocols in Pharmacology*, 86(1), e67–e67.

Beshah, F, Hunde, Y., Getachew, M., Bachheti, R.K., Husen, A. & Bachheti, A. (2020) Ethnopharmacological, phytochemistry and other potential applications of *Dodonaea* genus: A comprehensive review. *Current Research in Biotechnology*, 2, 103–119.

Chakraborthy, G. (2008a) Antidiabetic and antihyperlipidaemic effect of alcoholic extract of *Chlorophytum borivilianum* roots in alloxan induced diabetic albino rats. *Journal of Pharmacy Research*, 1(1), 29–33.

Chakraborthy, G. (2008) Antioxidant activity of the successive extracts of *Chlorophytum borivilianum* leaves. *Asian Journal of Chemistry*, 20, 5782–5784.

Chakraborthy, G. (2008b) Antioxidant activity of the successive extracts of Chlorophytum borivilianum leaves. *Asian Journal of Chemistry*, 20(7), 5782–5784.

Chauhan, R., Quraishi, A., Jadhav, S.K. & Keshavkant, S. (2016) A comprehensive review on pharmacological properties and biotechnological aspects of Genus Chlorophytum. *Acta Physiologiae Plantarum*, 38(5), 116.

Chen, Y. & Kirchmair, J. (2020) Cheminformatics in natural product-based drug discovery. *Molecular Informatics*, 39(12), e2000171–e2000171.

Das, S., Singhal, S., Kumar, N., Rao, C.M., Sumalatha, S., Dave, J., Dave, R. & Nandakumar, K. (2016) Standardised extract of safed musli (Chlorophytum borivilianum) increases aphrodisiac potential besides being safe in male Wistar rats. *Andrologia*, 48(10), 1236–1243.

Deore, S.L. & Khadabadi, S.S. (2009a) Larvicidal activity of the saponin fractions of Chlorophytum borivilianum santapau and Fernandes. *Journal of Entomology and Nematology*, 1(5), 064–066.

Deore, S.L. & Khadabadi, S.S. (2010) Antiproliferative activity of saponin fractions of Chlorophytum borivilianum. *Pharmacognosy Journal*, 2(16), 33–37.

Deore, S.L. & Khadabadi, S.S. (2008) Antiinflammatory and antioxidant activity of Chlorophytum borivilianum root extracts. *Asian Journal of Chemistry*, 20, 983–986.

Deore, S.L. & Khadabadi, S.S. (2009a) Larvicidal activity of the saponin fractions of Chlorophytum borivilianum santapau and Fernandes. *Journal of Entomology and Nematology*, 1(5), 064–066.

Deore, S.L. & Khadabadi, S.S. (2009b) A study of hypocholesterolemic effects of Chlorophytum borivilianum tubers in rats. *Pharmacol Online*, 1, 398–403.

Deore, S.L. & Khadabadi, S.S. (2010) Isolation and characterization of phytoconstituents from Chlorophytum borivilianum. *Pharmacognosy Research*, 2(6), 343–349.

Giribabu, N., Kumar, K.E., Rekha, S.S., Muniandy, S. & Salleh, N. (2014) Chlorophytum borivilianum (Safed Musli) root extract prevents impairment in characteristics and elevation of oxidative stress in sperm of streptozotocin-induced adult male diabetic Wistar rats. *BMC Complementary and Alternative Medicine*, 14(1), 291.

Goyal, M. (2017) Use of ethnomedicinal plants for prophylaxis and management of postpartum complications among the Marwari community of Jodhpur District of Rajasthan. *Food Quality and Safety*, 1(3), 203–210.

Grover, M. (2021) *Chlorophytum borivilianum* (Safed Musli): Nature's Wonder Gift. *Journal of Ayurveda and Integrated Medical Sciences*, 6(4), 93–102.

Joo, Y.E. (2014) Natural product-derived drugs for the treatment of inflammatory bowel diseases. *Intestinal Research*, 12(2), 103–109.

Katoch, M., Kumar, R., Pal, S. & Ahuja, v (2010) Identification of Chlorophytum species (*C. borivilianum, C. arundinaceum, C. laxum, C. capense and C. comosum*) using molecular markers. *Industrial Crops and Products*, 32(3), 389–393.

Kaur, R. & Kaur, S. (2020) Protective efficacy of *Chlorophytum borivilianum* root extract against murine visceral leishmaniasis by immunomodulating the host responses. *Journal of Ayurveda and Integrative Medicine*, 11(1), 53–61.

Kenjale, R., Shah, R. & Sathaye, S. (2008) Effects of *Chlorophytum borivilianum* on sexual behaviour and sperm count in male rats. *Phytotherapy Research*, 22(6), 796–801.

Kenjale, R.D., Shah, R.K. & Sathaye, S.S. (2007) Anti-stress and antioxidant effects of roots of *Chlorophytum borivilianum* (Santa Pau & Fernandes), *Indian J Exp Biol*. 45(11):974–979.

Khanam, Z., Singh, O., Singh, R. & Bhat, I.U.H. (2013) Safed musli (*Chlorophytum borivilianum*): A review of its botany, ethnopharmacology and phytochemistry. *Journal of Ethnopharmacology*, 150(2), 421–441.

Kumar, M., Meena, P., Verma, S., Kumar, M. & Kumar, A. (2010) Anti-tumour, anti-mutagenic and chemomodulatory potential of *Chlorophytum borivilianum. Asian Pacific Journal of Cancer Prevention*, 11(2), 327–334.

Kumar, M.P.M., Verma, S., Kumar, M. & Kumar, A. (2010) Anti-tumour, anti-mutagenic and chemomodulatory potential of *Chlorophytum borivilianum. Asian Pacific Journal of Cancer Prevention: APJCP*, 11(2), 327–334.

Kumar, S, Kalra, S., Kumar, S., Kaur, J. & Singh, K. (2012) Differentially expressed transcripts from leaf and root tissue of *Chlorophytum borivilianum*: A plant with high medicinal value. *Gene*, 511(1), 79–87.

Lande, A.A., Ambavade, S.D., Uma, U.S., Adkar, P., Ambavade, P.D. & Waghamare, A.B. (2015) Saponins isolated from roots of *Chlorophytum borivilianum* reduce acute and chronic inflammation and histone deacetylase. *Journal of Integrative Medicine*, 13(1), 25–33.

McCarthy, C.G., Alleman, R.J., Bell, Z.W. & Bloomer, R.J. (2012) A dietary supplement containing *Chlorophytum borivilianum* and velvet bean improves sleep quality in men and women. *Integrative Medicine Insights*, 7, 7–14.

Nakasha, J.J., Sinniah, U.R., Puteh, A.B. & Kumara, M. (2017) Influence of tuber weight and cutting on growth and yield of safed musli (*Chlorophytum borivilianum*). *Archives of Agronomy and Soil Science*, 63(5), 619–625.

Narasimhan, S., Govindarajan, R., Madhavan, V., Thakur, M., Dixit, V.K., Mehrotra, S., Madhusudanan, K.P. (2006) Action of (2→1)Fructo-oligopolysaccharide Fraction of *Chlorophytum borivilianum* against streptozotocin-induced oxidative stress. *Planta Medica*, 72(15), 1421–(1424). En.

Paine, M.F. (2020) Natural products: Experimental approaches to elucidate disposition mechanisms and predict pharmacokinetic drug interactions. *Drug Metabolism and Disposition*, 48(10), 956–962.

Panda, S.K., Das, G. & Tripathy, N.K. (2011) Studies on antiulcer activity of root tubers of *Chlorophytum borivilianum* santapau & Fernandes. *International Journal of Pharmaceutical Sciences Review and Research*, 9, 65–68.

Patil, S.M., Rane, N.R., Gholave, A.R., Yadav, S.R., Jyoti, J.P. & Sanjay, P. (2016) Study of molecular genetic diversity and evolutionary history of medicinally important endangered genus Chlorophytum using DNA barcodes. *Biochemical Systematics and Ecology*, 65, 245–252.

Prashar RaAK. (1995) Chemopreventive action of Ocimum sanctum on 2, 12-dimethylbenz (a) anthracene DMBA-induced papillomagenesis in the skin of mice. *International Journal of Pharmacognosy*, 33(3), 181–187.

Samaranayake, N. (2020) Chapter 2 – Leishmaniasis. In: Misra, G. & Srivastava, V. (Eds.) *Molecular advancements in tropical diseases drug discovery*. Academic Press, Cambridge University, pp. 21–46.

Sharma, G. & Kumar, M. (2012) Antioxidant and modulatory role of *Chlorophytum borivilianum* against arsenic-induced testicular impairment. *Journal of Environmental Sciences*, 24(12), 2159–2165.

Sharma, P., Singh, V., Maurya, S.K., Kamal, M. & Poddar, N. (2020) Antimicrobial and antifungal properties of leaves to root extracts and saponin fractions of *Chlorophytum borivilianum. Current Bioactive Compounds*, 9–18.

Sharma, S.K. & Kumar, M. (2011) Hepatoprotective effect of Chlorophytum borivilianum root extract against arsenic intoxication. *Pharmacologyonline*, 3, 1021–1032.

Singh BaSY. (2013) Antiviral and cytokine modulating potential of *Chlorophytum borivilianum* hot aqueous extract. *Cytokine*, 63(3), 301.

Singh, D., Pokhriyal, B., Joshi, Y. and Kadam, V. (2012) Phytopharmacological Aspects of *Chlorophytum borivilianum* (Safed Musli): A Review. *International Journal of Research in Pharmacy and Chemistry*, 2 (3), 853–859.

Singh, N.K., Yadav, A.K., Sirohi, P. Rani, M. Saraswat, S. Singh, M.P., Mani, A. & Srivastava, S. (2018) *Anticancer Activity of Herbal Medicine: Mechanism of Action. Anticancer Plants: Mechanisms and Molecular Interactions*, Springer Nature, Singapore, pp. 165–196.

Thakur, G., Bag, M., Sanodiya, B., Debnath, M., Joseph, A. Bhadauriya, P. Prasad, G. & Bisen, P. (2009) *Chlorophytum borivilianum*: A white gold for biopharmaceuticals and nutraceuticals. *Current Pharmaceutical Biotechnology*, 10, 650–666.

Thakur, M., Bhargava, S. & Dixit, V.K. (2007) Immunomodulatory Activity of Chlorophytum borivilianum Sant. *F. Evidence-Based Complementary and Alternative Medicine*, 4, 362693.

Thakur, M., Bhargava, S. & Dixit, V. (2008a) Immunomodulatory Activity of *Chlorophytum borivilianum* Sant. *F. Evidence-Based Complementary and Alternative Medicine: eCAM*, 4, 419–423.

Thakur, M., Bhargava, S. & Dixit, V. (2008b) Immunomodulatory Activity of *Chlorophytum borivilianum* Sant. *F. Evidence-Based Complementary and Alternative Medicine: eCAM*, 4(4), 419–423.

Tripathi, N., Saini, N., Nair, P. & Tiwari, S. (2012) Lack of genetic diversity of a critically endangered important medicinal plant Chlorophytum borivilianum in Central India revealed by AFLP markers. *Physiology and Molecular Biology of Plants: An International Journal of Functional Plant Biology*, 18(2), 161–167.

Truax, N.J. & Romo, D. (2020) Bridging the gap between natural product synthesis and drug discovery. *Natural Product Reports*, 37(11), 1436–1453.

Veeresham C. (2012) Natural products derived from plants as a source of drugs. *Journal of Advanced Pharmaceutical Technology and Research*, 3(4), 200–201.

12 A Comprehensive Review on the Medicinal Use of *Piper methysticum* (Kava)

Versha Parcha, Pankaj Bhandari, Sukanya Chhetri, Uday Kumar, and Deepak Kumar

12.1 INTRODUCTION

Piper methysticum (G. Forst), of the family Piperaceae, is an evergreen bush (Figures 12.1, 12.2), popularly referred to as kava or awa, meaning 'bitter', which is a favourable edible medicinal herb, indigenous to Oceania (Stevinson et al., 2002). It has been grown all over the areas of the islands of the Pacific Ocean, Micronesia, Polynesia, and Melanesia (Figure 12.3) and is considered to be by far the most important medicinal plant. Kava is used as a muscle relaxant, analgesic, and anti-anxiety agent (Bilia et al., 2002), for reducing weight, counteracting fatigue, nervousness, and insomnia. In earlier times, the conventional use of kava was ubiquitous, and it was consumed as a beverage, either as an anxiolytic agent or a nutritional supplement; these days, it is more likely to be consumed in kava bars as a recreational beverage.

In ancient times, it is said that the principal users of awa were the chiefs and priests but it has now become widely used. Particularly in more recent times, those who conduct tiresome, repetitive tasks value the characteristics of kava as a source of respite from stress and exhaustion of the muscles, especially for weary farmers, fishermen, hunters, and paddlers. Kawa is consumed by spiritual leaders during rituals such as canoe race-meets, a social rite undertaken to express appreciation to the divine both before and after events. In Fijian, Samoan, and Tongan societies, the conventional herbal drink made of kava was used as a ritual beverage and as a medicine, playing a key role in ceremonies performed to unite participants, greet visitors, and authenticate their community identities. In the South Pacific Islands, kawa is used as a pious herb used in prayer, as well as being particularly consumed for pleasure. In recreational and social gatherings, kava can be found as a popular drink for high-profile heads and elders, a welcome drink for visitors, for the commencement and final completion of an event or of a piece of work, to substantiate prestige, mark childbirth, marriage ceremony, and death, act as a stress buster, etc. To celebrate the naming of children, birth of male children, traditional hula of young girls, and chanting, kava is served as the chief drink in some other parts of Pacific countries. A prayer of gratitude is offered when awa cups are filled and gobbled up rather than sipped, leaving some liquid in the container to be poured upon the earth for paying thanks and gratitude to God. It is believed to be a sacred symbol and finds its vitality in religious libation instead of drinking – it is spread onto the ground in Tikopia. Kava is also taken to mark relationships and headship rites, for public redemption of offences. On Wallis Island, during the kava ceremony, official decisions are made and even enemies are reconciled and friendships are restored. Kava has a prominent role in community gatherings and can be visualized where you see Pope John Paul II and former first lady Mrs. Johnson drinking upon their visit to the Pacific. Not only serving and drinking, but kava is sometimes the only approach to mend broken relationships but kava root is presented to dignitaries as a symbol of salutation and harmony (Turner et al., 1986).

DOI: 10.1201/b22924-12

FIGURE 12.1 *Piper methysticum* plant.

FIGURE 12.2 (A) 'Roots' (rhizomes), (B) heart-shaped leaves, (C) black-coloured stem, (D) fruit, (E) small flowerbud of *Piper methysticum*.

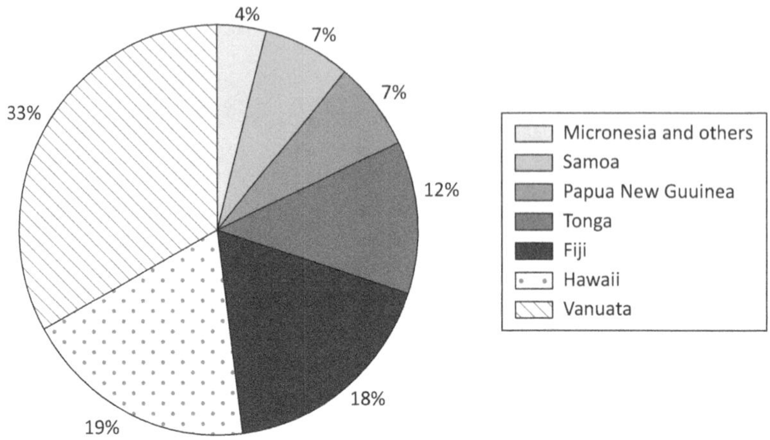

FIGURE 12.3 Botanical Distribution.

Over the past two decades in Western countries, kava has gained a lot of popularity, being promoted for various medicinal uses and is available in health food stores in the form of tablets, capsules, tinctures, and liquid extracts.

12.2 BOTANICAL DESCRIPTION

Botanical name: *Piper methysticum* [G. Forst]
Family: Piperaceae
Genus: *Piper*
Species: *methysticum*
Species authority: G. Forst

12.2.1 Vernacular Names

Kava; Kava-kava; Ava pepper shrub, Ava-ava; Ava root: Ava pepper, Awa; Gea; Gi; Grog; Intoxicating pepper; Kao; Fijian kava: Kava-kava: rhizome; Kava root; Kavapiper; Sakaua; Sakau; Kavapyrones; Kavarod; Pepe kava; Piperis Kavasporal forte; Kavekave; Kawa; Kawa kawa; Kawa pepper; Macropiper latifolium: Kawa Pfeffer; Kew; Malohu; Maluk; Maori kava; Meruk; Milik; Yagona; methystici rhizome; Rhizoma piperis methystici; Tonga; Yangona; Yaqona; Yongona

12.2.2 Botanical Illustration (Figure 12.3)

P. methysticum, an upright, sparsely branching, perennial shrub, is very hardy and grows up to nearly 12 feet high.

Roots: thick and softwood roots (actually, an underground stem or rhizome) when fresh, but harden later on. The colour of the rhizome varies from shades of white to dark yellow depending on the concentration of kavalactones in the light yellow-coloured exudate.–

Stems: are thick, erect, 1–3 cm in diameter, and succulent with heavily inflated nodes that differ in colour from greenish to blackish shades. The black awa is the rare one.

Leaves: are smooth, heart-shaped, light green in colour, and about 15 to 20 cm long.

Fruits: Fruits are rarely produced and are in the form of bone-seeded berries.

Flowers: Typically, 1.5 cm long, unisexual with no sepals, and having a male spike that is 3–9 cm high with two short stamens whereas the female spike bears flowers having a single basal ovule, with an ovary containing only one loculus chamber bearing a stigma at the top.

The plant spreads via the rhizome. The plant is usually harvested when it reaches a height of about 2–2.5 m. It needs to grow for at least 2–3 years to reach usable potency.

12.2.3 Part of Plant Used for Preparation of Kava

Almost all parts of the plants are used to prepare kava, i.e., the roots, the rhizomes, and sometimes the bark of the stems, but not the leaves. It is estimated that stem bark has a high kavalactone concentration (Coulter et al., 2007) so it has been used to produce kava. The possibility of cytotoxicity in the plant may be because of the higher concentration of alkaloids (Figure 12.4 and Figure 12.5).

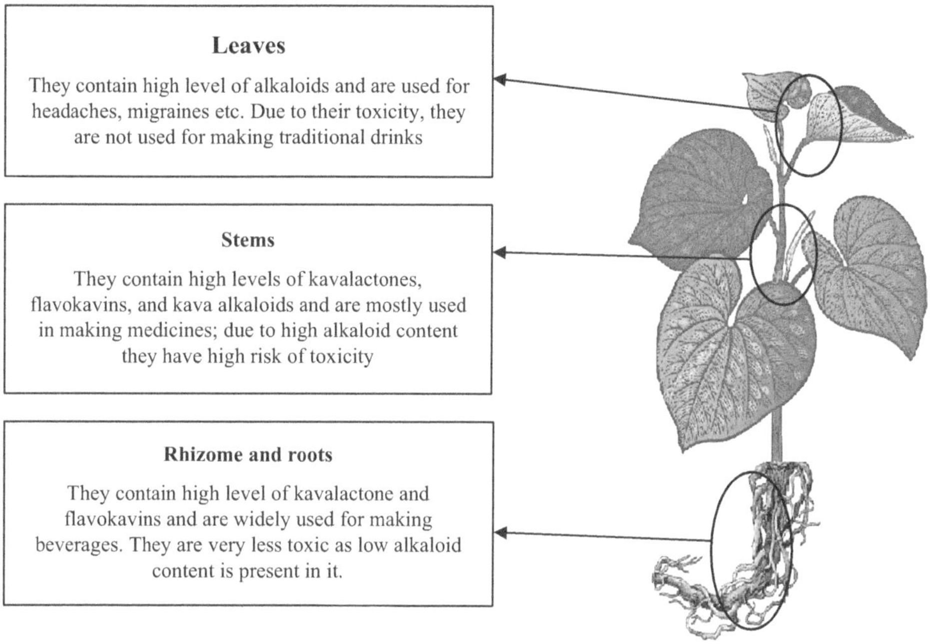

FIGURE 12.4 Importance of different parts of the kava plant.

(1) (2)

FIGURE 12.5 (1) Traditional preparation of kava, (2) kava as a nutritional supplement.

12.3 CHEMICAL CONSTITUENTS

Fresh kava rhizome consists of 80% water on average, whereas the dried rhizome consists of fibres, starch, proteins, water, sugars, minerals, and kavalactones at 20%, 4.3%, 3.6%, 12%, 3.2%, 3.2%, and 15%, respectively. The kavalactone component ranges from 3 to 20% of the rhizome dry weight, depending primarily on the plant's location and the variety. Fifty-six constituents have been reported from *P. methysticum*. These are divided into two groups, namely kavalactones and flavokavins. Table 12.1 summarizes the characteristics of each type of compound (Wang et al., 2021)

Kavalactones: Kavalactones are lipophilic, with a pyrone skeleton lactone, typically named as 4-methoxy-substituted 2-pyrones, and an aromatic styryl group or benzene ethyl substitution at position 6 (Bilia et al., 2004). It was estimated that about 29 different kavalactones have been isolated and identified. Few kavalactones were observed to be present in kava extracts using water, methanol, ethanol, chloroform, acetone, or hexane as solvent. The compounds were isolated and identified using chromatographic techniques like high-performance liquid chromatography (HPLC), gas chromatography-mass spectrometry (GC-MS), nuclear magnetic resonance (NMR), etc. Among all extracts, the acetone extract was found to contain the highest percentage of kavalactones.

Similar spectra and separation techniques on the rhizomes of kava helped in isolating multiple novel dimers of kavalactones, which are listed in Table 12.1.

Flavokavins: Similar studies, using methods of chromatographic separation, integrated with spectrum identification techniques on the roots of kava, helped in isolating and identifying another category named bis-kavalactones, listed in Table 12.1.

12.4 MEDICINAL USE

Kava is used to produce a calming medicated drink with proven medicinal effects, suitable for seeking relief from stress, to settle nerves or restlessness, simultaneously overcoming fatigue. Because of the presence of antidepressant components in kava, one can confront the 'blues' and become happy from an anxious state. Kava keeps the mind awake while the body relaxes, making it an excellent drink for treating illnesses such as headaches, migraines, and cramps. Kawa root can be used in a fresh or sun-dried form. Rhizomes of kava are washed, cleaned, cut into small pieces, and macerated with water in a blender to make an oral lipid-soluble suspension. Rhizomes are usually chewed by young people. Presently, powdered and packaged rhizome is also available on the market (Figure 12.6).

For the past 125 years, kava rhizomes have been used to treat severe gonorrhoea, fluor albus, vaginal infection, bedwetting, and other urogenital system disorders. Its local action is similar to pepper. Gonosan is used internally for gonorrhoea, which is kava exudate (20% [v/v]) in sandalwood oil. Because of its local anaesthetic properties, it is used for relieving pain and has an aphrodisiac effect. The capsules at a dose of 3 mg and two to four can be given at a time for antiseptic effects on urine. Being a strong diuretic, it is used to treat uric acid ailments, painful joints, respiratory tract sickness, and other ailments resulting from cardiac-related issues. (Grieve et al., 1931). Across the whole of Oceania, kava is used to promote calm, relaxation, and relief from sleep difficulties, as well as to combat weariness, unclog urinary tracts, ease asthma, and battle obesity. Taking kava is found to relieve pain in the head and convulsions, and to cure venereal disease and chlamydia. Kava also acts as a strength enhancer, relieves stomachache and also cures boils. In addition to rhizomes, some people use kava leaves for the treatment of different diseases. Another technique of treatment is to apply the masticated kava stump externally in addition to the conventional consumption of kava drink (Singh, 1992).

In Hawai'i, awa, when combined with lomi-lomi massage, serves as a calming treatment and induces calmness and sleep. Unless it is drunk in large quantities, it acts as an energy drink when feeling weak. Prolonged use of awa can adversely affect the skin and eyes. Children with an

TABLE 12.1
Different Chemical Constituents of *Piper methysticum*, Kavalactones [1–19] and Flavokavins [20–29]

1.

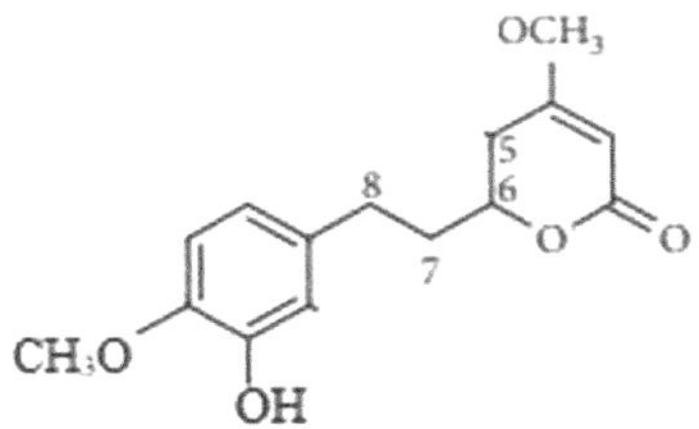

11-Hydroxy-12-methoxydihydrokavain

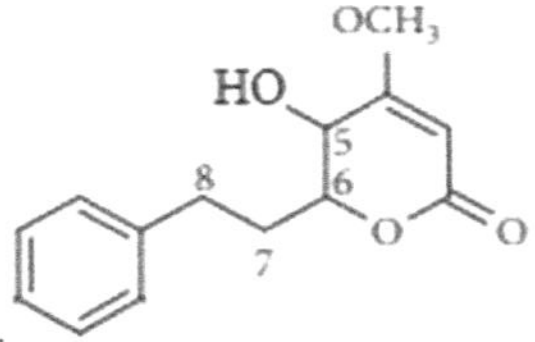

2.
7,8-Dihydro-5-hydroxykavain

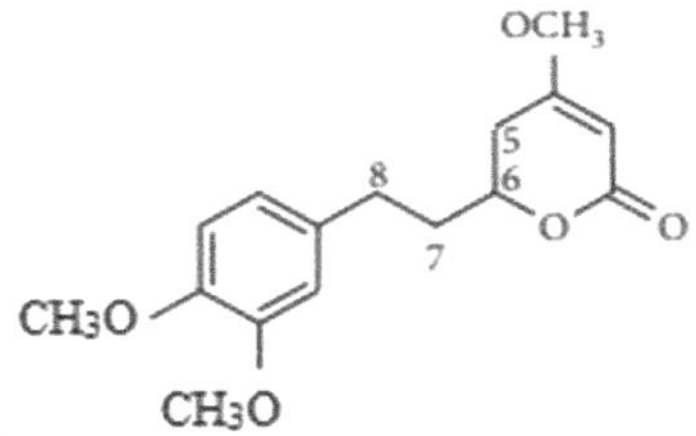

3.
11,12-Dimethoxydihydrokavain

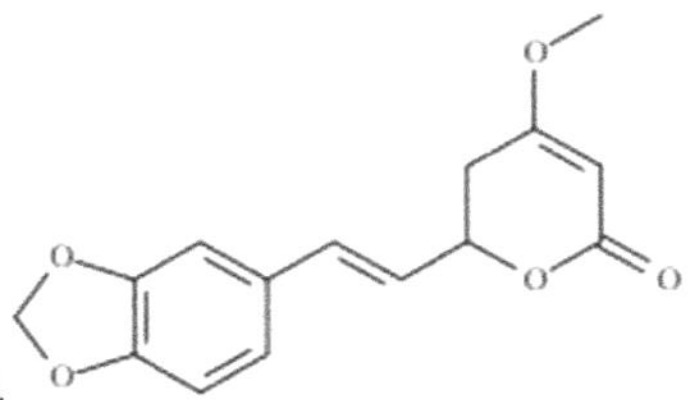

4.
Methysticin

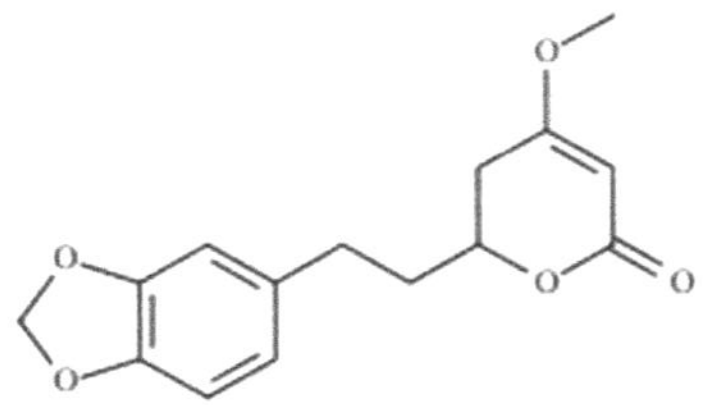

5.
Dihydromethysticin

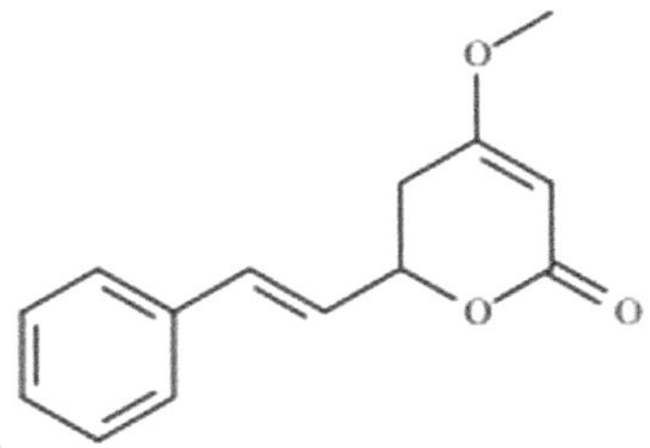

6.
Kavain

7.

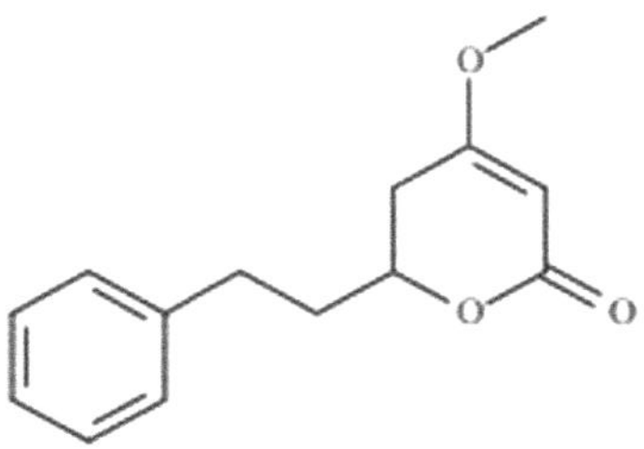

7,8-Dihydrokavain

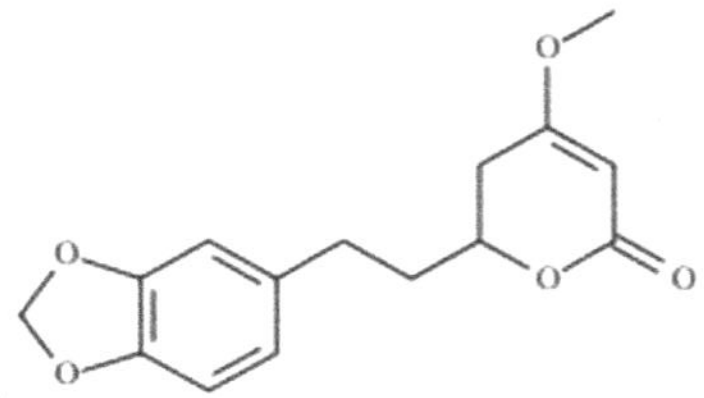

8.
1,2 anhydro 7,8-Dihydrokavain

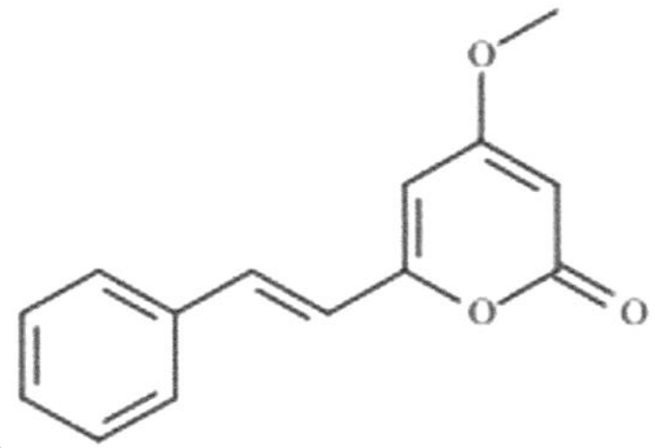

9.
Desmethoxy-yangonin

(Continued)

TABLE 12.1 (CONTINUED)
Different Chemical Constituents of *Piper methysticum*, Kavalactones [1–19] and Flavokavins [20–29]

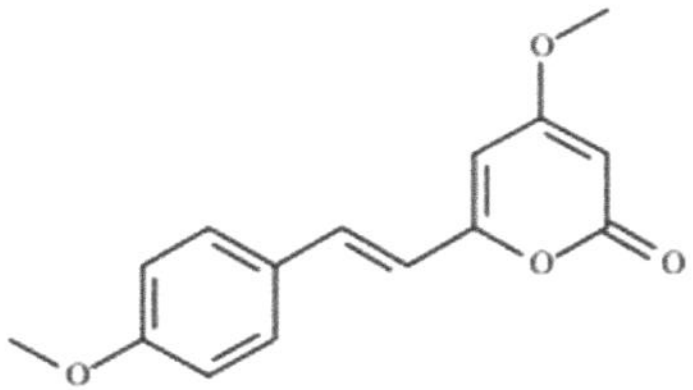
10.
Yangonin

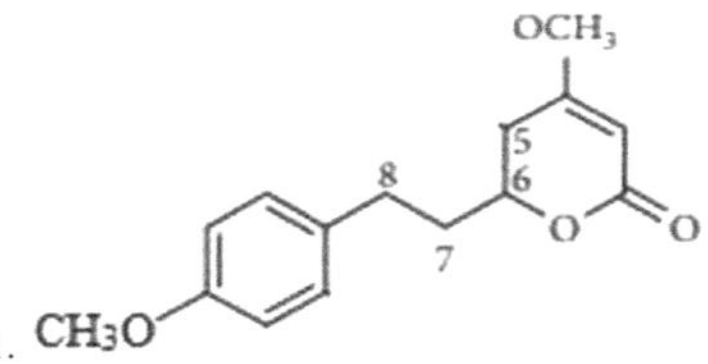

11.
5,6,7,8-Tetrahydro-yangonin

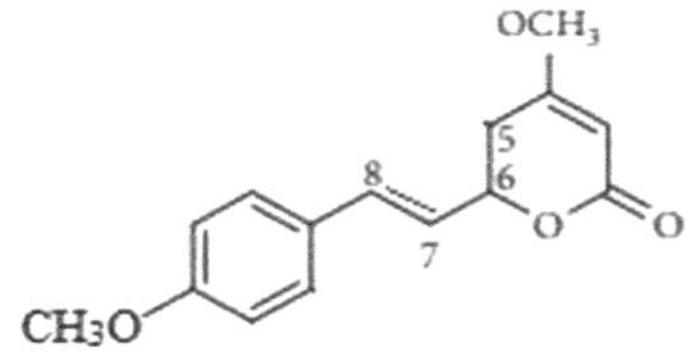

12.
5,6-Dihydro-yangonin

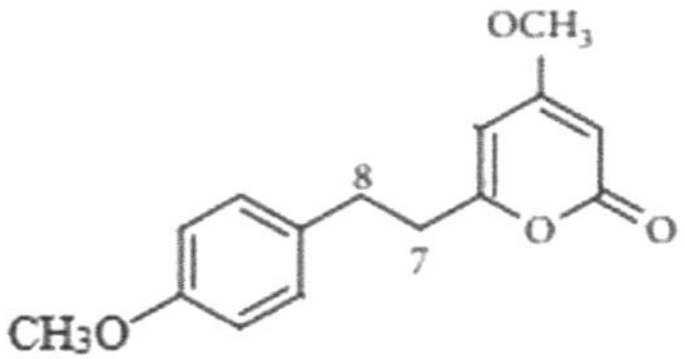

13.
7,8-Dihydro-yangonin

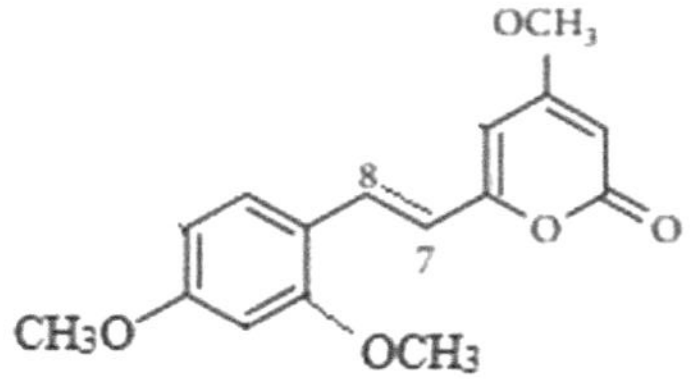

14.
10-Methoxy-yangonin

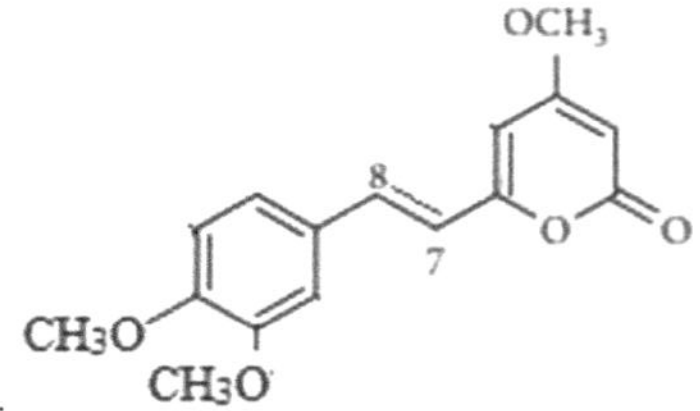

15.
11-Methoxy-yangonin

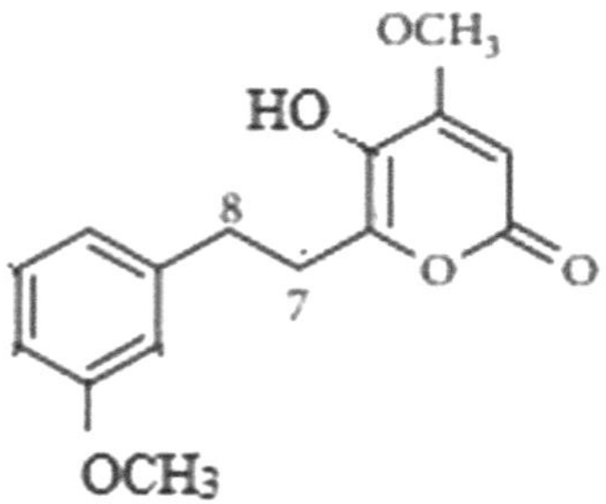

19.
5,6-Dihydro-5,6-dehydrokavain

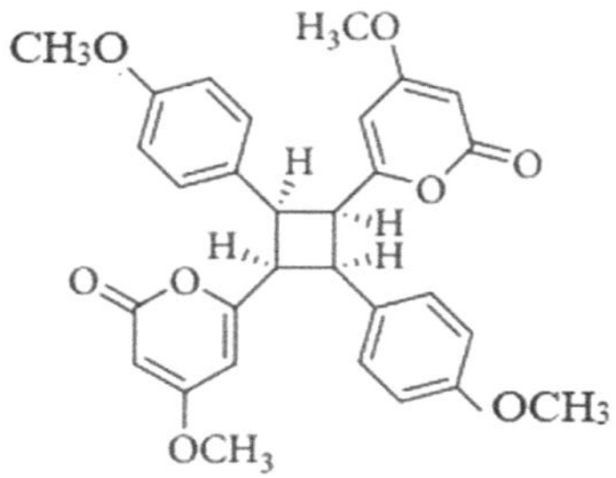

20.
Di-yangonins A

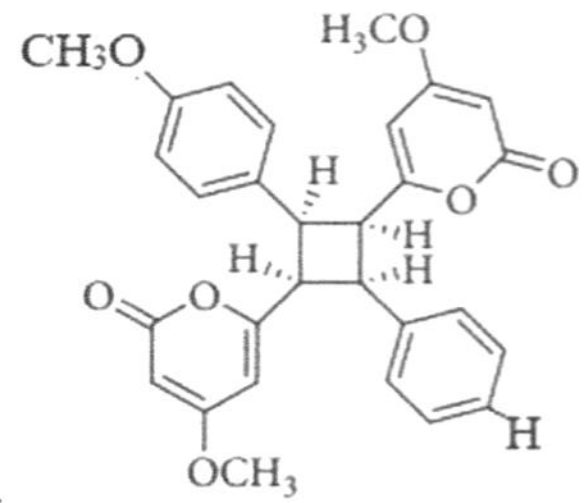

21.
Di-yangonins B

(*Continued*)

TABLE 12.1 (CONTINUED)
Different Chemical Constituents of *Piper methysticum*, Kavalactones [1–19] and Flavokavins [20–29]

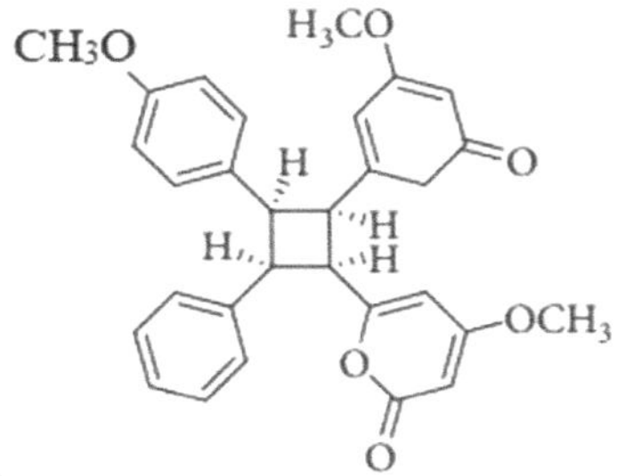

22.
Di-yangonins C

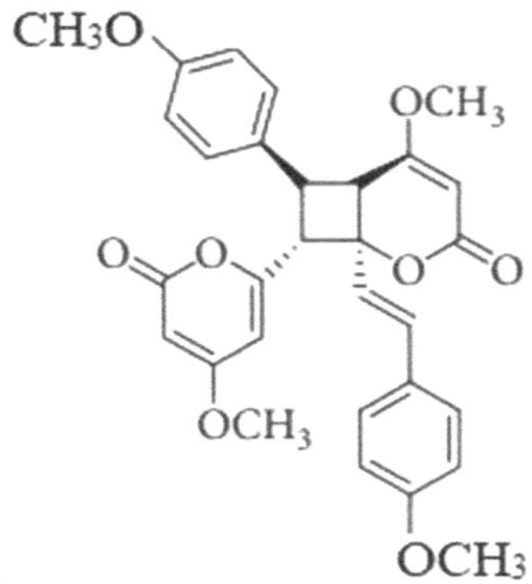

23.
Yangonin dimers A

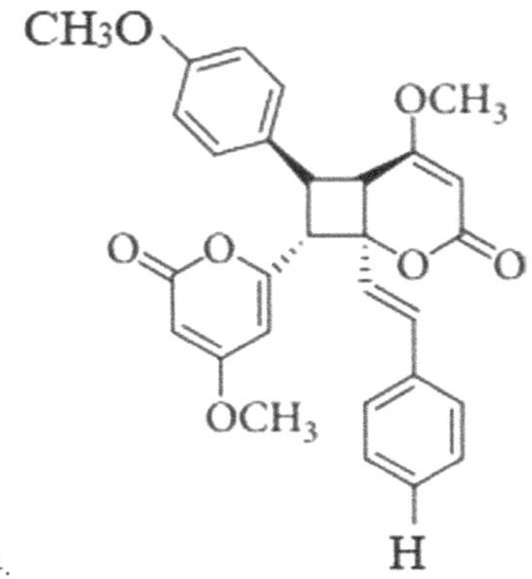

24.
Yangonin dimers B

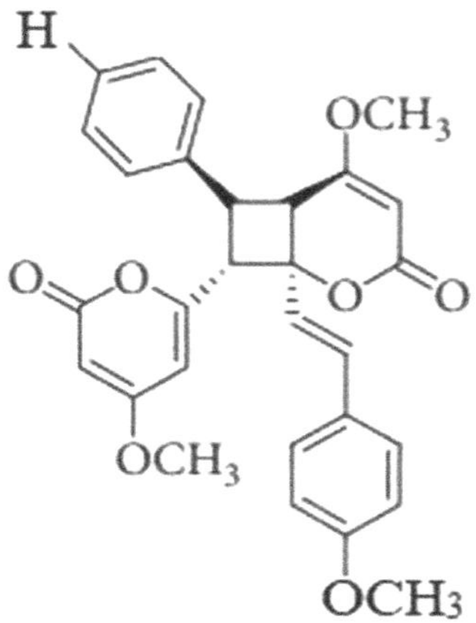

25.
Yangonin dimers C

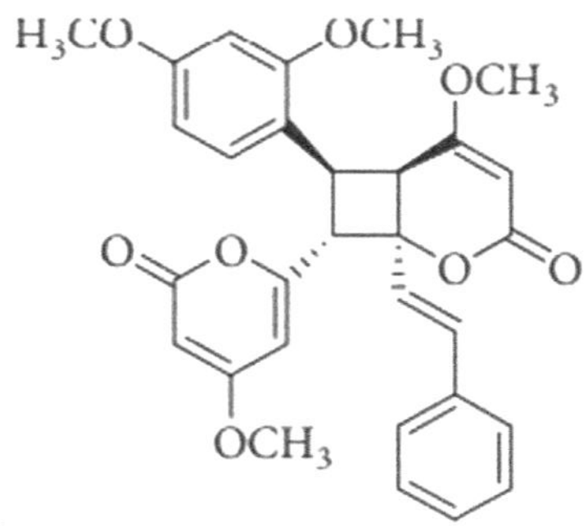

26.
Kavalactone A

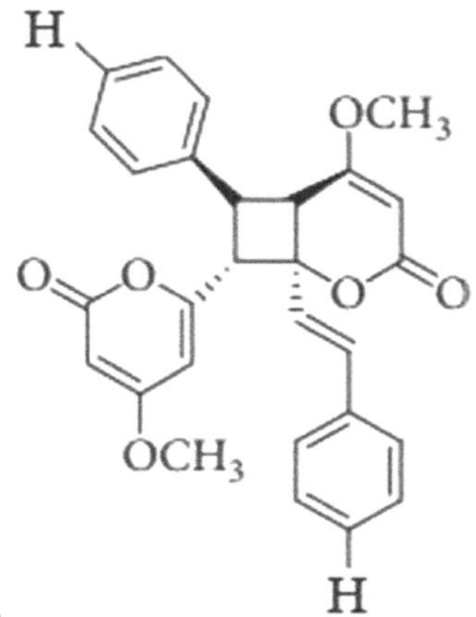

27.
Aniba-dimer A

(Continued)

TABLE 12.1 (CONTINUED)
Different Chemical Constituents of *Piper methysticum*, Kavalactones [1–19] and Flavokavins [20–29]

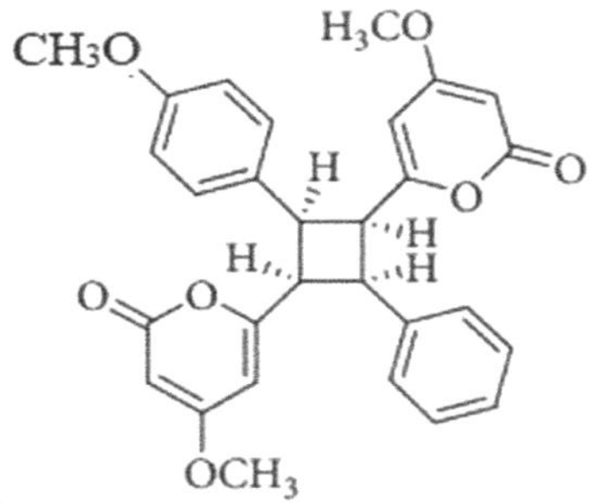 28. Rel-, *trans*-3-bis[6-(4-methoxy-2-pyronyl)]-cis-2, *trans*-4-diphenyl cyclobutane	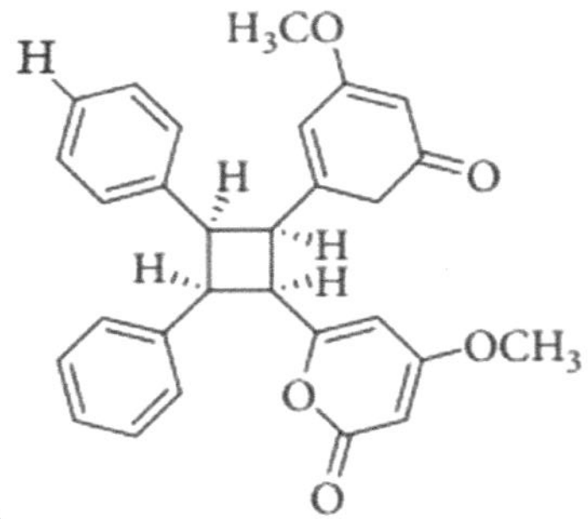 29. 6,6′-(3,4-Diphenylcyclobutane-1,2-diyl) bis (4-methoxy-2H-pyran-2-one)

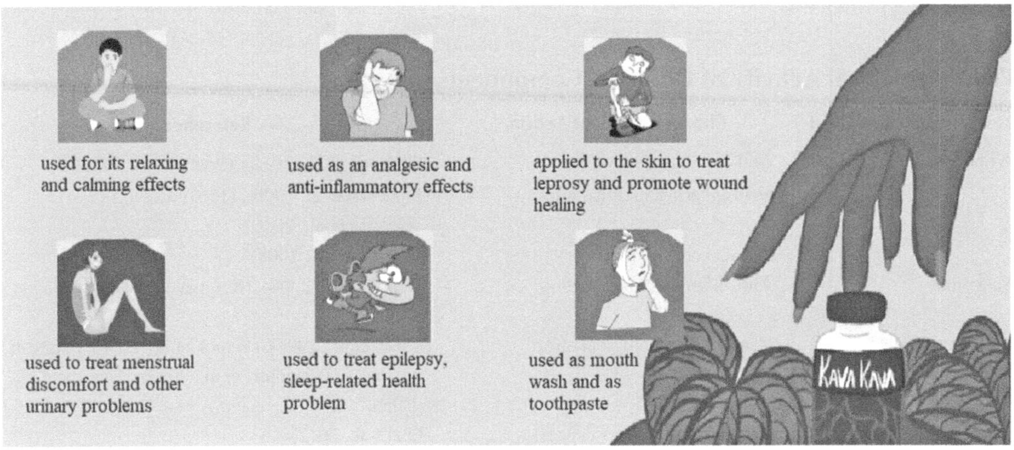

FIGURE 12.6 Common uses of *Piper methysticum* (kava).

unbalanced stomach are massaged with mixtures of kava and other plant ashes. Children are given buds to chew for general debility. The decoction of the whole plant and additional herbs is advised for lung and related issues. Additionally, kava combined with other herbs in coconut oil is used to treat urinating problems (Akana et al., 1922).

The rhizomes and roots of kava extracts are shown to possess numerous biological and pharmacological activities associated with them. Recent data on the anti-inflammatory pharmacological action, neurological function, and anticancer capabilities of kava and its components have been compiled in Table 12.2 and Figures 12.6 and 12.7.

12.4.1 Effects of Kava on Inflammation

Inflammation has an important role to play in many infectious diseases but epizootiology and molecular studies do reveal their relationship to a broad range of non-infectious diseases as well, ranging from cancer to neurological disorders (Hunter et al., 2012). Studies on kava concerning inflammation have been partially developed due to its painkilling potential (Jamieson et al., 1990) since anti-inflammatory drugs are also used for relieving pain (Nuki et al., 1990). Some of the earliest demonstrations of kava and compounds isolated from it in reducing inflammation have been reported long ago when studies were carried out on different inflammatory models. From the study, it could be concluded that kava and the natural kavalactones reduced inflammation as oedema in rat paws induced by formalin, serotonin, carrageenan, or dextran & UV light (Meyer et al., 1965).

There were earlier reports where kava had been used to treat infections related to kidneys, ureters, and related organs, immunity disorders, and the respiratory system (Meyer et al., 1965). In past developments, kava has been reported to have been used to treat infections of the urinary system, immune system disorders, and respiratory asthma (Folmer et al., 2006). Investigations have been conducted on kavain, dihydrokavain, methysticin, and dihydromethysticin for tumour necrosis factor-alpha (TNF-α) production in human acute monocytic leukaemia cells and for their responsiveness to lipopolysaccharide (LPS)-induced fatality in lab mice. All these isolated kavalactones have the potential to block LPS-induced TNF production, and kavain was found to be the best in the series which could also protect C57 mice from lethality dose of LPS.

Flavokavains A, B, and C particularly suppressed the expression of the pro-inflammatory inducible nitric oxide synthase (iNOS) and cyclooxygenase-2 (COX-2), as well as the subsequent production of nitric oxide (NO), PGE-2, and pro-inflammatory cytokines, such as TNF-, and the interleukins IL-1 and IL-6. This resulted in a reduction in NO production stimulated by LPS in murine macrophages. In addition, flavokavains A and B have been revealed to have immune regulatory functions,

TABLE 12.2
Pharmacological Activity of Different Components of Kava

Kava Component Used	Pharmacological Action		Reference
Whole plant	Anti-inflammation effect		(Shimoda et al., 2012), (Wang et al., 2021)
	Neuropharmacological effect		(Jamieson et al. 1990), (Jussofie et al., 1994), (Hegazy et al., 2019)
	Anti-cancer (lung cancer)		(Johnson et al., 2008)
Yangonin	Anti-inflammation effect		(Wruck et al., 2008), (Kong et al., 2018), (Dong et al., 2019)
	Neuropharmacological effect		(Baum et al., 1998), (Wruck et al., 2008), (Ligresti et al., 2012), (Fragoulis et al., 2017), (Hegazy et al., 2019),
	Anti-cancer (bladder cancer)		(Liu et al., 2017)
Desmethoxy-yangonin	Anti-inflammation		(Chou et al., 2013)
	Neuropharmacological effect		(Baum et al., 1998), (Tzeng et al., 2015)
Kavain	Anti-inflammation		(Wruck et al., 2008), (Pollastri et al., 2009), (Zhang et al., 2012), (Terazawa et al., 2013), (Huck et al., 2018), (Huck et al., 2019)
	Neuropharmacological effect		(Jamieson et al., 1990), (Gleitz et al., 1996), (Magura et al., 1997), (Walden et al., 1997), (Baum et al., 1998), (Wruck et al., 2008), (Chua et al., 2016), (Fragoulis et al., 2017), (Hegazy et al., 2019)
Methysticin	Anti-inflammation		(Wruck et al., 2008), (Pollastri 2009), (Zhang et al., 2012), (Fragoulis et al., 2017)
	Neuropharmacological effect		(Jamieson et al., 1990), (Magura et al., 1997), (Wruck et al., 2008), (Fragoulis et al., 2017),
Dihydromethysticin	Neuropharmacological effect		(Jamieson et al., 1990), (Tzeng et al., 2015)
	Anticancer	Osteosarcoma	(Dai et al., 2015)
		Colorectal cancer	(Pan et al., 2020)
Flavokavain	Anti-inflammation		(Kwon et al., 2013), (Abu et al., 2015), (Yang 2019)
	Anti-cancer		
		Prostate cancer	(Li et al., 2015),
		Lung cancer	(Li et al., 2020)
		Breast cancer	(Nadiah et al., 2015), (Jandial et al., 2017)
		Bladder cancer	(Zou et al., 2005), (Tang et al., 2008)
		Osteosarcoma	(Zhang et al., 2018)
		Flavokavain B	
		Lung cancer	(An et al., 2012), (Hseu et al., 2019), (Hua et al., 2020)
		Breast cancer	(Abu et al., 2015), (Abu et al., 2016), (Abu et al., 2018)
		Colon cancer	(Kuo et al., 2010), (Malami et al., 2017)
		Gastric cancer	(Chang et al., 2017)
		Thyroid cancer	(He et al., 2018)
		Oral cancer	(Hseu et al., 2012)
		Cervical cancer	(Yeap et al., 2017)
		Acute myeloid leukaemia	(Lee et al., 2018)

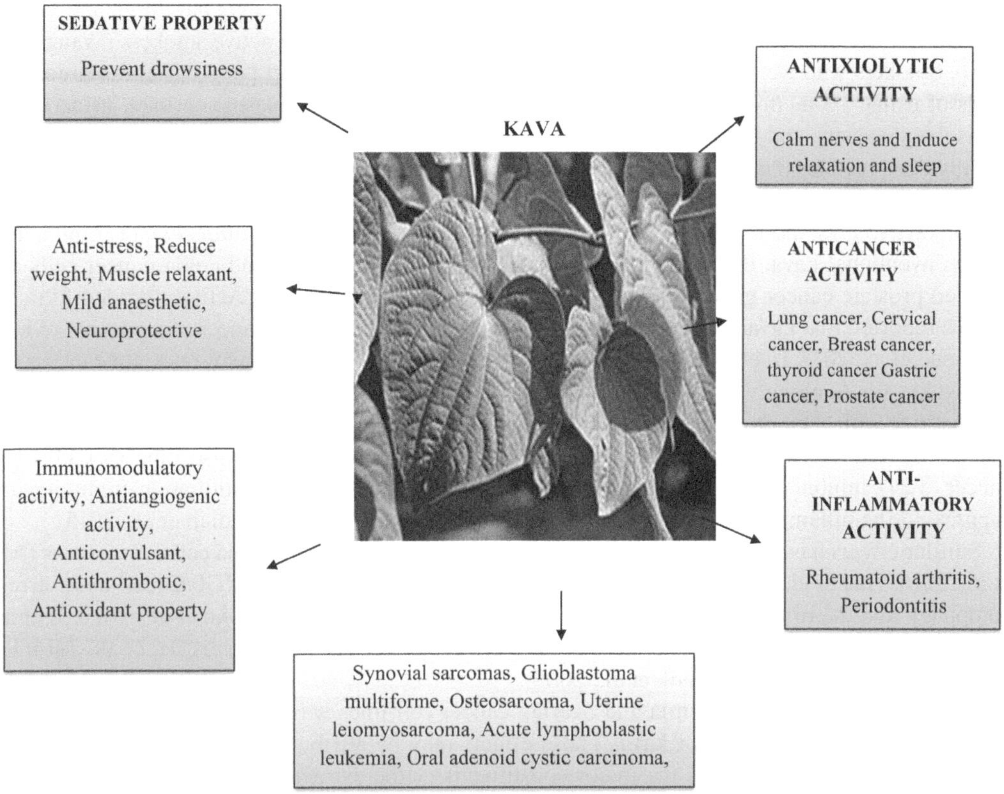

FIGURE 12.7 Detailed uses of *Piper methysticum* (kava).

like stimulated spleen cell proliferation and IL-2 and TNF-α secretion in sampled spleen cells as well as an additional increase in T-cell population subsets in mice used in cancer and immunology. Studies revealed that flavokavain B is more effective than flavokavain-A.

Studies showed that *P. methysticum* reduced the secretion of TNF-α and IL-17A significantly and upgraded the indications of Alzheimer's disease. Furthermore, the Nrf2 pathway in neuronal PC-12 and astroglial C6 cells were activated by methysticin, kavain, and yangonin.

On treatment with yangonin (20 mg/kg bodyweight), oestrogen-induced cholestasis was decreased because of activation of the farnesoid X receptor (FXR). Several studies stated that desmethoxyyangonin inhibited the rapid growth, T-cell activation, and activity of several in-vitro pro-inflammatory mediators. The studies on desmethoxyyangonin exhibited its protective effect on mice by regulating the IKK/NF-κB and Jak2/STAT3 inflammation signalling pathways.

12.4.2 Effect of Kava on Cancer

Epidemiological evidence-based research on kava's anti-neoplastic properties revealed that its use lessened the incidence of cancer. (Steiner et al., 2000). Countries with high kava consumption reported much lower incidences of cancer cases as compared to kava non-consuming countries. It can be further observed that the rate of cancer is higher in females than in males, the latter consuming more kava.

Kava components are reported to exhibit activity against the carcinogen-induced tumourigenesis model that kava can inhibit lung tumourigenesis, Johnson et al. (2008) reported that lung tumour multiplicity was reduced by an ethanolic kava extract at a dose of 10 mg/Kg. Later studies confirmed that treatment with a kavalactone-enriched fraction at doses of 5, 2.5, or 1.25 mg/g have high anti-tumour potential. A pilot clinical trial study was also carried out on twenty-one active smokers (Wang et al., 2020). Smokers who were fed kava capsules at the dose level of 225 mg per 3 capsules reduced the levels of urinary 3-methyladenine, urinary total nicotine equivalents, plasma cortisol, urinary total cortisol equivalents, and increased urinary excretion of total Nitrosonor Nicotine (NNA) which additionally demonstrates that kava has the impact of reducing lung cancer (Johnson et al., 2008).

Preventive effects on prostate cancer and colon cancer are also exhibited by the dietary consumption of kava (3 g/kg diet) (Yokoyama et al., 2016). Sea hibiscus (*Hibiscus tiliaceus*), in combination with conventional kava, displayed anti-cancer potential against breast and colon cancer cells and reduced prostate cancer growth in association with enhanced androgen. Activity-guided filtration studies reported that kavalactones and flavokavains are potentially anti-cancer components of kava (Einbond et al., 2017; Li et al., 2020); inhibition of cancer progression in taxol-resistant cancer cells by flavokavain-A, along with its ability to enhance antitumour immunity (Nadiah et al., 2015) and to synergize with chemotherapies, were also reported to induce G2/M arrest in a panel of cancer cell lines along with breast cancer cell lines (Tang et al., 2008). Flavokavain-A also suppressed bladder cancer (T24), inhibited lung metastasis of osteosarcoma cells, enhanced antitumour immunity, and suppressed the inflammatory process in breast cancer-challenged mice (Nadiah et al., 2015).

Similar effects have been reported for flavokavain B on several carcinoma cells, metastasis (Abu et al., 2016; Hua et al., 2020), angiogenesis, and to induce apoptosis, G2/MG0/G1cell cycle arrest, autophagy and the immune response (Kuo et al., 2010; Sakai et al., 2012; An et al., 2012; Ji et al., 2013; Rossette et al., 2017). Several kavalactones were reported to be bioactive (Dai et al., 2015; Liu et al., 2017; Pan et al., 2020; Shaik et al., 2009).

High activity against leukaemia and ovarian cancer cell lines was exhibited by crude dichloromethane and hexane extracts of kava. These effects may be attributed because *cis*-yangonin and flavokavain B were detected as the major constituents of the extracts. From the structure-activity standpoint, the disproportionately high activity of one of the isomeric forms of yangonin is exciting. *Cis*-yangonin is more potent than its structural isomer and its activity is closely comparable to that of the dimethoxy-yangonine compound reported earlier (Sotheeswaran et al., 2002) to block tumour necrosis factor-alpha (TNF-α) release. In summary, studies using both animal and cell culture models have shown that kava and several of its constituents have chemopreventive and anticancer properties. To establish the relevance of the 'mechanisms', detailed studies of ADME (Absorption, distribution diffusion, metabolism excretion) and pharmacokinetic data of kava and along its individual components, studies need to be carried out with *in-vitro* models and cancer-induced models which will render support and provide evidence for human clinical trials. Very prominent activity against malignant ovarian and leukaemia cancer cell lines was exhibited by crude dichloromethane and hexane extracts of kava. These effects may be attributed to the compounds *cis*-yangonin and flavokavain B present as major constituents of the extracts.

12.4.3 Effect of Kava on Anxiety

Kava use is very well known for its psychoactive effects which popularized it as a non-prescription treatment for multiple neurological conditions (Singh et al., 2002; Volgin et al., 2020). Many pharmacological actions, such as anxiolytic, sedative, anticonvulsant, local anaesthetic, convulsive, anti-inflammatory, and analgesic activities have been confirmed, but the mechanism of these actions is unknown.

Kavalactones, bioactive constituents of kava, are sold in the market today as herb-based medicines and as a dietary nutrient for anxiety disorders and With regard to response times and cognitive processing, kava has been shown to have both the least negative effects and the greatest potential for positive effects. In addition, it reduces nervousness without losing mental balance. At therapeutic doses, kava has been established as a powerful well-recognized, and non-addictive drink but long-term consumption of very high doses can have potential side-effects, such as headaches, weakness, restlessness, and tremors (Cawte et al., 1986; Warnecke et al., 1990, 1991). Kava, a popular drink made from *P. methysticum* in the Pacific Island nations, may be used as a supplement to treat anxiety disorders. As discussed above, kavalactone (α-pyrone) has been identified as the ingredient responsible for the reported anxiolytic and sedative effects of kava. There is evidence that kavalactone may mediate anxiolytic and sedative effects through positive regulation of gamma-aminobutyric acid A (GABA A) receptor function, which acts outside the classical benzodiazepine binding site. Individual kava lactones have been shown to increase the binding of bicuculline methochloride and muscimol to brain membranes without influencing the binding of diazepam or flunitrazepam, demonstrating that they interact with the agonist/competitive antagonist binding sites on GABA A receptors. Enhanced muscimol binding indicates that local selectivity is highest in the amygdala and medulla oblongata, with minimal effect on the cerebellum and frontal lobe, and suggesting that kavalactone may have subtype specificity for specific GABA A receptors. Functional studies on recombinant α1β2γ2L-GABA A receptors have shown that kavalactone acts as a positive modulator, but such studies need to be extended to explore different combinations of subunits. However, the latest study on the mechanisms of action of isolated kavalactones has revealed other neuroactive effects which are not directly related to the GABA receptor function. Thus, kava may be useful in the medication of a variety of degenerative illnesses and nervous system conditions. One of the neuroprotective effects of kavalactones is mediated by the P38/nuclear factor-kappaB (nuclear factor-κB)/cyclooxygenase-2 (COX-2) signalling pathway. P38 mitogen-activated protein kinase (MAPK) is considered to be a stress-regulating kinase. Kavalactone activates p38, which then activates the downstream kinase, with phosphorylation and degradation of IB inhibitor of NF-kB, leading to nuclear translocation of NF-κB. One of the pro-inflammatory reactions downstream of NF-B binding to DNA is the activation of cyclooxygenase enzymes, particularly COX-2. COX converts arachidonic acid into prostaglandins, and pharmacological inhibition of COX can alleviate inflammation and pain symptoms. It has been found that dihydro-5,6-dehydrokavain (DDK) and 5,6-dehydrokavain (or desmethoxy-yangonin) inhibit H_2O_2-induced P38 phosphorylation in this pathway's signalling molecule (Rao et al., 2014). Methysticin (Shaik et al., 2009), a kavain derivative with a ring B with an additional cyclopentane ring on top, inhibited NF-B activity in cells modelled by a luciferase construct. In a cell-free system, dihydrokawain and yangonin were shown to inhibit COX-2 activity (Wu et al., 2002). Other than anxiety, kava finds use in various psychopharmacological advantages, e.g., epilepsy in (Cairney et al., 2002). Kava reduces cravings associated with substance abuse, such as alcohol, tobacco, cocaine, and heroin, as reported by Steiner et al. (2001). New Zealand has used kava as part of addiction rehabilitation programmes, with a 90% success rate (Aporosa, 2019). Kava consumption thus appears to be clinically non-addictive (Bilia et al., 2002), coupled with showing improvements in recognition memory tasks and increased accuracy in performance with respect to visual attention and working memory tasks (Münte et al., 2004; Sarris et al., 2011). No safety issues were reported with anxiolytic and anti-depressant activities (Sarris et al., 2009). Sleep spindle, sleep-inducing and -improving effects and increase in deep sleep time have been reported as well (Emser et al., 1991). Treatment with a dose of 200 mg kava/day achieved excellent results in patients with non-psychotic anxiety-related sleep disorders (Lehrl, 2004). A thorough characterization of kava composition is required before establishing it as a drug candidate.

12.4.4 Effect of Kava on Menstrual Problems and Menstrual Wellness

Empirical evidence suggests that kava helps with symptoms like menstrual cramps or changes in mood, but with no solid supportive research. Older studies suggest that kava may help with menopausal symptoms, but it is not currently a valid therapeutic option. However, kava is commonly used by herbalists to treat anxiety and insomnia caused by Premenstrual Syndrome (PMS) (Romm et al., 2010). According to a comprehensive review and analysis of the evidence for kava having anxiolytic action, it is found to be more effective than a placebo. Several small studies have been reported about effective of kava in menopause-related restlessness (304 cases). However, no studies have looked at kava's effectiveness in treating PMS symptoms. Concerns about the safety of kava have recently arisen as a result of approximately 80 international case reports of hepatotoxicity allegedly linked to kava consumption, prompting several countries to prohibit kava sales.

12.4.5 Effect of Kava on Insomnia

Research studies related to the use of kava for the treatment of insomnia are mainly on animal models. Kavalactone, called kevain, causes the drug's sedative effect. However, in Germany, research on human beings concluded that people given a dose of 200 mg kava extract daily showed tremendous relief from sleeping disorders after 14 days (Lehrl, 2004). Despite the promising results, the conclusions were not clear, as even people provided with a placebo showed significant improvements in sleep (Lehrl, 2004).

12.4.6 Other Effects of Kava

The effect of kava on insomnia is related to its effectiveness in anxiolytic action. Kava is used as a solution for scratchiness of the throat and also for the treatment of chest colds (Fern et al., 2019)

12.5 ADVERSE EFFECTS OF KAVA

Over the past decade, adverse effects of kava have been reported, like four cases of drivers in Iowa, USA, who were not able to operate a motor vehicle after consumption of kava. Detailed information on the use of kava products or dosage was lacking so studies could not be supported. However, the presence of yangonin and methysticin in the urine in the two of above cases confirmed the consumption of kava by the drivers. A number of clinical trials on kava's potential in modulating the central nervous system and its effects on driving have been carried out but without promising findings. The prominent adverse effect so far is the hepatotoxicity issue, which is discussed below.

12.5.1 Hepatotoxicity

In the past, very few safety issues with reference to the traditional use of kava were reported. But, in the early 2000s, Western countries reported cases related to liver toxicity. Kava use as a herbal anxiolytic drug has been banned in the United Kingdom and other European countries. Worldwide, above 100 cases of liver toxicity have been reported which were associated with the consumption of kava. The hepatotoxic risk cannot be ignored, although the debate about the causal relationship between side-effects and kava use cannot be established with sufficient certainty. There are reports suggesting that the hepatotoxicity may be due to the mixing of other parts of the plant in the rhizome preparations, improper storage of kava materials, misidentification of other plants for kava, the use of low-quality cultivars, improper handling, drug–herb interactions, overdosing, etc.

Aqueous extracts of kava were used by Pacific Islanders but with no apparent adverse effects. *In-vitro* and *in-vivo* studies recently have confirmed that the water fraction of kava is less cytotoxic than the organic solvent fraction (Jhoo et al., 2006; Sorrentino et al., 2006). The latest commercial products are based on the alcohol or acetone extraction processes that can release toxic compounds (e.g., alkaloids) from kava plants. Therefore, some experts suggested that the extraction method could be a prime factor in understanding liver damage observed with modern kava preparations (Whitton et al., 2003).

An urgent reason for the ban on the use of kava in Germany was because of the uncertainty regarding the anxiolytic effect of kava. The prohibition on kava in the UK and different nations (for example Austria, Germany, and France) caused expected harm of US$1.2 billion to the business (Gruenwald, 2006). A few specialists now believe that, taking everything into account, the ban is no longer reasonable. To explain possible toxicity mechanisms, several theories have emerged. However, there is still debate about whether kava is hepatotoxic or safe. As a result, prohibition was found to be unjustified and, recently, the UK licensing authority upheld the ban.

12.6 CONCLUSIONS AND FUTURE PERSPECTIVES

P. methysticum, commonly known as kava, has a long history of use in traditional medicinal systems for various purposes. It is primarily known for its anxiolytic properties, helping to combat anxiety. However, it has also been used for the treatment of other conditions such as inflammation, insomnia, cancer, gonorrhea, asthma, headaches, menstrual issues, urinary tract infections, and skin problems. Additionally, it has been reported to act as a muscle relaxant and an anti-obesity agent. The main constituents of kava are divided into two groups: kavalactones and flavokavins. These two major constituents are responsible for most of the biological activities associated with kava. So far, 56 constituents have been identified from P. methysticum. The composition of kava varies depending on whether it is fresh or dried. Fresh kava rhizome consists of approximately 80 percent water on average. Dried rhizome, on the other hand, contains fibers, starch, proteins, water, sugars, minerals, and kavalactones, with concentrations of approximately 20%, 4.3%, 3.6%, 12%, 3.2%, 3.2%, and 15%, respectively. The kavalactone component can range between 3 and 20% of the weight of the dry rhizome, depending on factors such as the age of the plant and the variety used. In the past, there have been very few safety issues reported regarding the traditional use of kava. However, the use of kava as an herbal anxiolytic drug has been banned in the United Kingdom and some other European countries due to concerns about potential side effects. The causal relationship between these side effects and kava use is still a matter of controversy. Some reports suggest that the side effects may be attributed to factors such as the mixing of other parts of the plant with rhizomes, improper storage of kava materials, low-quality cultivars, improper handling, drug-herb interactions, or overdosing.

In the future, further research is needed to better understand the potential toxicity mechanisms associated with kava use. This would allow for a clearer understanding of its safety profile and enable the optimal utilization of its medicinal properties for the benefit of mankind.

REFERENCES

Abu Bakar, A., Akhtar, M.N., Mohd Ali, N., Yeap, S.K., Quah, C.K., Loh, W.S. et al. (2018) Design, synthesis and docking studies of flavokawain B type chalcones and their cytotoxic effects on MCF-7 and MDA-MB-231 cell lines. *Molecules*, 23(3), 616.

Abu, N., Mohamed, N.E., Yeap, S.K., Lim, K.L., Akhtar, M.N., Zulfadli, A.J. et al. (2015) In vivo antitumor and antimetastatic effects of flavokawain B in 4T1 breast cancer cell-challenged mice. *Drug design, Development and Therapy*, 9, 1401.

Abu, N., Akhtar, M.N., Yeap, S.K., Lim, K.L., Ho, W.Y., Abdullah, M.P. et al. (2016) Flavokawain B induced cytotoxicity in two breast cancer cell lines, MCF-7 and MDA-MB231 and inhibited the metastatic potential of MDA-MB231 via the regulation of several tyrosine kinases In vitro. *BMC Complementary and Alternative Medicine*, 16(1), 1–14.

Akana, A. (1922) *Hawaiian herbs of medicinal value: Found among the mountains and elsewhere in the Hawaiian Islands, and known to the Hawaiians to possess curative and palliative properties most effective in removing physical ailments.* Honolulu: Board of Health of the Territory of Hawaii.

An, J., Gao, Y., Wang, J., Zhu, Q., Ma, Y., Wu, J. et al. (2012) Flavokawain B induces apoptosis of non-small cell lung cancer H460 cells via Bax-initiated mitochondrial and JNK pathway. *Biotechnology Letters*, 34(10), 1781–1788.

Aporosa, A.S., Atkins, M. & Brunton, R. (2020) Kava drinking in traditional settings: Towards understanding effects on cognitive function. *Human Psychopharmacology*, 35(2), e2725. doi: 10.1002/hup.2725.

Baum, S.S., Hill, R. & Rommelspacher, H. (1998) Effect of kava extract and individual kavapyrones on neurotransmitter levels in the nucleus accumbens of rats. *Progress in Neuro-Psychopharmacology and Biological Psychiatry*, 22(7), 1105–1120.

Bilia, A.R., Gallori, S. & Vincieri, F.F. (2002) Kava-kava and anxiety: Growing knowledge about the efficacy and safety. *Life Sciences*, 70(22), 2581–2597.

Bilia, A.R., Scalise, L., Bergonzi, M.C. & Vincieri, F.F. (2004) Analysis of kavalactones from Piper methysticum (kava-kava). *Journal of Chromatography B*, 812(1–2), 203–214.

Cairney, S., Maruff, P. & Clough, A.R. (2002) The neurobehavioural effects of kava. *Australian & New Zealand Journal of Psychiatry*, 36(5), 657–662.

Cawte, J. (1986) Parameters of kava used as a challenge to alcohol. *Australian & New Zealand Journal of Psychiatry*, 20(1), 70–76.

Chang, C.T., Hseu, Y.C., Thiyagarajan, V., Lin, K.Y., Way, T.D., Korivi, M. et al. (2017) Chalcone flavokawain B induces autophagic-cell death via reactive oxygen species-mediated signaling pathways in human gastric carcinoma and suppresses tumor growth in nude mice. *Archives of Toxicology*, 91(10), 3341–3364.

Chou, T.W., Feng, J.H., Huang, C.C., Cheng, Y.W., Chien, S.C., Wang, S.Y. et al. (2013) A plant kavalactone desmethoxyyangonin prevents inflammation and fulminant hepatitis in mice. *PLOS One*, 8(10), e77626.

Chua, H.C., Christensen, E.T., Hoestgaard-Jensen, K., Hartiadi, L.Y., Ramzan, I., Jensen, A.A. et al. (2016) Kavain, the major constituent of the anxiolytic kava extract, potentiates GABAA receptors: *Functional characteristics and molecular mechanism. PLoS One*, 11(6), e0157700.

Coulter D., Tamayo C., Sotheeswaran S., Ulbricht C. (2007) *Assessment of the risk of hepatotoxicity with Kava products.* Geneva: World Health Organization.

Dai, J.Q., Huang, Y.G. & He, A.N. (2015) Dihydromethysticin kavalactone induces apoptosis in osteosarcoma cells through modulation of PI3K/Akt pathway, disruption of mitochondrial membrane potential and inducing cell cycle arrest. *International Journal of Clinical and Experimental Pathology*, 8(5), 4356.

Dong, R., Wang, J., Gao, X., Wang, C., Liu, K., Wu, J. et al. (2019) Yangonin protects against estrogen-induced cholestasis in a farnesoid X receptor-dependent manner. *European Journal of Pharmacology*, 857, 172461.

Einbond, L.S., Negrin, A., Kulakowski, D.M., Wu, H.A., Antonetti, V., Jalees, F. et al. (2017) Traditional preparations of kava (Piper methysticum) inhibit the growth of human colon cancer cells in vitro. *Phytomedicine*, 24, 1–13.

Emser W. & Bartylla, K (1991) Improvement in quality of sleep: Effect of kava extracts WS 1490 on the sleep patterns in healthy people. *Neurol. Psychiatr*, 5, 636–642.

Fern, K. (2019) Tropical plants database. *Fecha de consulta: Entre febrero y marzo de consulta: Entre febrero y marzo de.* http://tropical.theferns.info/ (Accessed 12 March 2019).

Folmer, F., Blasius, R., Morceau, F., Tabudravu, J., Dicato, M., Jaspars, M. et al. (2006) Inhibition of TNFα-induced activation of nuclear factor κB by kava (Piper methysticum) derivatives. *Biochemical Pharmacology*, 71(8), 1206–1218.

Fragoulis, A., Siegl, S., Fendt, M., Jansen, S., Soppa, U., Brandenburg, L.O. et al. (2017) Oral administration of methysticin improves cognitive deficits in a mouse model of Alzheimer's disease. *Redox Biology*, 12, 843–853.

Johnston, G.A.R., Chebib, M., Duke, R.K., Fernandez, S.P., Hanrahan, J.R., Hinton, T. & Mewett, K.N. (2009) Herbal product receptors. *Encyclopedia of Neuroscience*, 4, 1095–1101.

Gleitz, J., Beile, A. & Peters, T. (1996) (±)-Kavain inhibits the veratridine-and KCl-induced increase in intracellular Ca2+ and glutamate-release of rat cerebrocortical synaptosomes. *Neuropharmacology*, 35(2), 179–186.

Grieve, M. & Leyel, H. (1931) *Book: 'A Modern Herbal'*, vol.-1. Hafner Publishing.

Gruenwald J. (2006) Welcome back Kava? *Nutraceuticals World*, 26–27. [Google Scholar]

Ji, T., Lin, C. & Krill, L.S. (2013) Flavokawain B, a kava chalcone, inhibits growth of human osteosarcoma cells through G2/M cell cycle arrest and apoptosis. *Molecular Cancer*, 12, 55–59 doi: 10.1186/1476-4598-12-55.

He, Q., Liu, W., Sha, S., Fan, S., Yu, Y., Chen, L. & Dong, M. (2018) Adenosine 5′ -monophosphate-activated protein kinase-dependent mTOR pathway is involved in flavokawain B-induced autophagy in thyroid cancer cells. *Cancer Science*, 109(8), 2576–2589.

Hegazy, N.H., Breitinger, H.G. & Breitinger, U. (2019) Kavalactones from Kava (Piper methysticum) root extract as modulators of recombinant human glycine receptors. *Biological Chemistry*, 400(9), 1205–1215.

Hseu, Y.C., Huang, Y.C., Thiyagarajan, V., Mathew, D.C., Lin, K.Y., Chen, S.C. et al. (2019) Anticancer activities of chalcone flavokawain B from Alpinia pricei Hayata in human lung adenocarcinoma (A549) cells via induction of reactive oxygen species-mediated apoptotic and autophagic cell death. *Journal of Cellular Physiology*, 234(10), 17514–17526.

Hseu, Y.C., Lee, M.S., Wu, C.R., Cho, H.J., Lin, K.Y., Lai, G.H. et al. (2012) The chalcone flavokawain B induces G2/M cell-cycle arrest and apoptosis in human oral carcinoma HSC-3 cells through the intracellular ROS generation and downregulation of the Akt/p38 MAPK signaling pathway. *Journal of Agricultural and Food Chemistry*, 60(9), 2385–2397.

Hua, R., Pei, Y., Gu, H., Sun, Y. & He, Y. (2020) Antitumor effects of flavokawain-B flavonoid in gemcitabine-resistant lung cancer cells are mediated via mitochondrial-mediated apoptosis, ROS production, cell migration and cell invasion inhibition and blocking of PI3K/AKT Signaling pathway. *Journal of BUON*, 25(1), 262–267.

Huck, O., Han, X., Mulhall, H., Gumenchuk, I., Cai, B., Panek, J. et al. (2019) Identification of a Kavain analog with efficient anti-inflammatory effects. *Scientific Reports*, 9(1), 1–10.

Huck, O., You, J., Han, X., Cai, B., Panek, J. & Amar, S. (2018) Reduction of articular and systemic inflammation by Kava-241 in a Porphyromonas gingivalis-induced arthritis murine model. *Infection and Immunity*, 86(9), e00356-18.

Hunter, P. (2012) The inflammation theory of disease: The growing realization that chronic inflammation is crucial in many diseases opens new avenues for treatment. *EMBO Reports*, 13(11), 968–970.

Jamieson, D.D. & Duffield, P.H. (1990) The antinociceptive actions of kava components in mice. *Clinical and Experimental Pharmacology and Physiology*, 17(7), 495–507.

Jandial, D.D., Krill, L.S., Chen, L., Wu, C., Ke, Y., Xie, J. et al. (2017) Induction of G2M arrest by flavokawain A, a kava chalcone, increases the responsiveness of HER2-overexpressing breast cancer cells to herceptin. *Molecules*, 22(3), 462.

Jhoo, J.W., Freeman, J.P., Heinze, T.M., Moody, J.D., Schnackenberg, L.K., Beger, R.D. et al. (2006) In vitro cytotoxicity of nonpolar constituents from different parts of kava plant (Piper methysticum). *Journal of Agricultural and Food Chemistry*, 54(8), 3157–3162.

Johnson, T.E., Kassie, F., O'Sullivan, M.G., Negia, M., Hanson, T.E., Upadhyaya, P. et al. (2008) Chemopreventive Effect of Kava on 4-(Methylnitrosamino)-1-(3-pyridyl)-1-butanone plus Benzo [a] pyrene–Induced Lung Tumorigenesis in A/J Mice. *Cancer Prevention Research*, 1(6), 430–438.

Kwon, D.J., Ju, S.M., Youn, G.S., Choi, S.Y. & Park, J. (2013) Suppression of iNOS and COX-2 expression by flavokawain A via blockade of NF-kappaB and AP-1 activation in RAW 264.7 macrophages. *Food and Chemical Toxicology*, 58, 479–486. doi: 10.1016/j.fct.2013.05.031.

Jussofie, A., Schmiz, A. & Hiemke, C. (1994) Kavapyrone enriched extract from Piper methysticum as modulator of the GABA binding site in different regions of rat brain. *Psychopharmacology*, 116(4), 469–474.

Kong, Y., Gao, X., Wang, C., Ning, C., Liu, K., Liu, Z. et al. (2018) Protective effects of yangonin from an edible botanical Kava against lithocholic acid-induced cholestasis and hepatotoxicity. *European Journal of Pharmacology*, 824, 64–71.

Kuo, Y.F., Su, Y.Z., Tseng, Y.H., Wang, S.Y., Wang, H.M. & Chueh, P.J. (2010) Flavokawain B, a novel chalcone from Alpinia pricei Hayata with potent apoptotic activity: Involvement of ROS and GADD153 upstream of mitochondria-dependent apoptosis in HCT116 cells. *Free Radical Biology and Medicine*, 49(2), 214–226.

Lee, J.J., Koh, K.N., Park, C.J., Jang, S., Im, H.J. & Kim, N. (2018) The combination of flavokawain B and daunorubicin induces apoptosis in human myeloid leukemic cells by modifying NF-κB. *Anticancer Research*, 38(5), 2771–2778.

Lehrl, S. (2004) Clinical efficacy of kava extract WS® 1490 in sleep disturbances associated with anxiety disorders: Results of a multicenter, randomized, placebo-controlled, double-blind clinical trial. *Journal of Affective Disorders*, 78(2), 101–110.

Li, J., Zheng, L., Yan, M., Wu, J., Liu, Y., Tian, X. et al. (2020) Activity and mechanism of flavokawain A in inhibiting permeability Pglycoprotein expression in paclitaxel resistance of lung cancer. *Oncology Letters*, 19(1), 379–387.

Li, X., Yokoyama, N.N., Zhang, S., Ding, L., Liu, H.M., Lilly, M.B. et al. (2015) Flavokawain A induces deNEDDylation and Skp2 degradation leading to inhibition of tumorigenesis and cancer progression in the TRAMP transgenic mouse model. *Oncotarget*, 6(39), 41809.

Ligresti, A., Villano, R., Allarà, M., Ujváry, I. & Di Marzo, V. (2012) Kavalactones and the endocannabinoid system: The plant-derived yangonin is a novel CB1 receptor ligand. *Pharmacological Research*, 66(2), 163–169.

Liu, Z., Ha, U.S., Yu, K., Wu, C., Yokoyama, N. & Zi, X. (2017) Kavalactone yangonin induces autophagy and sensitizes bladder cancer cells to flavokawain A and docetaxel via inhibition of the mTOR pathway. *Journal of Biomedical Research*, 31(5), 408.

Magura, E.I., Kopanitsa, M.V., Gleitz, J., Peters, T. & Krishtal, O.A. (1997) Kava extract ingredients,(+)-methysticin and (±)-kavain inhibit voltage-operated Na+-channels in rat CA1 hippocampal neurons. *Neuroscience*, 81(2), 345–351.

Malami, I., Abdul, A.B., Abdullah, R., Kassim, N.K.B., Rosli, R., Yeap, S.K. et al. (2017) Crude extracts, flavokawain B and alpinetin compounds from the rhizome of alpinia mutica induce cell death via UCK2 enzyme inhibition and in turn reduce 18S rRNA biosynthesis in HT-29 cells. *PLoS One*, 12(1), e0170233.

Meyer, H.J. (1965) Antagonistic effects of geunine kawa pyrones in experimental inflammations and fever. *Klin. Wochenschr*, 43, 469–470.

Nadiah, A., Nurul Elyani, M., Swee Keong, Y., Kian Lam, L., Akhtar, M.N., Aimi Jamil, Z. et al. (2015) *In Vivo* anti-tumor effects of flavokawain A in 4T1 breast cancer cell-challenged mice. *Anti-Cancer Agents in Medicinal Chemistry*, 15, 905–915.

Nuki, G. (1990) Pain control and the use of non-steroidal analgesic anti-inflammatory drugs. *British Medical Bulletin*, 46(1), 262–278.

Pan, H., Liu, F., Wang, J., Zhao, M., Wang, D., Jia, C. et al. (2020) Dihydromethysticin, a natural molecule from Kava, suppresses the growth of colorectal cancer via the NLRC3/PI3K pathway. *Molecular Carcinogenesis*, 59(6), 575–589.

Pollastri, M.P., Whitty, A., Merrill, J.C., Tang, X., Ashton, T.D. & Amar, S. (2009) Identification and characterization of Kava-derived compounds mediating TNF- α suppression. *Chemical Biology & Drug Design*, 74(2), 121–128.

Rao, Y.K., Shih, H.N., Lee, Y.C., Cheng, W.T., Hung, H.C., Wang, H.C. et al. (2014) Purification of kavalactones from Alpinia zerumbet and their protective actions against hydrogen peroxide-induced cytotoxicity in PC12 cells. *Journal of Bioscience and Bioengineering*, 118(6), 679–688.

Romm, A., Clare, B., Stansbury, J.E., Ryan, L., Trickey, R., Lee, L. & Hywood, A.J. (2010) Menstrual wellness and menstrual problems. *Botanical Medicine for Women's Health*, 97–185. A. Romm, Ed.

Rossette, M.C., Moraes, D.C., Sacramento, E.K., Romano-Silva, M.A., Carvalho, J.L., Gomes, D.A. et al. (2017) The in vitro and in vivo antiangiogenic effects of flavokawain B. *Phytotherapy Research*, 31(10), 1607–1613.

Sakai, T., Eskander, R.N., Guo, Y., Kim, K.J., Mefford, J., Hopkins, J. et al. (2012) Flavokawain B, a kava chalcone, induces apoptosis in synovial sarcoma cell lines. *Journal of Orthopaedic Research*, 30(7), 1045–1050.

Sarris, J., Kavanagh, D.J., Byrne, G., Bone, K.M., Adams, J. & Deed, G. (2009) The Kava Anxiety Depression Spectrum Study (KADSS): A randomized, placebo-controlled crossover trial using an aqueous extract of Piper methysticum. *Psychopharmacology*, 205(3), 399–407.

Sarris, J., LaPorte, E. & Schweitzer, I. (2011) Kava: A comprehensive review of efficacy, safety, and psychopharmacology. *Australian & New Zealand Journal of Psychiatry*, 45(1), 27–35.

Shaik, A.A., Hermanson, D.L. & Xing, C. (2009) Identification of methysticin as a potent and non-toxic NF-κB inhibitor from kava, potentially responsible for kava's chemopreventive activity. *Bioorganic & Medicinal Chemistry Letters*, 19(19), 5732–5736.

Shimoda, L.M., Park, C., Stokes, A.J., Gomes, H.H. & Turner, H. (2012) Pacific island Awa (Kava) extracts, but not isolated kavalactones, promote proinflammatory responses in model mast cells. *Phytotherapy Research*, 26, 1934–1941.

Singh, Y.N. (1992) Kava: An overview. *Journal of Ethnopharmacology*, 37(1), 13–45.

Singh, Y.N. & Singh, N.N. (2002) Therapeutic potential of kava in the treatment of anxiety disorders. *CNS Drugs*, 16(11), 731–743.

Sorrentino, L., Capasso, A. & Schmidt, M. (2006) Safety of ethanolic kava extract: Results of a study of chronic toxicity in rats. *Phytomedicine*, 13(8), 542–549.

Sotheeswaran, S. (2002) Anticancer activity studies on kava (Piper Methysticum) extracts. In Pacific Kava Research Symposium, Suva.

Steiner, G.G. (2000) The correlation between cancer incidence and kava consumption. *Hawaii Med. J.* 59, 420–422.

Steiner, G.G. (2001) Kava as an anticraving agent: Preliminary data. *Pacific Health Dialog*, 8(2), 335–339.

Stevinson, C., Huntley, A. & Ernst, E. (2002) A systematic review of the safety of kava extract in the treatment of anxiety. *Drug Safety*, 25(4), 251–261.

Tang, Y., Simoneau, A.R., Xie, J., Shahandeh, B. & Zi, X. (2008) Effects of the kava chalcone flavokawain a differ in bladder cancer cells with wild-type versus mutant p53. *Cancer Prevention Research*, 1, 439.

Terazawa, R., Akimoto, N., Kato, T., Itoh, T., Fujita, Y., Hamada, N. et al. (2013) A kavalactone derivative inhibits lipopolysaccharide-stimulated iNOS induction and NO production through activation of Nrf2 signaling in BV2 microglial cells. *Pharmacological Research*, 71, 34–43.

Turner, J.W. (1995) Substance, symbol and practice: The power of kava in Fijian society. *Canberra Anthropology*, 18(1–2), 97–118.

Turner, J.W. (1986) 'The water of life': Kava ritual and the logic of sacrifice. *Ethnology*, 25(3), 203–214.

Tzeng, Y.M. & Lee, M.J. (2015) Neuroprotective properties of kavalactones. *Neural Regeneration Research*, 10(6), 875.

Volgin, A., Yang, L., Amstislavskaya, T., Demin, K., Wang, D., Yan, D. et al. (2020) DARK classics in chemical neuroscience: Kava. *ACS Chemical Neuroscience*, 11(23), 3893–3904.

Walden, J., Von Wegerer, J., Winter, U., Berger, M. & Grunze, H. (1997) Effects of kawain and dihydromethysticin on field potential changes in the hippocampus. *Progress in Neuro-Psychopharmacology & Biological Psychiatry*, 21(4), 697–706.

Wang, Y., Su, C., Zhang, B., Niu, Y., Ren, R., Zhao, X. et al. (2021) Biological activity, hepatotoxicity, and structure-activity relationship of kavalactones and flavokavins, the two main bioactive components in Kava (Piper methysticum). *Evidence-Based Complementary and Alternative Medicine*, 2021.

Wang, Y., Su, C., Zhang, B., Niu, Y., Ren, R., Zhao, X. et al. Ma, X. (2021) Biological activity, hepatotoxicity, and structure-activity relationship of kavalactones and flavokavins, the two main bioactive components in Kava (Piper methysticum). *Evidence-Based Complementary and Alternative Medicine*, 2021.

Wang, D., Yang, L., Wang, J., Hu, G., Liu, Z., Yan, D. et al. (2020) Behavioral and physiological effects of acute and chronic kava exposure in adult zebrafish. *Neurotoxicology and teratology*, 79, 106881.

Warnecke G. (1991) Psychosomatic dysfunctions in the female climacteric. Clinical effectiveness and tolerance of Kava Extract WS 1490. Fortschritte der Medizin, 109, 119–122.

Warnecke, G., Pfaender, H., Gerster, G. & Gracza, E. (1990) Wirksamkeit von Kawa-kawa-Extrakt beim klimakterischen syndrom. *Z Phytother*, 11, 81–86.

Whitton, P.A., Lau, A., Salisbury, A., Whitehouse, J. & Evans, C.S. (2003) Kava lactones and the kava-kava controversy. *Phytochemistry*, 64(3), 673–679.

Wruck, C.J., Götz, M.E., Herdegen, T., Varoga, D., Brandenburg, L.O. & Pufe, T. (2008) Kavalactones protect neural cells against amyloid β peptide-induced neurotoxicity via extracellular signal-regulated kinase 1/2-dependent nuclear factor erythroid 2-related factor 2 activation. *Molecular Pharmacology*, 73(6), 1785–1795.

Wu, D., Yu, L., Nair, M.G., De Witt, D.L. & Ramsewak, R.S. (2002) Cyclooxygenase enzyme inhibitory compounds with antioxidant activities from Piper methysticum (kava kava) roots. *Phytomedicine*, 9(1), 41–47.

Yang, J. (2019) X. A new kavalactone dimer from piper methysticum. *Chemistry of Natural Compounds*, 55(4), 606–609. doi: 10.1007/s10600-019-02759-8.

Yeap, S.K., Abu, N., Akthar, N., Ho, W.Y., Ky, H., Tan, S.W. et al. (2017) Gene expression analysis reveals the concurrent activation of proapoptotic and antioxidant-defensive mechanisms in flavokawain B–treated cervical cancer HeLa cells. *Integrative Cancer Therapies*, 16(3), 373–384.

Yokoyama, Y., Nishi, M., Murayama, H., Amano, H., Taniguchi, Y., Nofuji, Y. et al. (2016) Association of dietary variety with body composition and physical function in community-dwelling elderly Japanese. *The Journal of Nutrition, Health & Aging*, 20(7), 691–696.

Zhang, L., Rowe, A., Braet, F. & Ramzan, I. (2012) Macrophage depletion ameliorates kavalactone damage in the isolated perfused rat liver. *The Journal of Toxicological Sciences*, 37(2), 447–453.

Zhang, Y., Zvi, Y.S., Batko, B., Zaphiros, N., O'Donnell, E.F., Wang, J. et al. (2018) Down-regulation of Skp2 expression inhibits invasion and lung metastasis in osteosarcoma. *Scientific Reports*, 8(1), 1–13.

Zou, L., Harkey, M.R. & Henderson, G.L. (2005) Synthesis, in vitro, reactivity, and identification of 6-phenyl-3-hexen-2-one in human urine after kava-kava (Piper methysticum) ingestion. *Planta Medica* 71(2), 142–146. doi: 10.1055/s-2005-837781.

13 Medical Benefits and Side-effects of *Lepidium meyenii* Root

Deepak Kumar Verma, Vinay Pathak, Sapna Yadav, Sadiya Sameer, Navneet Kumar, Aashna Sinha, Kundan Kumar Chaubey, Krishan Raj Singh, Gaurav Bhardwaj, Alazar Essayas, and Rakesh Kumar Bachheti

13.1 INTRODUCTION

Lepidium meyenii (Maca), also called Peruvian ginseng, is a cruciferous root vegetable (Brassicaceae family), related to turnips and radishes. It was first cultivated in Peru and is available in various hues, including red, black, yellow, and brown (Gonzales, 2012), and different Maca variants have been found to have diverse biological characteristics. Maca 'root' is actually a swollen hypocotyl, and, in the wild, it grows only between 4,000 and 4,500 m above sea level in the central Peruvian Andes, notably on the Junin plateau. It resembles a big radish and is resilient to adverse condition weather (Figure 13.1). The Maca plant's resilience in the face of adversity shows potential health benefits for humans if consumed in the recommended dosage (Ybañez-Julca et al., 2020). Typically, the producers provide Maca as a powder, in capsule form, or as a tincture (Figure 13.2). Although it is frequently used as a supplement, it is crucial to confirm that it has been tested for toxins before consuming it. Although Maca has many remarkable health advantages, some experts have also found that the root can have a few adverse effects (Orellana Mendoza et al., 2021).

13.2 CONSTITUENTS OF *LEPIDIUM MEYENII* ROOT

In terms of its basic ingredients, Maca root is primarily composed of 59% carbohydrates, 10 to 14% proteins, 10% water, 8.5% dietary fibre, and 2.2% fat (Gonzales, 2012). In addition to being rich in copper, manganese, zinc, vitamin C, and vitamins B_1 and B_2, Maca is also high in concentration of potassium, calcium, iron, and iodine. It contains 19 amino acids, 20 fatty acids, a significant amount of choline, and gamma-aminobutyric acid (GABA) (Gonzales-Arimborgo et al., 2016). The main active component of the Maca root is the alkaloid macaridine, which has not been found in any other plant. A distinctive fatty acid found in Maca is called macamide. The bitter taste of Maca is a result of its active ingredients, which also include glucosinolates (Wang & Zhu, 2019). Fresh Maca has ten times the amount of glucosinolates as other cruciferous vegetables. Red Maca has the highest concentration of glucosinolates, followed by black and yellow Maca roots (Zevallos-Concha et al., 2016). Furthermore, the colour of the maca root affects the number and concentration of dietary polyphenols it contains.

13.3 RECOMMENDED DOSAGE OF *LEPIDIUM MEYENII* ROOT

There has yet to be an established formula for the ideal Maca root dosage, although some health professionals advise starting with 3 g daily and gradually increasing it to up to 9 g daily. When taken with

DOI: 10.1201/b22924-13

FIGURE 13.1 Typical roots of *Lepidium meyenii* (adopted from Tafuri et al. 2019).

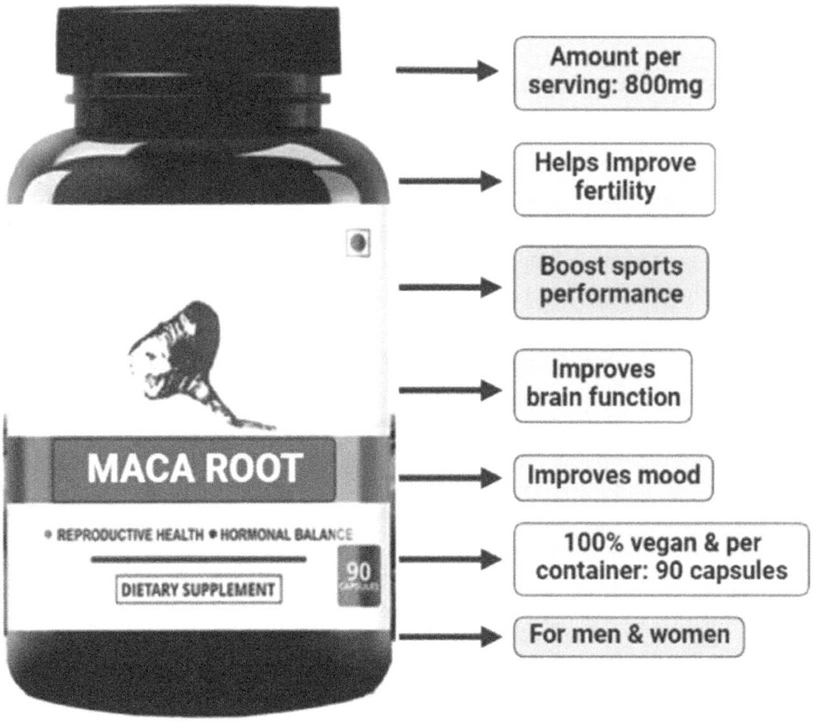

FIGURE 13.2 Maca root as a supplement.

meals for up to four months, 3 g per day is considered safe (Brooks et al., 2008; Gonzales-Arimborgo et al., 2016). A nutritionist should be consulted before starting Maca root supplements since dosages vary depending on the type of therapy received or the illness the person suffers from (Sprouse & van Breemen, 2016). A person can add up to two or three teaspoons of Maca per serving to ordinary meals or beverages, whether in root or powder form. Juices, smoothies, soups, and tea may all be made with Maca as an ingredient. Consuming Maca in the morning is recommended since it is an energy booster. Regardless of how it is taken, Maca has several health benefits for our bodies (Shin et al., 2010). It is often recommended to take it for six to sixteen weeks before expecting to notice any visible changes.

13.4 MEDICAL BENEFITS OF *LEPIDIUM MEYENII* ROOT

Numerous studies suggest that Maca root is used to treat several medical issues, including erectile dysfunction, hair loss, depression, and indications of low libido and menopause (Beharry & Heinrich, 2018). Additionally, it is employed in Peruvian folk medicine to increase energy levels. In the following paragraphs, we will describe a few of the health advantages of Maca root that have been supported by research.

13.4.1 Improves Sexual Function and Libido

According to a group of experts who examined several previously published clinical trials, taking Maca improves sexual performance (Dording et al., 2015). They claim that, although some researchers have suggested that Maca could be beneficial for conditions including erectile dysfunction and low libido, the overall number of studies, the size of the sample, and the quality of the studies made it impossible to draw firm conclusions (Shin et al., 2010). Researchers gave patients either 1.5 g or 3 g of Maca or a placebo (an inert tablet) for 12 weeks in a prior trial, and they found that both dosages of Maca improved sexual desire more than the placebo but had no impact on testosterone or oestrogen levels (Gonzales-Arimborgo et al., 2016).

13.4.2 Treats Sexual Dysfunction Caused by Antidepressants

Some adverse effects of certain antidepressants include low libido, dry vagina, and difficulty achieving orgasm. Women with sexual dysfunction brought on by antidepressants may benefit from Maca. In a 12-week experiment, women using antidepressant drugs known as SSRIs (selective serotonin reuptake inhibitors) or SNRIs (serotonin-norepinephrine reuptake inhibitors) or placebo were given Maca root or a placebo. After that, individuals taking Maca saw decreased sexual dysfunction (Dording et al., 2008, 2015). Figure 13.3 shows the benefits of *Lepidium meyenii* to male and female reproductive health.

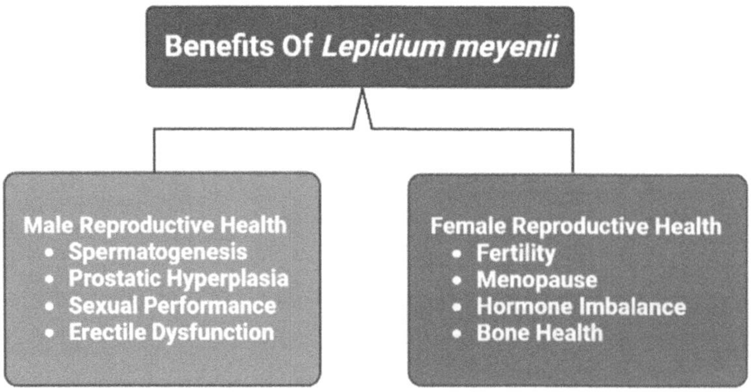

FIGURE 13.3 Benefits of *Lepidium meyenii* in male and female reproductive health.

13.4.3 Improves Fertility

Semen quality and hormone levels were examined after a group of males were given Maca or a placebo in a research study. There were no significant changes in sperm counts or hormone levels after 12 weeks, but sperm concentration and motility improved (Gonzales et al., 2002; Stone et al., 2009; Zenico et al., 2009; Melnikovova et al., 2015). Compared with yellow and red Maca, prior research indicated that black Maca had the strongest impact on spermatogenesis (Gonzales et al., 2006). Maca is historically used to increase male and female fertility. In reality, Father Bernabe Cobo, a chronicler of the Spanish conquest of Peru, referred to the fertility-enhancing properties of Maca in the first part of the seventeenth century (Cobo et al., 1956).

13.4.4 Helps to Reduce Size of Enlarged Prostate

According to certain studies, Maca extract can help to treat mice with enlarged prostrates (Gonzales et al., 2008; Fano et al., 2017). Furthermore, rats' prostate size was dramatically decreased by red Maca but not by yellow or black Maca (Gonzales et al., 2005).

13.4.5 Reduces Depression and Anxiety

Maca may aid in enhancing mood in cases of depression. Researchers gave Maca to postmenopausal women for six weeks and observed significant reductions in depression when compared with those who took a placebo (Rubio et al., 2006; Brooks et al., 2008; Stojanovska et al., 2015).

13.4.6 Reduces Signs of Inflammation and Oxidative Stress

A short study showed that the subjects who took Maca root for some time had lower levels of inflammatory markers (e.g., IL-6) than those who did not take it (Gonzales et al., 2013). Another study on mice showed that black Maca reduced the level of markers of oxidative stress (Rubio et al., 2011; Zha et al., 2014).

13.4.7 Lowers High Blood Pressure

An observational study of 50 people indicated that Maca consumption is linked to lower blood pressure. The study found that six weeks of Maca supplementation of the diet dramatically reduced blood pressure in postmenopausal women, although additional human studies with larger sample sizes are required to confirm Maca's effects on blood pressure (Figure 13.4) (Gonzales et al., 2013). Maca lowers the availability of angiotensin, a hormone that raises blood pressure by lowering angiotensin-converting enzyme (ACE) activity.

13.4.8 Helps Bone Growth

Maca supplementation showed enhanced bone density in menopausal women following four months of ingestion in a pilot trial. Reparagen, a Maca-derived chemical, has effectively reduced pain and stiffness and improved physical function in patients with osteoarthritis (Mehta et al., 2007). Maca has also been shown to decrease bone loss in rats in response to lowered oestrogen concentrations (Meissner et al., 2006).

13.4.9 Reduces Menopause Symptoms

In postmenopausal women, Maca helps to improve menopausal symptoms (irritability, pain from hot flashes, and nocturnal sweating). In postmenopausal women, Maca powder reduces levels of

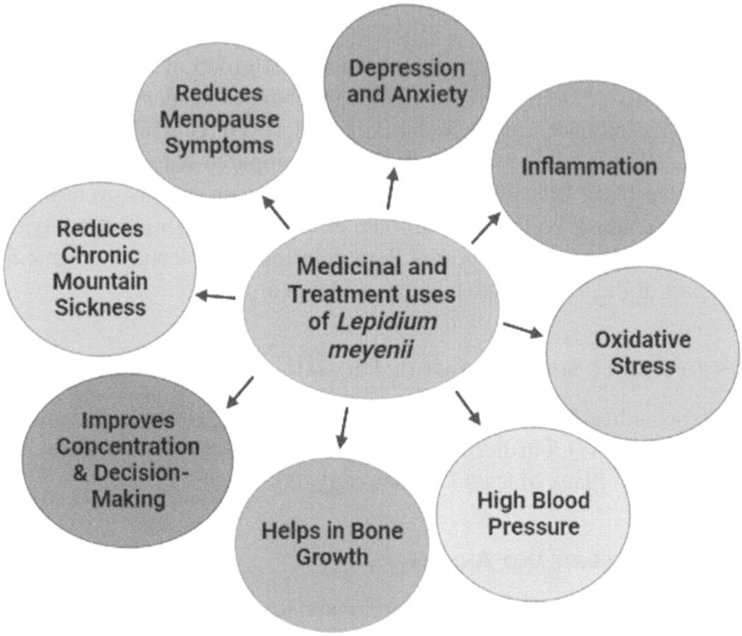

FIGURE 13.4 Medicinal uses of *Lepidium meyenii.*

follicle-stimulating hormone and luteinizing hormone while increasing oestrogen and progesterone levels (Meissner et al., 2005, 2006).

13.4.10 Improves Exercise Performance and Reduces Chronic Mountain Sickness

A pilot study showed that consuming Maca root improved sportsmen's performance during training or exercise (Stone et al., 2009). It has also been observed that consuming Maca root reduces chronic mountain sickness (Penaloza & Arias-Stella, 2007; Gonzales-Arimborgo et al., 2016). Chronic mountain sickness is linked to elevated haemoglobin levels at high altitudes, and black Maca is known to lower haemoglobin levels in such people.

13.4.11 Improves Concentration and Decision-making

Several kinds of research have demonstrated that Maca root helps to reduce negative feelings and optimize hormone production. It can also be utilized to help with weight loss. Although it does not enhance metabolism or fat burning, it may increase energy levels, motivating people to exercise more regularly and adhere to a diet plan (Rubio et al., 2006, 2007, 2011).

13.5 MECHANISMS OF ACTION OF COMPONENTS OF *LEPIDIUM MEYENII* ROOT

Maca is high in macamides, fatty acids that influence the endocannabinoid system. Macamides work by inhibiting the enzyme fatty acid amide hydrolase (FAAH), which breaks down anandamide (Kasprzak et al., 2018). Anandamide binds to the CB1 receptor in the cannabinoid system, causing feelings of bliss. Maca protects against oxidative stress by neutralizing free radicals (Tafuri et al., 2019). By raising IGF-1 levels in human cartilage, Maca improves bone health (Chen et al., 2021).

13.6 TOXICITY

After being used as food in the Andes for many years, there is minimal data available on the toxicity of Maca. Maca has minimal *in-vitro* and *in-vivo* acute oral toxicity in animals (Valerio & Gonzales, 2005). The 3-(4,5-dimethylthiazol-2-yl)-2,5-diphenyltetrazolium bromide viability assay revealed that, up to 10 mg/mL, the aqueous and methanolic extracts of Maca did not exhibit cytotoxicity in primary hepatocyte cultures. According to this discovery, Maca does not exhibit *in-vitro* hepatotoxicity. In contrast, a little cytoprotective impact, which could not possibly be mediated by antioxidant capacity, was observed (Valentová et al., 2006). On three T3 cell and cancer cell lines, aqueous extracts of yellow, red, and black Maca did not exhibit cytotoxic action. At 1 g/kg bodyweight, it had no impact on the regular growth of preimplantation mice embryos. LD_{50} for rats at a single dosage of Maca extract (17 g/kg bodyweight) did not show acute toxicity. Oral administration of Maca extract at doses of 1–2 g per kg of bodyweight for mice and 1 g per kg of bodyweight for humans is safe (Darrigo et al., 2004). Establishing a scientific method is important to assess the promising biological activity in humans and address its safe usage (Gonzales & Valerio, 2006).

13.6.1 Side-effects of *Lepidium meyenii* Root

There is little information about the safety and hazardous effects of Maca usage. It is considered safe in large dosages because it is a natural food. Also, the impact of Maca on hormone levels is unknown. For example, although some research found no influence on sex hormones, animal studies indicated greater levels of luteinizing hormone, progesterone, and testosterone in response to Maca root (Ybañez-Julca et al., 2020; Azad et al., 2021). If an individual has breast cancer, uterine cancer, ovarian cancer, endometriosis, uterine fibroids, or thyroid problems, one should not use Maca without consulting a physician (Oshima et al., 2003). Like many other supplements, Maca has not been evaluated for safety in pregnant women, nursing mothers, individuals with medical conditions, people on medical prescriptions, and children. Maca root is generally considered safe to consume, but it is recommended to buy it from a laboratory that has tested it for toxicity. Pregnant or breastfeeding women are suggested not to consume Maca since it has not been scientifically tested on such populations. Although Maca is considered to be well tolerated, with rare and minimal side-effects, it is still recommended to start with a very small amount and slowly increase the dose to the full rate.

Residents of Peru's Andes Mountains take up to 100 g of Maca each day without experiencing any adverse effects, but they recommend only eating dehydrated or boiled Maca root because raw Maca might create health problems. Maca is well tolerated up to 3 g/day in clinical investigations, whereas, in a small trial, consuming 0.6 g of Maca per day for 90 days raised levels of a liver damage marker (aspartate aminotransferase) and diastolic blood pressure in 95 patients with metabolic syndrome (Valentová et al., 2008). Maca contains (1R,3S)-1-methyl-1,2,3,4-tetrahydro-beta-carb oline-3-carboxylic acid (MTCA) that might cause DNA mutations, and certain food safety agencies have warned against Maca uses because of this risk (Piacente et al., 2002). However, some researchers contradict this, claiming that boiling MTCA renders it inactive (Gonzales & Gonzales-Castañeda, 2009). Some Maca users have reported menstruation irregularities, such as an altered menstrual cycle, stomach cramps, moodiness, and insomnia as side-effects (Corazza et al., 2014). Before taking Maca as a supplement, it is advised to speak with a doctor to prevent these and other unforeseen adverse events and interactions (Sprouse & van Breemen, 2016).

13.7 CONCLUSION

Due to extensive marketing campaigns for the plant, the interest in and demand for Maca have increased globally during the past 20 years. The use of Peruvian Maca in the treatment of various illnesses and dysfunctions is noted in the literature. Peruvian Maca is superior to other hypocotyls like potatoes, carrots, and turnips in composition, and has a high nutritional value comparable to cereal grains (Korkmaz et al., 2018). This root contains 80% water when it is fresh, leaving just a modest of the nutrients. There is minimal information available on the risks associated with Maca use. Due to the fact that it is a natural food, it is believed to be safe even when taken in large amounts. Additionally, it is still being determined how Maca affects hormone levels. However, because various bioactives are present, extrinsic and intrinsic variables associated with culture and extraction methods might vary its pharmacological effect. Based on this, several methods may be investigated to generate extracts with high pharmacological potential.

REFERENCES

Azad, P. et al. (2021) Protective role of estrogen against excessive erythrocytosis in Monge's disease. *Experimental & Molecular Medicine*, 53(1), 125–135. https://doi.org/10.1038/s12276-020-00550-2.

Beharry, S. & Heinrich, M. (2018) Is the hype around the reproductive health claims of Maca (*Lepidium meyenii*Walp.) justified. *Journal of Ethnopharmacology*, 211, 126–170. https://doi.org/10.1016/j.jep.2017.08.003.

Brooks, N.A. et al. (2008) Beneficial effects of *Lepidium meyenii* (Maca) on psychological symptoms and measures of sexual dysfunction in postmenopausal women are not related to estrogen or androgen. *Menopause: The Journal of The North American Menopause Society*, 15(6): 1157–1162.

Chen, R., Wei, J. & Gao, Y. (2021) A review of the study of active components and their pharmacology value in *Lepidium meyenii* (Maca). *Phytotherapy Research*, 35(12), 6706–6719. https://doi.org/10.1002/ptr.7257.

Cobo, B. (1956) *History of the new world.* de Mateos, F. (Ed). Madrid: Ediciones Atlas.

Corazza, O. et al. (2014) Sexual enhancement products for sale online: raising awareness of the psychoactive effects of yohimbine, Maca, horny goat weed, and Ginkgo biloba. *BioMed Research International*, 841798–841798. https://doi.org/10.1155/2014/841798.

Darrigo, G., Benavides, V. & Pino, J. (2004) Preliminary evaluation effect of *Lepidium meyenii*Walp on the embryonic development of mouse. *Revista Peruana de Biología*. 11(1)103–106.

Dording, C.M. et al. (2008) A double-blind, randomized, pilot dose-finding study of maca root (L. meyenii) for the management of SSRI-induced sexual dysfunction. *CNS Neuroscience & Therapeutics*, 14(3), 182–191. https://doi.org/10.1111/j.1755-5949.2008.00052.x.

Dording, C.M. et al. (2015) A double-blind placebo-controlled trial of maca root as treatment for antidepressant-induced sexual dysfunction in women. *Evidence-Based Complementary and Alternative Medicine*. https://doi.org/10.1155/2015/949036.

Fano, D. et al. (2017) N-butanol and aqueous fractions of red maca methanolic extract exerts opposite effects on androgen and oestrogens receptors (Alpha and beta) in rats with testosterone-induced benign prostatic hyperplasia. *Evidence-Based Complementary and Alternative Medicine, ECAM*, 9124240–9124240. https://doi.org/10.1155/2017/9124240.

Gonzales, C., Rubio, J., Gasco, M., Nieto, J., Yucra, S. & Gonzales, G.F. (2006) Effect of short-term and long-term treatments with three ecotypes of *Lepidium meyenii* (Maca) on spermatogenesis in rats. *Journal of Ethnopharmacology*, 103, 448–454. https://doi.org/10.1016/j.jep.2005.08.035.

Gonzales, G.F. et al. (2002) Effect of *Lepidium meyenii* (MACA) on sexual desire and its absent relationship with serum testosterone levels in adult healthy men. *Andrologia*, 34(6), 367–372. https://doi.org/10.1046/j.1439-0272.2002.00519.x.

Gonzales, G.F. et al. (2008) Antagonistic effect of *Lepidium meyenii* (red Maca) on prostatic hyperplasia in adult mice. *Andrologia*, 40(3), 179–185. https://doi.org/10.1111/j.1439-0272.2008.00834.x.

Gonzales, G.F. and Gonzales-Castañeda, C. (2009) The Methyltetrahydro-{beta}-Carbolines in Maca (Lepidium meyenii). Evidence-based complementary and alternative medicine :eCAM. 2008/06/19 edn, 6(3), 315–316. doi:10.1093/ecam/nen041.

Gonzales, G.F. (2012) Ethnobiology and Ethnopharmacology of *Lepidium meyenii* (Maca), a Plant from the Peruvian Highlands. *Evidence-Based Complementary and Alternative Medicine: ECAM*, 193496–193496. https://doi.org/10.1155/2012/193496.

Gonzales, G.F., Gasco, M. & Lozada-Requena, I. (2013) Role of Maca (*Lepidium meyenii*) consumption on serum interleukin-6 levels and health status in populations living in the Peruvian Central Andes over 4000 m of altitude. *Plant Foods For Human Nutrition (Dordrecht, Netherlands)*, 68(4), 347–351. https://doi.org/10.1007/s11130-013-0378-5.

Gonzales, G.F., Miranda, S., Nieto, J., Fernandez, G., Yucra, S., Rubio, J. et al. (2005) Red, Maca (*Lepidium meyenii*) reduced prostate size in rats. *Reproductive Biology and Endocrinology*, 3, 5. https://doi.org/10.1186/1477-7827-3-5.

Gonzales, G.F. & Valerio, L.G., Jr. (2006) Medicinal plants from Peru: A review of plants as potential agents against cancer. *Anti-Cancer Agents in Medicinal Chemistry*, 6(5), 429–444.

Gonzales, G.F. & Gonzales-Castañeda, C. (2009) 'The Methyltetrahydro-{beta}-Carbolines in Maca (Lepidium meyenii). *Evidence-based complementary and alternative medicine* :eCAM. 2008/06/19 edn, 6(3), pp. 315–316. doi:10.1093/ecam/nen041.

Gonzales-Arimborgo, C. et al. (2016) Acceptability, safety, and efficacy of oral administration of extracts of black or red Maca (*Lepidium meyenii*) in adult human subjects: A randomized, double-blind, placebo-controlled study. *Pharmaceuticals (Basel, Switzerland)*, 9(3), 49. https://doi.org/10.3390/ph9030049.

Kasprzak, D. et al. (2018) (Maca) – multidirectional health effects – Review. *Current Issues in Pharmacy and Medical Sciences*, 31(3), 107–112. https://doi.org/10.1515/cipms-2018-0021.

Korkmaz, S. ED1, Shalaby, E. ED2 & Azzam, G.M. (2018) Antioxidants in Maca (*Lepidium meyenii*) as a supplement in nutrition. In: *Antioxidants in Foods and Its Applications*. Rijeka: IntechOpen, https://doi.org/10.5772/intechopen.75582.

Mehta, K. et al. (2007) Comparison of glucosamine sulfate and a polyherbal supplement for the relief of osteoarthritis of the knee: A randomized controlled trial [ISRCTN25438351]. *BMC Complementary and Alternative Medicine*, 7, 34–34. https://doi.org/10.1186/1472-6882-7-34.

Meissner, H.O. et al. (2005) Use of gelatinized Maca (*Lepidium peruvianum*) in early postmenopausal women. *International Journal of Biomedical Science : IJBS*, 1(1), 33–45.

Meissner, H.O. et al. (2006) Hormone-balancing effect of pre-gelatinized Organic Maca (*Lepidium peruvianum* Chacon): (III) Clinical responses of early-postmenopausal women to Maca in double blind, randomized, Placebo-controlled, crossover configuration, outpatient study. *International Journal of Biomedical Science : IJBS*, 2(4), 375–394.

Melnikovova, I. et al. (2015) Effect of Lepidium meyenii Walp. on Semen parameters and serum hormone levels in healthy adult men: A double-blind, randomized, placebo-controlled pilot study. *Evidence-Based Complementary and Alternative Medicine*, 1–6. https://doi.org/10.1155/2015/324369.

Orellana Mendoza, E. et al. (2021) Heavy metals in soils and edible tissues of *Lepidium meyenii* (Maca) and health risk assessment in areas influenced by mining activity in the central region of Peru. *Toxicology Reports*, 8, 1461–1470. https://doi.org/10.1016/j.toxrep.2021.07.016.

Oshima, M., Gu, Y. & Tsukada, S. (2003) Effects of *Lepidium meyenii*Walp and *Jatropha macrantha* on blood levels of estradiol-17 β, progesterone, testosterone and the rate of embryo implantation in mice. *Journal of Veterinary Medical Science*, 65(10), 1145–1146. https://doi.org/10.1292/jvms.65.1145.

Penaloza, D. & Arias-Stella, J. (2007) The heart and pulmonary circulation at high altitudes. *Circulation*, 115(9), 1132–1146. https://doi.org/10.1161/CIRCULATIONAHA.106.624544.

Piacente, S. et al. (2002) Investigation of the tuber constituents of Maca (*Lepidium meyenii*Walp.). *Journal ofAgricultural and Food Chemistry*, 50(20), 5621–5625. https://doi.org/10.1021/jf020280x.

Rubio, J. et al. (2006) Effect of three different cultivars of *Lepidium meyenii* (Maca) on learning and depression in ovariectomized mice. *BMC Complementary and Alternative Medicine*, 6, 23–23. https://doi.org/10.1186/1472-6882-6-23.

Rubio, J. et al. (2007) Aqueous and hydroalcoholic extracts of Black Maca (*Lepidium meyenii*) improve scopolamine-induced memory impairment in mice. *Food and Chemical Toxicology*, 45(10), 1882–1890. https://doi.org/10.1016/j.fct.2007.04.002.

Rubio, J. et al. (2011) Aqueous Extract of Black Maca (*Lepidium meyenii*) on Memory Impairment Induced by Ovariectomy in Mice. *Evidence-Based Complementary and Alternative Medicine :ECAM*, 253958–253958. https://doi.org/10.1093/ecam/nen063.

Shin, B.-C. et al. (2010) Maca (*L. meyenii*) for improving sexual function: A systematic review. *BMC Complementary and Alternative Medicine*, 10(1), 44. https://doi.org/10.1186/1472-6882-10-44.

Sprouse, A.A. & van Breemen, R.B. (2016) Pharmacokinetic interactions between drugs and botanical dietary supplements. *Drug Metabolism and Disposition: The Biological Fate of Chemicals*, 44(2), 162–171. https://doi.org/10.1124/dmd.115.066902.

Stojanovska, L. et al. (2015) Maca reduces blood pressure and depression, in a pilot study in postmenopausal women. *Climacteric*, 18(1), 69–78. https://doi.org/10.3109/13697137.2014.929649.

Stone, M. et al. (2009) A pilot investigation into the effect of maca supplementation on physical activity and sexual desire in sportsmen. *Journal of Ethnopharmacology*, 126(3), 574–576. https://doi.org/10.1016/j.jep.2009.09.012.

Tafuri, S. et al. (2019) Chemical analysis of *Lepidium meyenii* (Maca) and its effects on redox status and on reproductive biology in stallions. *Molecules*, 24(10). https://doi.org/10.3390/molecules24101981.

Valentová, K. et al. (2008) Maca (*Lepidium meyenii*) and yacon (Smallanthussonchifolius) in combination with silymarin as food supplements: In vivo safety assessment. *Food and Chemical Toxicology*, 46(3), 1006–1013. https://doi.org/10.1016/j.fct.2007.10.031.

Valentová, K., Buckiová, D., Křen, V., Pe˘knicová, J.., Ulrichová, J. &Šimánek, V. (2006) The in vitro biological activity of *Lepidium meyenii* extracts. *Cell Biology and Toxicology*, 22(2), 91–99.

Valerio, L. & Gonzales, G.F. (2005) Toxicological aspects of South American herbs: *Uncaria tomentosa* (cat's claw) and *Lepidium meyenii* (Maca): A critical synopsis. *Toxicological Reviews*, 24(1), 11–35.

Wang, S. & Zhu, F. (2019) Chemical composition and health effects of Maca (*Lepidium meyenii*). *Food Chemistry*, 288, 422–443. https://doi.org/10.1016/j.foodchem.2019.02.071.

Ybañez-Julca, R.O. et al. (2020) Maca extracts and estrogen replacement therapy in ovariectomized rats exposed at high altitude. *Reproductive Medicine and Biology*, 20(1), 88–95. https://doi.org/10.1002/rmb2.12357.

Zenico, T. et al. (2009) Subjective effects of *Lepidium meyenii* (Maca) extract on well-being and sexual performances in patients with mild erectile dysfunction: A randomised, double-blind clinical trial. *Andrologia*, 41(2), 95–99. https://doi.org/10.1111/j.1439-0272.2008.00892.x.

Zevallos-Concha, A. et al. (2016) Effect of gamma irradiation on phenol content, antioxidant activity and biological activity of black maca and red maca extracts (*Lepidium meyenii*walp). *Toxicology Mechanisms and Methods*, 26(1), 67–73. https://doi.org/10.3109/15376516.2015.1090512.

Zha, S. et al. (2014) Extraction, purification and antioxidant activities of the polysaccharides from Maca (*Lepidium meyenii*). *Carbohydrate Polymers*, 111, 584–587. https://doi.org/10.1016/j.carbpol.2014.05.017.

14 Phytopharmacological Aspects of *Valeriana officinalis* Root

Pratibha Kumari, Anupam Bhatt, and Dipika Rana

ABBREVIATIONS

Ac = acetyl; Iv = isovaleryl; Iv-Iv = isovaleryloxyisovaleryl; MV = β-methylvaleryl.

14.1 INTRODUCTION

The valerian plant, *Valeriana officinalis* L., belongs to the family Caprifoliaceae. For thousands of years, plants from this family have been used as medicinal herbs. *V. officinalis,* commonly called valerian, Indian valerian, and garden heliotrope, is called Tagara and Nata in Ayurveda, and in folk medicine, it is known as Sugandhbala (Khare, 2008; McCabe et al., 2002; Carlini et al., 2003; Houghton et al., 1999; Miyasaka et al., 2006; Grünwald, 1995; Morazzoni, 1995; Gupta and Virmani 1968; Rücker et al., 1978; Violon et al., 1984). *Valeriana officinalis* var. *latifolia* is a perennial herb distributed throughout Asia, Europe, and North America (Houghton, 1999). *V. officinalis* is geographically distributed in the Himalayas, Kashmir, Afghanistan, Europe, China, Nepal, and Pakistan (Figure 14.1). *V. officinalis* has pinnately dissected leaves bearing 6–10 pairs of lance-shaped leaflets, with numerous small pink or white flowers in dense clusters. It grows up to 1.5 m in height, and can be glabrous or more-or-less pubescent. It has a solitary, short, sub-erect stem with furrowed margins; the stem is not thicker than the roots, and the plant has stoloniferous rootstocks. The mature seeds are small, tapered, 5 mm in diameter, and almost hairless (Nandhini et al., 2018) (Figure 14.2).

The therapeutic properties attributed to valerian include anxiety relief, sedation, and insomnia treatment (Patocka et al., 2010). Valerian essential oils, representing approximately 0.2–2.8% of the plant by weight, are obtained by steam distillation of the roots and rhizomes of *V. officinalis* and contain sesquiterpenes, including valerenic acid, isovaleric acid, valeric acid, valerenal, valeranone, and other compounds (Circosta et al., 2007). According to Patocka & Jakl (2010), the root extract of *V. officinalis* contains alkaloids of between 0.01 and 0.05% by weight, volatile essential oil from 0.02 to 0.08%, and iridoid valepotriates from 0.5% to 2.0% by weight. Alpha-methyl, skyanthine, valeranine, chatinine, and pyrylketone are the major alkaloids (Patocka et al., 2010). Similarly, extracts of aboveground tissues contain 0.5% to 2.0% of iridoid valepotriates, triterpenes, lignans, and flavonoids (Murti et al., 2011). Similarly, the dry extract of *V. officinalis* contained 0.020 to 0.075% valeric acid (Dugaheh et al., 2013).

The plant *V. officinalis* is native to Europe and Asia but has now spread to North America. This plant species grows best in wet wooded areas and has been widely grown in northern Europe. The Netherlands produces the majority of Europe's supply. It is best grown in low-lying, limed wet sandy soil. It is harvested and dried in the late autumn between August and September (Huang, 1999).

14.2 MEDICINAL USES

Roots of *V. officinalis* have long been used in traditional herbal medicine for relaxing and antispasmodic treatments (Oshima et al., 1995; Houghton et al., 1999), and to treat heart arrhythmia (Jia et al., 1999). Valerian root has been used as a soothing or sleep-inducing product for over a century (Becker et al., 1985; McCabe et al., 2002). Hysteria, hypochondriasis, nervous unrest, sleeplessness,

DOI: 10.1201/b22924-14

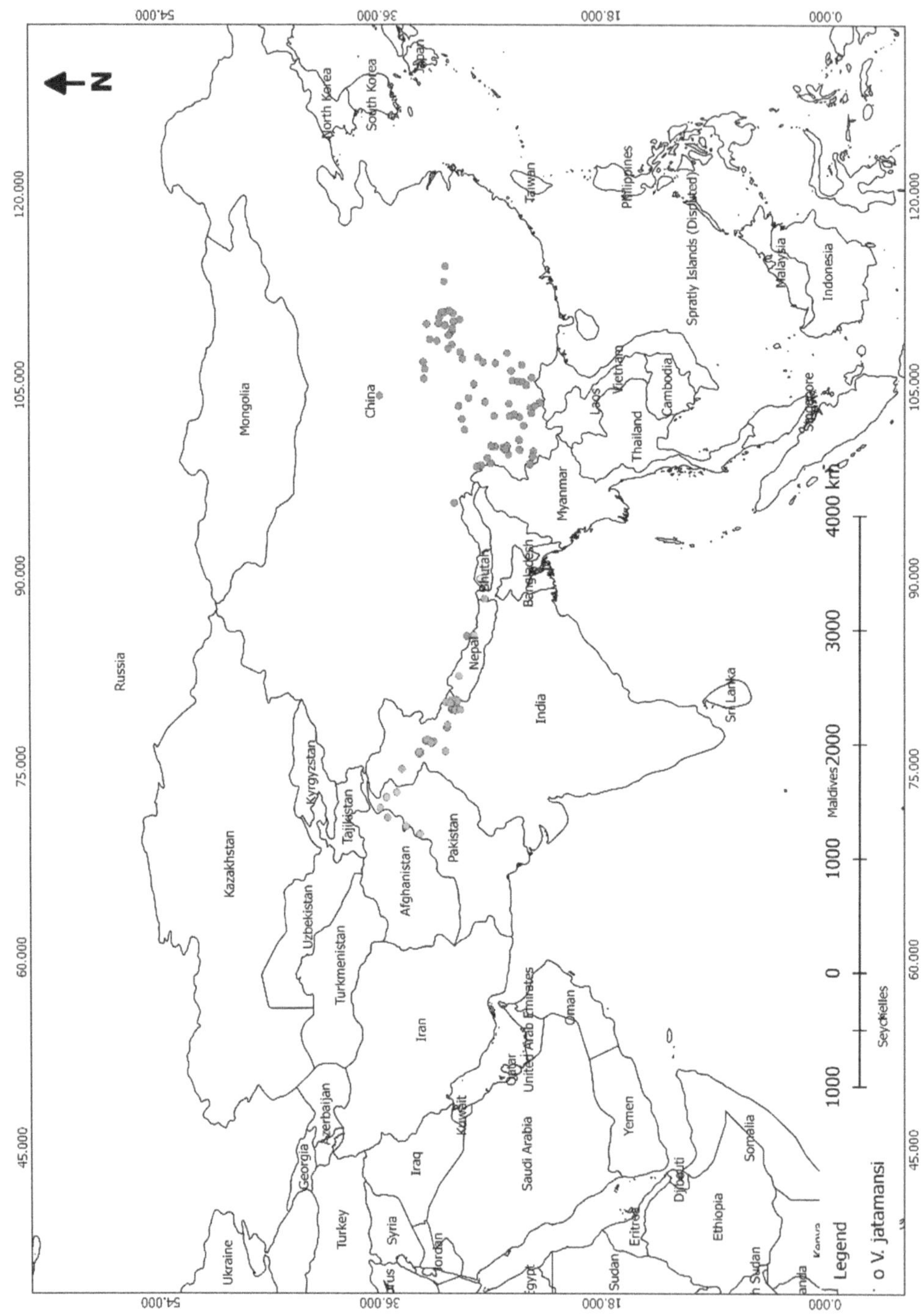

FIGURE 14.1 Map depicting the distribution of *Valeriana officinalis* around the world.

FIGURE 14.2 *Valeriana officinalis* plant, bearing flowers.

and other emotional disorders are treated with it (Council of Industrial and Clinical Research, 2009; Nithya et al., 2016). For at least 1,000 years, valerian has been used to treat stomach issues and urinary infections. Its bioactivities include sedative, anxiolytic, depressive, antispasmodic, and anti-HIV properties (Hazelhoff et al., 1982; Morazzoni et al., 1995; Murakami et al., 2002; Carlini et al., 2003; Miguel et al., 2008). According to research, valerian has a sedative effect on the brain that invigorates stressed individuals and energizes exhausted individuals, putting the system into balance (Muller et al., 2006). It is unclear which elements of *V. officinalis* preparations are responsible for the relaxing and anxiolytic properties of valerian extracts. Components of valerian extracts include valepotriates (iridoids), valerenic acid (a sesquiterpenoid molecule), and valerenic acid derivatives. It is generally believed that, among these constituents, valepotriates and valerenic acids provide the calming effects of valerian extracts, but these properties need further investigation (Hendricks et al., 1981). It is mainly sold as a sleeping aid in the United States, but it is also used to treat uneasiness, anxiety, and shivering in Europe (Houghton et al., 1999). In Brazil, it is used as a traditional drug for its relaxing, antidepressant, and neuropsychiatric properties (Hendrick et al., 1981). The rhizome of *V. officinalis* has been used as a sedative in Europe for centuries. It acts as a mild sedative or drug to treat insomnia; thus, it is frequently used as a milder substitute for harsher, synthetic drugs that induce sedation. Benzodiazepines are used to treat anxiety disorders and sleep problems caused by anxiety (Miyasaka et al., 2006), but are addictive. Valerian extracts are readily accessible supplements for dietary intake, mainly consisting of dried root or root extracts encapsulated in gelatin or tablets. A single dose contains between 50 mg and 1 g of dried root extract. These supplements are widely used; approximately 210 million doses are sold in the United States and 125 million doses in Europe each year (Grunwald, 1995). It has been prescribed to treat insomnia, although studies don't support this use (Sipnella, 2001). Studies have also shown that restlessness, anxiety, agitation, and stress are valerian's most common harmful side-effects (Tariq & Pulisetty, 2008). It is known to cause side-effects when taken in large amounts, and it, like most sleep aids, can lead to dependency (Garges et al., 1998). It is not recommended to use valerian if you have liver issues (Cohen & Del Torro, 2008). Although it has been reported to be a useful treatment for anxiety, it has also been linked to headaches and sleeping problems in some people. Valerian can potentially irritate some people as they may not possess the digestive conversion characteristics (Dennehy et al., 2005).

14.3 FUNDAMENTAL USES OF *V. OFFICINALIS*

Ayurveda describes valerian as a versatile medicine that can treat various illnesses and diseases. It was given to soldiers who had suffered 'shellshock' during World War I. Lotions containing valerian extract can also be used to soothe skin irritations and swollen joints, relieve menstrual cramps, and act as a carminative, easing gas in the stomach and intestines. Valerian may be an effective sleep aid and tranquilizer, although researchers disagree on its effectiveness. According to another study in the United States, valerian is used as a calming agent; findings by a Swiss team showed that valerian and lemon balm mixtures induced sleep significantly better than a placebo. Multiple studies currently available have not been conclusive, so further research may help clarify the question (National Health Institute Report, 2017). Currently, it is not used as a food source. The European Council has listed *Valerianas* pp. as a natural food flavourant, and valerian is 'generally recognized as safe', GRAS (Lanje et al., 2020).

14.4 PHYTOCHEMISTRY OF *VALERIANA OFFICINALIS*

Rhizomes of *Valeriana* spp. are commonly used as supplements, consumed as tablets, liquid extracts, and sometimes as teas. The root is used as a medicine in the form of fresh juice or dried powder (Hendricks, 1981). Even though the calming properties of the root were explored years ago, the chemical components responsible for these effects have yet to be discovered and confirmed (Morazzoni, 1995). There is little relationship between the volatile oil concentration of the plant and its clinical effects (Hendricks, 1985; Holzl, 1998). Valepotriates (baldrinals), valerenic acid, valerenal, and valeranone, and other compounds present in valerian essential oil have all been linked at some point to valerian's impact on the central nervous system (Franck et al., 1970). Valerian contains around 150 chemical compounds, many of which have physiological effects (Jiang et al., 2007). Numerous active pharmacological compounds have been evaluated in valerian extracts, including terpenes, alkaloids, organic acids, and their derivatives, and valepotriates and flavones. Generally, it is accepted that the valepotriate compounds are responsible for the relaxing effect of *V. officinalis* (Patocka et al., 2010).

14.4.1 Alkaloids

Valerian contains alkaloids (0.01–0.05%), particularly terpene alkaloids (Patocka et al., 2010). Valerianine, chatinine, valerianine, naphthyridin, alpha-methyl pyrryl ketone, and actinidine methyl ketone are the most prominent valerian alkaloids (Duke, 1985).

14.4.1.1 Actinidine

The essential oil of valerian root contains a monoterpene pyridine alkaloid that is cyclopenta [c] pyridine skeleton-derived (Johnson et al., 1971) (Figure 14.3). The compound actinidine, which attracts cats, is found in valerian. Quinolinic acid and lysine are the precursors in the biosynthesis of actinidine, (Auda et al., 1967). B Actinidine inhibits the activity of GABA-ergic signalling by interfering with its metabolism and gamma-aminobutyric acid (GABA) and acts as an agonist on Benzodiazepine receptors. As a result, it triggers the allosteric modulation of GABA receptors (Patocka et al., 2010).

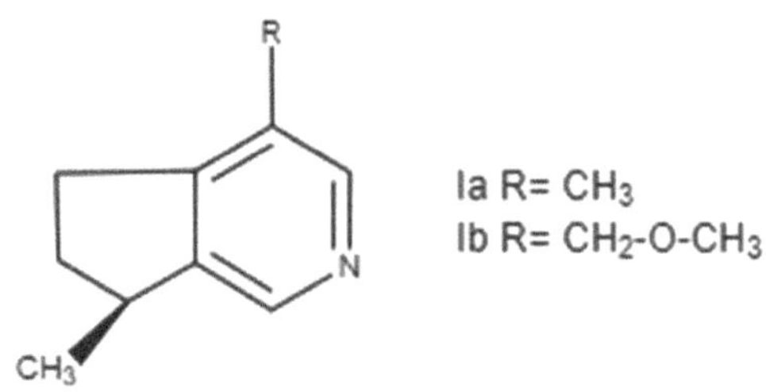

FIGURE 14.3 The chemical structure of major compounds present in essential oil of *Valeriana officinalis*. Valerian alkaloids actinidine (Ia) and valerianine (Ib).

14.4.1.2 Chatinine

Chatinine can be isolated from valerian, though no studies have been carried outon its biological properties (Baby et al., 2005). In 1970, a study was conducted in Germany on alpha-methyl pyrryl ketone for its effects on the central nervous system (Pilerood et al., 2013).

14.4.2 Amino Acids

Tyrosine, γ-amino butyric acid (GABA), glutamine were among the free amino acids found in water extracts of *Valeriana* spp. GABA is a well-studied neurotransmitter that has an inhibitory role in various metabolic processes (Huang et al., 2006; Wu et al., 2008; Chen et al., 2013).

14.4.3 Flavonoids

Flavonoids are present in very low concentrations in *Valeriana* spp., such as acacetin, linarin, diosmetin, apigenin, luteolin, quercetin, kaempferol, and luteolin (Nazarova et al., 1981; Santos et al., 1994; Piccinelli et al., 2004).

14.4.4 Terpenes and Organic Acids

The volatile valerian essential oil contains organic acids and terpenes and accounts for 0.2 to 2.8% of its dry weight of root. The plant's roots and aerial parts are the source of essential oils (Funke et al., 1975). Monoterpenes and sesquiterpenes are two chemical groups of terpenes. The most important valerian organic compounds include valeric acid, bornyl acetate, valerenic acid, isovalerenic acid, acetoxyvalerenic acid, bornyl isovalerenate, isovaleric-1-pinene, 1-camphene, 1-borneol, terpineol, valeranone, and cryptofauronol. Few oil constituents are thought to have sedative characteristics. The bornyl isovalerate and isovaleric acid are the sources of valerian's distinctive fragrance. The volatiles 3-methylbutanoic acid and isovaleric acid have no pharmacological or toxic qualities but share the drug's odour. On the other hand, it was discovered in 2007 that isovaleric acidaemic patients have lower ATPase activity in the synaptic membranes of the cerebral cortex, which may contribute to their neurological impairments (Ribeiro et al., 2007).

14.4.4.1 Valeric Acid

Isovaleric acid, and 3-methylbutanoic acid, is used in perfumes and cosmetics as a pleasant odour alternative to n-pentanoic acid (Figure 14.4). N-pentanoic acid has a very unpleasant odour, whereas its volatile esters tend to have pleasant odours. That is also the case for 3-methylbutanoic acid. The two acids share only a common odour, but have no significant pharmacological or toxicological effects, although 3-methylbutanoic acid is thought to be responsible for the pathophysiology of neurological dysfunction due to its reduction of ATPase activity (Ribeiro et al., 2007).

14.4.4.2 Valerenic Acid

Valerenic acid (Figure 14.5IIa) and its aldehyde, valerenal (Figure 14.5IIb), are monoterpenes that have pharmacological properties (Figure 14.5). Valerian is thought to work through the gamma-aminobutyric acid (GABA) pathway. Research has shown that a valerian extract binds to GABA receptors, but the effect of such a binding is still to be proven (Cavadas et al., 1995). The medicinal

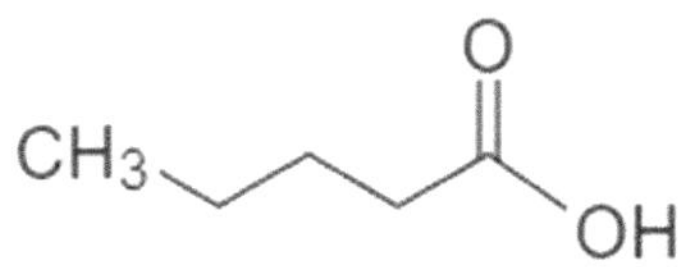

FIGURE 14.4 Valeric acid.

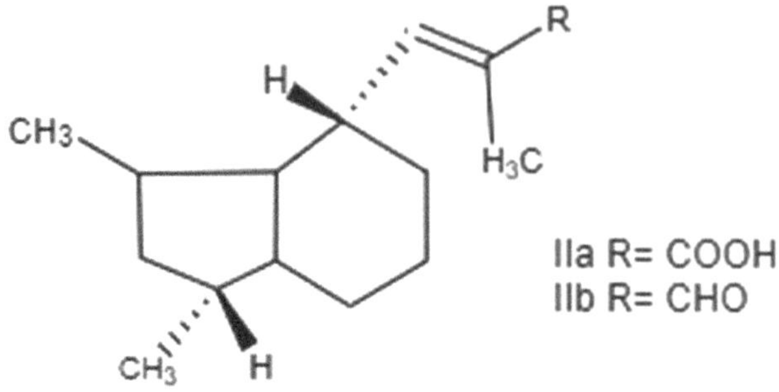

FIGURE 14.5 Valerenic acid (IIa) and its aldehyde, valerenal (IIb).

effects of valerenic acid and valerian extract are due to the regulation of GABAA receptor activity, according to data from the study. Actions of valerian extract and valerenic acid mediate the GABAA receptor function modification. Thus, valerian increases the relaxing effects of anaesthetics and other drugs that operate via GABA receptors, meaning that valerian used prior to surgery could produce valerian–anaesthetic interactions (Yuan et al., 2004). Valerenic acid has recently been found to influence GABAA receptors (Trauner et al., 2008); this is also known to diffuse into the nervous system transcellularly (Neuhaus et al., 2008). According to Dietz et al. (2005), valerenic acid is a receptor agonist of the 5-HT region, with a high affinity for the 5-HT(5a) receptor but only a low affinity for the 5-HT(2b) receptor and the serotonin transporter. Acetylvalerenolic acid, valerenic acid, and valerenal were found to be NF-B prohibitors at a concentration of 100 g/mL. NF-B activity was reduced by 4% with acetylvalerenolic acid, while NF-B activity was lowered by 25% with valerenic acid (3) (Jacobo-Herrera et al., 2006).

14.4.4.3 Valeranone

Valeranone was tested as a medical drug to address hyperkinetic behaviour disorders (Gupta & Virmani 1968). Valeranone (Figure 14.6) was pharmacologically investigated in animal experiments for its sedative, tranquilizing, and antihypertensive properties, but the activity of valeranone was lower than those of the standard substances used (Rücker et al., 1978).

14.4.5 Valepotriates

Valepotriates are iridoid monoterpenes that have been esterified. The name derives from the valeriana-epoxy-triester of polyhydroxycyclopenta-(c)-pyrans with carboxylic acids: á-isovaleroxy-isovaleric, acetic, âacetoxy-â-methyl valeric acid, valeric, isovaleric, â-methyl valeric, â-acetoxy-isovaleric, and â-hydroxyisovaleric (Thies, 1969). This is a significant element, with 50–80%active chemicals. There are the following types of valepotriates: monoene and diene

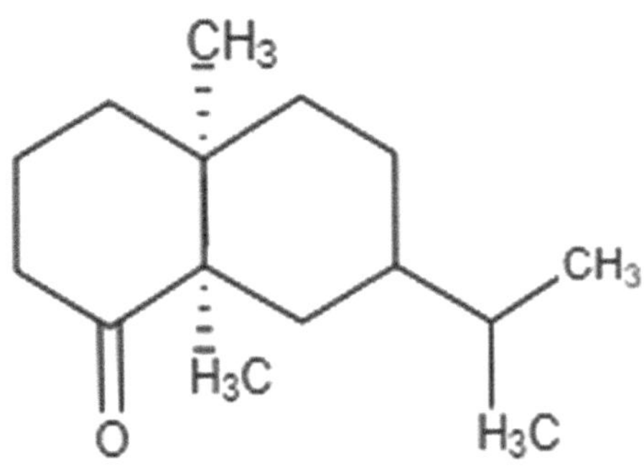

FIGURE 14.6 The terpene valeranone.

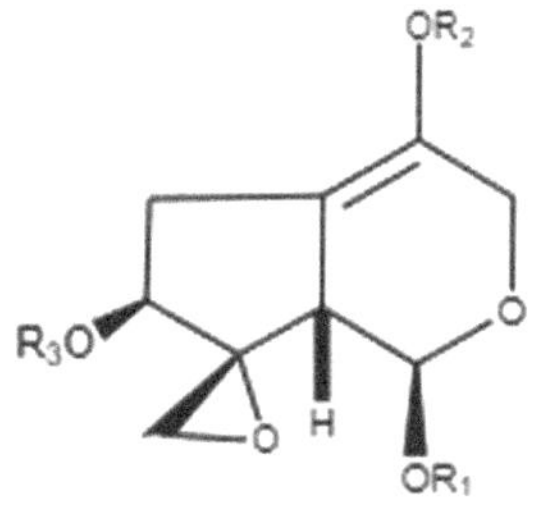

IVa R_1= Iv, R_2 = Ac, R_3 = Iv
IVb R_1= Iv, R_2 = Iv, R_3 =Ac
IVc R_1 = Iv, R_2 = Ac, R_3 = Ac
IVd R_1 = Iv, R_2 = Ac, R_3 = MV

FIGURE 14.7 The structure of diene valepotriates from *Valeriana officinalis*: valtrate (IVa), isovaltrate (IVb), 7-desisovaleroyl-7-acetyl-valtrate (IVc). and homovaltrate (IVd).

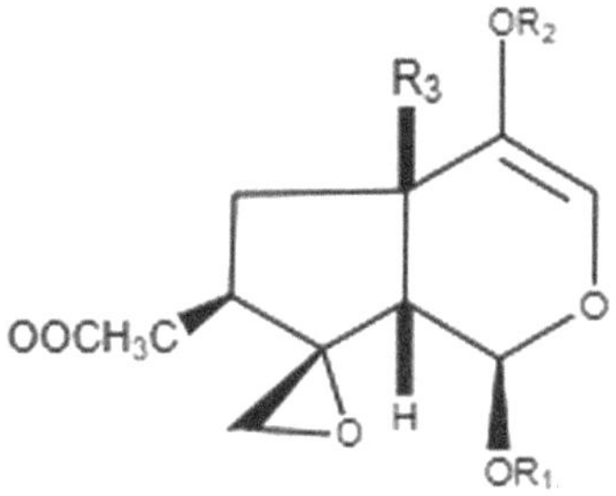

Va R_1= Iv, R_2 = Iv, R_3 = H
Vb R_1= Iv, R_2 = Iv-Iv, R_3 = OH

FIGURE 14.8 The structure of monoene valepotriates from *Valeriana officinalis*: didrovaltrate (Va), isovaleoxyhydrooxydidrovaltrate (Vb).

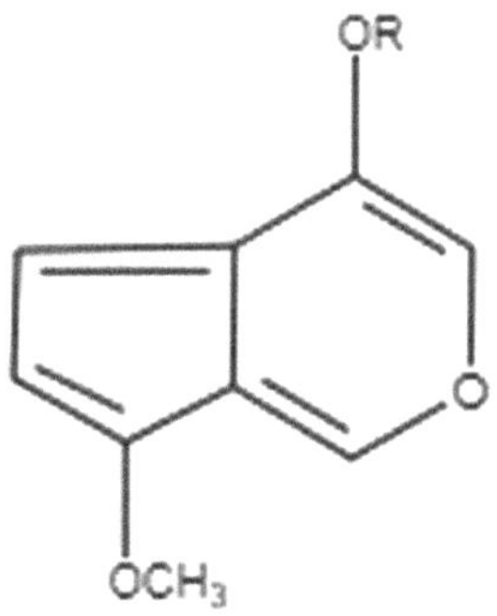

VIa R= Ac
VIb R= Iv

FIGURE 14.9 Degradation products of valtrates are baldrinal (VIa) and homobaldrinal (VIb).

derivatives. 7-desisovaleroyl-7-acetylvaltrate (Ivc, Figure 14.7), valtrate (Iva, Figure 14.7), isovaltrate (Ivb, Figure 14.7), and 7-homovaltrate (Ivd, Figure 14.7) are the most common diene valepotriates, whereas didrovaltrate (Va, Figure 14.8) and is ovaleroxyhydroxydidrovaltrate (Vb, Figure 14.8) are the most common monoene derivatives (Figure 14.8). The amount of valepotriates found in different animals varies greatly. The underground organs of valerian have a higher valepotriate concentration than does the rest of the plant (Violon et al., 1984). Valepotriates are generally thermolabile, indicating that they dissolve quickly in acidic or alkaline water and alcoholic solutions. In anhydrous methanol, the diene valepotriates were relatively stable when kept at 20°C. When dissolved in methanol or ethanol with a small amount of aqua and kept at ambient temperature for a few weeks, it decomposes by 90% (Bos et al., 1996). The yellow-coloured baldrinals (Via, Figure 14.9) are the major degradation products of the valepotriates (Bos et al., 2002). As a result of their chemical reactivity, baldrinals can form polymers (Bos et al., 1996).

14.5 PHARMACOLOGICAL ASPECTS

The history of valerian is a long one. Hippocrates first described it and later prescribed it to treat insomnia, so valerian has been used for medicine for centuries. It is used in pharmaceuticals because of its sedative, anticonvulsant, depressive, antihypertensive, narcotic, antispasmodic, and anxiolytic properties (Plushner, 2000). Recently, it was shown that the *V. officinalis* aqueous extract has cyto-protective effects towards human neuroblastoma cells. A study in 1998 found that valerian was the tenth most popular medicinal herb in the United States. Valerian extracts are now widely used as dietary supplements, and the US sells 210 million doses of the drug annually, compared with 125 million doses in Europe. They usually consist of either dry root or extracts packaged into pills or gelatin capsules (Lanje et al., 2020). Other pharmacological properties of *V. officinalis* are cardio-vascular (Zhang et al., 1982), anti-anxiety (Hattesohl et al., 2008), sedative (Fernandez et al., 2004), gastrointestinal, attention-enhancing, and other neurological activities (Criscosta et al., 2007).

14.6 CONCLUSION

V. officinalis is known for its broad medicinal properties and usage in folk medicines. Ayurveda describes valerian as a versatile medicine that can treat various illnesses and diseases. The relaxing and tranquilizing properties of valerian roots make them popular for treating insomnia and serving as an antidepressant. Studies have shown that it has been widely used for various functions like curing skin irritations and swollen joints, as a carminative, and for relieving menstrual cramps. Though there have been multiple studies on the effectiveness and usage of *V. officinalis*, more precise and in-depth investigations are required to achieve further bioprospection and product development. Furthermore, the Indigenous mountain people's large-scale cultivation of the species can be a valuable source of income. It is essential that proper cultivation training is imparted to the farmers for popularization and maximum output of the product.

REFERENCES

Auda, H., Waller, G.R. & Eisenbraun, E.J. (1967) Biosynthesis of methylcyclopentane monoterpenoids: III. ACTINIDINE. *Journal of Biological Chemistry*, 242(18), 4157–4160.

Baby, R., Cabezas, M., Castro, E., Filip, R. & de Reca, N.W. (2005) Quality control of medicinal plants with an electronic nose. *Sensors and Actuators B: Chemical*, 106(1), 24–28.

Becker, H. & Chavadej, S. (1985) Valepotriate production of normal and colchicine-treated cell suspension cultures of Valeriana wallichii. *Journal of Natural Products*, 48(1), 17–21.

Bos, R., Woerdenbag, H.J. & Pras, N. (2002) Determination of valepotriates. *Journal of Chromatography A*, 967(1), 131–146.

Bos, R., Woerdenbag, H.J., Hendriks, H., Zwaving, J.H., De Smet, P.A., Tittel, G. et al. (1996) Analytical aspects of phytotherapeutic valerian preparations. *Phytochemical Analysis*, 7(3), 143–151.

Carlini, E.A. (2003) Plants and the central nervous system. *Pharmacology Biochemistry and Behavior*, 75(3), 501–512.

Cavadas, C., Araujo, I., Cotrim, M.D., Amaral, T., Cunha, A.P., Macedo, T. et al. (1995) In vitro study on the interaction of Valeriana officinalis L. extracts and their amino acids on GABAA receptor in rat brain. *Arzneimittel-Forschung*, 45(7), 753–755.

Chen, J.S., Wu, J.K., Liu, L., Zhang, Y., Wang, F.J. & Du, X.W. (2013) Studies on improving sleep function and relative mechanism of mice by petroleum extract of Valeriana amurensis. *Chinese Journal of Experimental Traditional Medical Formulae*, 19(24), 245–249.

Circosta, C., De Pasquale, R., Samperi, S., Pino, A. & Occhiuto, F. (2007) Biological and analytical characterization of two extracts from Valeriana officinalis. *Journal of Ethnopharmacology*, 112(2), 361–367.

Cohen, D.L. & Del Toro, Y. (2008) A case of valerian-associated hepatotoxicity. *Journal of Clinical Gastroenterology*, 42(8), 961–962.

Council of Scientific and Industrial Research. (2009) *Valeriana officinalis* Linn. The wealth of India: A dictionary of Indian materials and industrial products. *New Delhi: Council of Scientific and Industrial Research*, 9, 426.

Dennehy, C.E., Tsourounis, C. & Horn, A.J. (2005) Dietary supplement-related adverse events reported to the California poison control system. *American journal of health-system pharmacy*, 62(14), 1476–1482.

Dietz, B.M., Mahady, G.B., Pauli, G.F. & Farnsworth, N.R. (2005) Valerian extract and valerenic acid are partial agonists of the 5-HT5a receptor in vitro. *Molecular Brain Research*, 138(2), 191–197.

Dugaheh, M.A., Meisami, F., Torabian, Z. & Sharififar, F. (2013) Antioxidant effect and study of bioactive components of Valeriana sisymbriifolia and Nardostachys jatamansii in comparison to Valeriana officinalis. *Pakistan Journal of Pharmaceutical Sciences*, 26(1), 53–59.

Duke, J.A. (1985) *CRC Handbook of medicinal herbs*. Boca Raton, FL: CRC.

Fernández, S., Wasowski, C., Paladini, A.C. & Marder, M. (2004) Sedative and sleep-enhancing properties of linarin, a flavonoid-isolated from *Valeriana officinalis*. *Pharmacology Biochemistry and Behavior*, 77(2), 399–404.

Franck, B., Petersen, U. & Huper, F. (1970) Valerianine, a tertiary monoterpene alkaloid from valerian. *Angewandte Chemie International Edition*, 9, 891.

Funke, E.D. and Friedrich, H. (1975) Valepotriates in the aerial parts of some more valerianaceae species (author's transl). *Planta Medica*, 28(3), 215–224.

Gupta, P.D. & Virmani, V. (1968) Clinical trial of jatamansone (syn: valeranone) in hyperkinetic behavior disorders. *Neurol India*, 16, 168–173.

Garges, H.P., Varia, I. & Doraiswamy, P.M. (1998) Cardiac complications and delirium associated with valerian root withdrawal. *Jama*, 280(18), pp.1566–1567.

Grünwald, J. & Buettel, K. (1995) The European phytomedicines market: Figures, trends, analysis. *Herbal Gram*, 34, 60–65.

Hattesohl, M., Feistel, B., Sievers, H., Lehnfeld, R., Hegger, M. & Winterhoff, H. (2008) Extracts of Valeriana officinalis L. sl show anxiolytic and antidepressant effects but neither sedative nor myorelaxant properties. *Phytomedicine*, 15(1–2), 2–15.

Hazelhoff, B., Malingré, T.M. & Meijer, D.K. (1982) Antispasmodic effects of valeriana compounds: An in-vivo and in-vitro study on the guinea-pig ileum. *Archives internationales de pharmacodynamie et de thérapie*, 257(2), 274–287.

Hendriks, H., Bos, R., Allersma, D.P., Malingré, T.M. & Koster, A.S. (1981) Pharmacological screening of valerenal and some other components of essential oil of Valeriana officinalis. *Planta Medica*, 42(05), 62–68.

Hendriks, H., Bos, R., Woerdenbag, H.J. & Koster, A.S. (1985) Central nervous depressant activity of valerenic acid in the mouse. *Planta Medica*, 51(01), 28–31.

Hendriks, H., Geertsma, H.J. & Malingre, T.M. (1981) The occurrence of valeranone and crypto-fauronol in the essential oil of Valeriana officinalis L. sl collected in the northern part of The Netherlands. *Pharmaceutisch weekblad*, 3(1), 1316–1320.

Holzl, J. (1998) Valerian-Valerian officinalis. *Zeitschrift fur Phytotherapie*, 19, 47–54.

Houghton, P.J. (1999) The scientific basis for the reputed activity of Valerian. *Journal of Pharmacy and Pharmacology*, 51(5), 505–512.

Huang, B., Zheng, H. & Qin, L. (2006) Material basis of the sedative and hypnotic activities of Valeriana officinalis. *Pharmaceutical Care and Research*, 6(3), 165.

Huang, K.C. (1999) Tonics and supporting herbs. *The pharmacology of Chinese herbs*. CRC Press, Boca Raton. 263–264.

Jacobo-Herrera, N.J., Vartiainen, N., Bremner, P., Gibbons, S., Koistinaho, J. & Heinrich, M. (2006) F- κB modulators from Valeriana officinalis. *Phytotherapy Research: An International Journal Devoted to Pharmacological and Toxicological Evaluation of Natural Product Derivatives*, 20(10), 917–919.

Jia, J.N. & Zhang, B.H. (1999) Effect of valerian extract (V3d) on cardiovascular system [J]. *J Guangxi Coll Tradit Chin Med*, 16, 40–42.

Jiang, X., Zhang, J.C., Liu, Y.W. & Fang, Y. (2007) Studies on chemical constituents of *Valeriana officinalis. Zhong yao cai= Zhongyaocai= Journal of Chinese Medicinal Materials*, 30(11), 1391–1393.

Johnson, R.D. & Waller, G.R. (1971) Isolation of actinidine from Valeriana officinalis. *Phytochemistry*. 10:3334.

Khare, C.P. (2008) *Indian medicinal plants: An illustrated dictionary*. Springer Science & Business Media; Springer New York, pp. 1–692.

Lanje, C.N., Patil, S.R. & Wankhade, A.M. (2020) Medicinal natural drug of Valerian (Valerina Officinalis): an-over review. *American Journal of PharmTech Research*, 10(01), 148–173.

McCabe, S. (2002) Complementary herbal and alternative drugs in clinical practice. *Perspectives in Psychiatric Care*, 38(3), 98–107.

Miguel, H., Feistel, B., Hartwig, S., Romanus, L., Mirjam, H., & Hilke, W. (2008) Extracts of Valeriana officinalis L. S.L. show anxiolytic and antidepressant effects but neither sedative nor myorelaxant properties. *Phytomedicine*, 15, 2–15.

Miyasaka, L.S., Atallah, Á.N. & Soares, B. (2006) Valerian for anxiety disorders. *Cochrane Database of Systematic Reviews* (4), CD004515.

Morazzoni, P. & Bombardelli, E. (1995) Valeriana officinalis: Traditional used and recent evaluation of activity. *Fitoterapia (Milano)*, 66(2), 99–112.

Müller, S.F. & Klement, S. (2006) A combination of valerian and lemon balm is effective in the treatment of restlessness and dyssomnia in children. *Phytomedicine*, 13(6), 383–387.

Murakami, N., Ye, Y., Kawanishi, M., Aoki, S., Kudo, N., Yoshida, M. et al. (2002) New Rev-transport inhibitor with anti-HIV activity from Valerianae Radix. *Bioorganic & Medicinal Chemistry Letters*, 12(20), 2807–2810.

Murti, K., Kaushik, M., Sangwan, Y. & Kaushik, A. (2011) Pharmacological properties of Valeriana officinalis-a review. *Pharmacologyonline*, 3, 641–646.

Nandhini, S., Narayanan, K.B. & Ilango, K. (2018) Valeriana officinalis: A review of its traditional uses, phytochemistry and pharmacology. *Asian Journal of Pharmaceutical and Clinical Research*, 11(1), 36–41.

National Institutes of Health and National Institutes of Health. (2017) Office of dietary supplements. *Department of Health and Human Services. ods. od. nih. gov/factsheets/iron-HealthProfessional* (1).

Nazarova, I.P., Glushenkova, A.I. & Umarov, A.U. (1981) Gossypol-like compounds of the cotton plant. Methods of determining gossypol. *Chemistry of Natural Compounds*, 17(2), 87–102.

Neuhaus, W., Trauner, G., Gruber, D., Oelzant, S., Klepal, W., Kopp, B. et al. (2008) Transport of a GABAA Receptor Modulator and Its Derivatives from Valeriana officinalis L. s. l. Across an in Vitro Cell Culture Model of the Blood-Brain Barrier. *Planta Medica*, 74(11), 1338–1344.

Nithya, S. & Muthuraman, M.S. (2016) An overview on the biological perspectives of Nardostachys Jatamansi. *International Journal of Pharmacy and Pharmaceutical Sciences*, 8(2), 31–36.

Oshima, Y., Matsuoka, S. & Ohizumi, Y. (1995) Antidepressant principles of Valeriana fauriei roots. *Chemical and Pharmaceutical Bulletin*, 43(1), 169–170.

Patočka, J. & Jakl, J. (2010) Biomedically relevant chemical constituents of Valeriana officinalis. *Journal of Applied Biomedicine*, 8(1), 11–18.

Piccinelli, A.L., Arana, S., Caceres, A., di Villa Bianca, R.D.E., Sorrentino, R. & Rastrelli, L. (2004) New lignans from the roots of Valeriana p rionophylla with antioxidative and vasorelaxant activities. *Journal of Natural Products*, 67(7):1135–1140.

Pilerood, S.A. & Prakash, J. (2013) Nutritional and medicinal properties of valerian (Valeriana officinalis) herb: A review. *International Journal of Food Sciences and Nutrition Diet*, 1, 25–33.

Plushner, S.L. (2000) Valerian: Valeriana officinalis. *American Journal of Health-System Pharmacy*, 57(4), 328–335.

Ribeiro, C.A., Balestro, F., Grando, V. & Wajner, M. (2007) Isovaleric acid reduces Na+, K+-ATPase activity in synaptic membranes from the cerebral cortex of young rats. *Cellular and Molecular Neurobiology*, 27(4), 529–540.

Rücker, G., Tautges, J., Sieck, A. Wenzl, H. & Graf, E. (1978) Isolation and pharmacodynamic activity of the sesquiterpene aleranone from Nardostachys jatamansi DC (in German). *Arzneimittelforschung*, 28:7–13.

Santos, M.S., Ferreira, F., Faro, C., Pires, E., Carvalho, A.P. & Cunha, A.P. (1994) The amount of GABA present in aqueous extracts of valerian is sufficient to account for [3H] GABA release in synaptosomes. *Planta Medica*, 60(5), 475–476.

Spinella, M. (2001) Herbal medicines and epilepsy: The potential for benefit and adverse effects. *Epilepsy & ehavior*, 2(6), 524–532.

Tariq, S.H. & Pulisetty, S. (2008) Pharmacotherapy for insomnia. *Clinics in Geriatric Medicine*, 24(1), 93–105.

Thies, P.W. (1969) On the chromomgenic behavior of valepotriate. 5. Report on the active substances of Valerian. *Arzneimittel-forschung*, 19(3), 319–322.

Trauner, G., Khom, S., Baburin, I., Benedek, B., Hering, S. & Kopp, B. (2008) Modulation of GABAA receptors by valerian extracts is related to the content of valerenic acid. *Planta Medica*, 74(01), 19–24.

Violon, C., Dekegel, D. & Vercruysse, A. (1984) Relation between valepotriate content and differentiation level in various tissues from Valerianeae. *Journal of Natural Products*, 47(6), 934–940.

Wu, B., Fu, Y.-M., Huang, A.-H. & Ma, Y.-J. (2008) Changes of GABA and Glu content in hippocampus of PTZ-induced epileptic rats treated with volatile oil of *Valeriana*. *Chinese Archives of Traditional Chinese Medicine*, 26(11), 2476–2477.

Yuan, C.S., Mehendale, S., Xiao, Y., Aung, H.H., Xie, J.T. & Ang-Lee, M.K. (2004) The gamma-aminobutyric acidergic effects of valerian and valerenic acid on rat brainstem neuronal activity. *Anesthesia & Analgesia*, 98(2), 353–358.

Zhang, B.H., Meng, H.P., Wang, T., Dai, Y.C., Shen, J., Tao, C. et al. (1982) Effects of Valeriana officinalis L extract on cardiovascular system. *Yao xue xue bao= Acta pharmaceutica Sinica*, 17(5), 382–384.

15 Photochemistry, Medicinal and Commercial Applications of *Beta vulgaris* L.

M.C. Purohit, Anuj Kandwal, Sonali Purohit, Archana Bachheti, and Arun Kumar Khajuria

15.1 INTRODUCTION

Plants and plant-based preparations have been of interest in human communities since antiquity. Each community develops its traditions with its native flora and fauna to serve its objectives. In ancient literature, it is also clearly mentioned that a person of an area/zone/locality can be cured by local flora present in their surroundings; in other words, one can understand the source plant for medicines to cure an ailment by observing the habit and shape of the plant as it resembles the different human organs (Dhiman et al., 2006; Khajuria et al., 2021; Bachheti et al., 2022; Asfaw et al., 2022).

There has been a swing in the increasing global demand for herbal products for curing ailments and in diet (Bisht & Khajuria, 2014; Khajuria et al., 2021). People are looking for alternatives to improve their quality of life by changing food habits or consuming more plant-based food, such as 'functional foods,' which fulfil both objectives, i.e., providing nourishment and protecting people from general diseases. With minor changes in food habits and shifting towards plant-based diets, the population is supposed to have benefits such as improved cardiovascular health (Adams et al., 2006), improved gut health, protection from constipation or irritable bowel syndrome, diabetes, and obesity (Morris et al., 2018; Mengstu et al., 2021), several forms of cancers and less production of reactive oxygen species (ROS) (Edziri et al., 2019; Ceclu & Nistor, 2020; Gamba et al., 2021; Abate et al., 2022).

One such important plant is *Beta vulgaris* or beetroot, which is in high demand due to its considerable health benefits (due to its protective role against cancer, cardiovascular, hypertension, diabetes, hepatic diseases, immune dysfunction, and many others) and diverse phytochemicals. One important class of phytoconstituent in beetroot is the betalains, water-soluble pigments derived from the amino acid L-tyrosine. Ingestion of beetroot provides a good means of increasing nitric oxide (NO) concentration *in vivo*. This increased blood NO concentration promotes vasodilation through different cellular mechanisms, significantly reducing blood pressure (Kapil et al., 2010). In addition, the plant has good pharmacological properties too. This chapter aims to provide in-depth information about the medicinal and commercial uses of *B. vulgaris*.

15.2 COMMON NAMES AND SYNONYMS

Beta vulgaris L.; *Beta orientalis* L.; *Beta vulgaris* subsp. *orientalis* Aellen; *Beta vulgaris* var. *cicla* L.; **Vernacular and Hindi** Chakundar; **English** Beet.

DOI: 10.1201/b22924-15

15.3 BOTANY AND CYTOLOGY OF THE PLANT

15.3.1 Botany

B. vulgaris (Amaranthaceae), commonly known as beetroot or chakundar, is an annual, biennial, or rarely perennial plant, originating from *B. vulgaris* subsp. *maritima*, also called 'sea beet'. Beet is an economically important plant consisting of several distinct cultivated lineages of *B. vulgaris* subsp. *Vulgaris*, namely sugar beet, table beet, fodder beet, and chard. The plant is originally native to the Middle East but is presently distributed around the globe. Plants are consumed raw, converted to juice, used in salads, cooked in some dishes, and processed as pickles at home. The plant is a glabrous, erect herb with a tuberous or very rarely cylindrical root; its colour ranges from white to yellow to purplish or sometimes violet; the stem is branched in the upper part (Figure 15.1). Leaves are mostly basal, ovate to oblong-ovate, passing into linear bracts. Flowers are greenish, with many axillary or terminal, 2- to 6-flowered clusters, forming panicles. There are five tepals, oblong, obtuse, and with a tuberculate base to the fruit. Utricle is adnate to the disc and base of tepals; seeds are smooth, reddish-brown (Shishkin, 1936; Gaur, 1999). Stigmas are 2–3.

15.3.2 Germination

The plant grows well in temperatures ranging between 12 and 19°C, whereas the maximum growing temperature for the plant is 35°C. During seed germination, the preferable soil temperature is > 7°C, and the ideal pH for beetroot production is between 6.0 and 8.0. Further, beetroot prefers deep, friable, well-drained, sandy to silt loams for growth. Thus, the crop is generally cultivated during the winter months in the Tarai-Bhabar belt in northern India.

15.3.3 Cytology

Cytological studies of *Beta vulgaris* at the global level suggested that plants are diploid, with 2n=18 chromosomes (Galewski & McGrath, 2020).

15.4 PHYTOCHEMISTRY OF *BETA VULGARIS*

New drug discovery from plants is primarily based on traditional knowledge or folk use of that plant, thus involving the combined approach of phytochemistry and ethnobotany, which include the isolation of active compounds from the preparation used in ethnobotany. The beginning of this started with the isolation of morphine from opium in the early nineteenth century, which opened

FIGURE 15.1 (a) *Beta vulgaris*; (b) *Beta vulgaris* showing secondary growth.

the door for the isolation and characterization of other important medicinal compounds from plants, some of which are still in use, i.e., codeine, digitoxin, and quinine, to cure different ailments. In India, drug discovery programmes started with the combined efforts of the Council of Scientific and Industrial Research along with other R&D institutes in 1996 to discover new bioactive molecules from natural sources like plants, fungi, microbes, etc. The Central Drug Research Institute (CDRI), Lucknow and the Regional Research Laboratory (RRL), Jammu has taken the lead in drug discovery endeavours (Bhutani & Ghoil 2010). *B. vulgaris* has an excellent reputation for containing different phytochemicals and is composed of 4–12% sugar, 0.8% fibre, 1.5% protein, 0.1% fat, 12 to 20% dry matter, and many minerals such as sodium, potassium, phosphorus, calcium, iron, zinc, copper, boron, silica, and selenium along with small amounts of vitamins A, C, K, E and B (Babarykin et al., 2019; Lechner & Stoner, 2019; Abdo et al., 2020). Due to its diverse phytochemicals, the plant is ranked tenth among all vegetables in terms of its natural antioxidant potential. *B. vulgaris* is a rich source of phytochemicals exhibiting antioxidant activity, like triterpenes, flavonoids, coumarins, carotenoids, and sesquiterpenoids (Bangar et al., 2022b).

The plant is also known to contain phenolics (caffeic acid, betalains, ellagic acid, syringic acid, vanillic acid, and ferulic acid), flavonoids (myricetin, kaempferol, betagarin, betavulgarin, cochliophilin A, dihydro-isorhamnetin, and quercetin) (as shown in Figure 15.2), terpenoids, steroids, tannins, saponins, glycosides, and inorganic nitrates (NO_3) (Bailey et al., 2009; Chhikara et al., 2019; Kazimierczak et al., 2016). In addition, beetroot also exhibits great variation in organic acids, such as shikimic acid, malic acid, citric acid, fumaric acid, glycolic acid, quinic acid, oxalic acid, succinic acid, glutaric acid, adipic acid, tartaric acid; amino acids, such as leucine, asparagine, aspartic acid, glutamine, glutamic acid, tyrosine; and sugars, such as saccharose, invertase, raffinose, pectins, fructose, glucose, arabinose, galactose, xylan, dextran, levulan, saccharin, and coniferin (Zhang et al., 2013; Ceclu & Nistor, 2020).

The intense red colour of beetroot is due to the presence of the specific secondary plant metabolites, the betalains (phenolics). Betalains are water-soluble plant pigments that makeup up to 70% of the total phenolic content of the beetroot, and their estimated limits may range up to 0.8–1.3 g/L of fresh beetroot juice. Betalains mainly contain betacyanins (up to 60%) and betaxanthins (up to 40%) (Tesoriere et al., 2008; Ravichandran et al., 2013; Sporna et al., 2015).

Because of these phytochemicals, beetroot exhibits various pharmacological activities, i.e., antibacterial (John et al., 2017; El-Beltagi et al., 2018), antifungal (Citores et al., 2016), antioxidant (Spórna et al., 2015; Guldiken et al., 2016), antistress and anti-anxiety (Sulakhiya et al., 2016), anti-inflammatory (El Gamal et al., 2014; Martinez et al., 2015), hepatoprotective (Kassem et al., 2020), anticancer (Reddy et al., 2005; Nyirády et al., 2010; Kapadia et al., 2011), hyperglycaemic (Olumese & Oboh, 2016; Oztay et al., 2015; Mirmiran et al., 2020), cardioprotective (Lundberg et al., 2011; Dos et al., 2020; Mirmiran et al., 2020), antihypertensive (Wong et al., 2014) and fertility-restorative activities (Kitazaki et al., 2015).

15.5 COMMERCIAL AND MEDICINAL USE OF *BETA VULGARIS*

During antiquity, the beet was primarily consumed for its leaves, and the literature from some old civilizations confirmed the same. Ancient Greek, Roman, and Egyptian civilizations describe this plant's use in many cuisines. However, the use of the root was missing, and the use of beetroot for medicinal purposes cannot be traced back to these civilizations (Wrusset al., 2015; Chhikara et al., 2019; Abdo et al., 2020; Goldman & Janick, 2021).

Lugi Squalermo (1561), in his work 'De simplicibus', described a variety of beet from Greece with bright red roots like turnip, which at that time was popularly known by the name of Cochinoguglia in Greece (Biancardi et al., 2012; Goldman & Janick, 2021). The dates associated with the development of this root crop (beet with a swollen root) are unknown, but evidence suggests that it occurred only in the 1500s.

The root was not much enlarged in the early wild forms, and the beet was primarily a leaf crop.

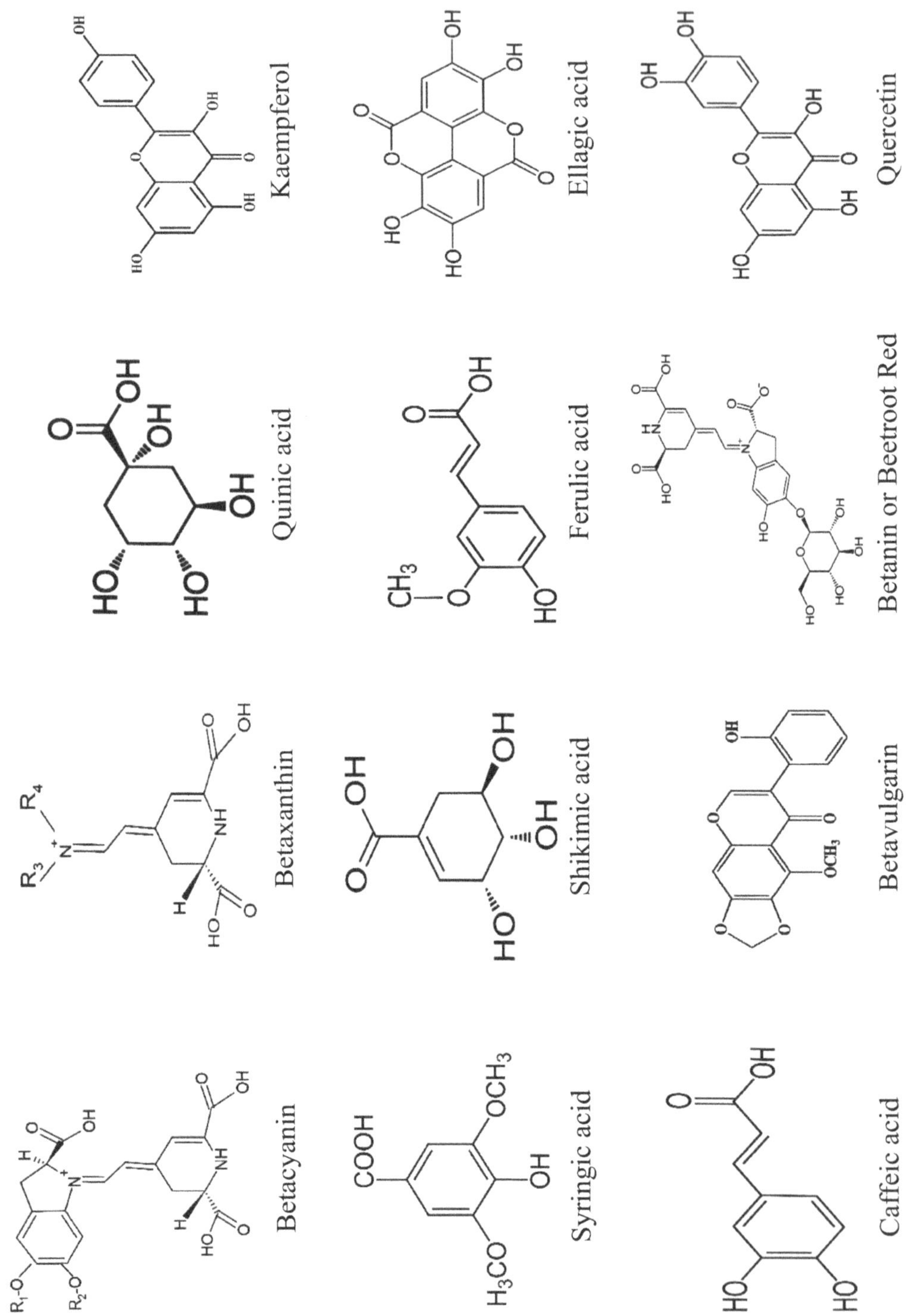

FIGURE 15.2 Phytochemicals present in different parts of *Beta vulgaris*.

The earliest evidence for lineages with expanded roots occurs in Egypt around 3500 BC. Today's enlarged beets are possible because of the presence of supernumerary cambium in the root, which allows root expansion. The nutritional constituents of the beet root are soluble and insoluble dietary fibres, antioxidants, a significant amount of vitamins C, B_1, B_2, B_3, B_6, and B_{12}, whereas vitamin A is found in the leaves (Wootton-Beard & Ryan 2011; Ravichandran et al., 2013; Singh et al., 2014; Morgado et al., 2016).

Consumption of beetroot is beneficial in treating diseases like obesity, high blood pressure, heart disease, and diabetes (Halvorsen et al., 2002; Bailey et al., 2009; Baiao et al., 2017). Regular consumption of beetroot reduces high blood pressure and improves cardiac health (Larsen et al., 2006; Lundberg et al., 2011; Olumese & Oboh 2016; Mirmiran et al., 2020). Furthermore, its high iron content, coupled with a high copper content, regenerates and reactivates the blood and makes the iron more available to the body, helping anaemic patients. Several workers also document the use of beetroot in digestive disorders. According to some, consuming the root after mixing it with honey on an empty stomach helps to treat peptic ulcers and irregular bowel syndrome; in some parts of England, the juice is advocated as an easily digested food for the aged, weak, or infirm (Cermak et al., 2012; Clifford et al., 2015; Guldiken et al., 2016; Dominguez et al., 2018).In some parts of Africa, the roots are used as an antidote to cyanide poisoning.

Furthermore, beetroot juice mixed with fruits of *Emblica officinalis* (amla, Indian gooseberry), the rhizome of ginger, and roots of carrot help to improve stomach health and, at the same time, increase the libido. In addition to this, in some communities, beetroot extract is used to cure different skin ailments, i.e., boils, skin inflammation, and outbreaks of pimples and pustules. In mythology, Aphrodite is said to have eaten beet to retain her beauty. Regular raw use of beetroot in salad increases the production of male sex hormones, which might be due to its high amount of boron. Furthermore, the juice of the root is considered to be one of the most powerful cleansers of the body. In folk magic, it is advocated that, if a woman and man eat from the same beet, they will fall in love. Mahto (2014) concluded that beetroot fodder could be incorporated in place of maize in the concentrated mixture for growing children up to 50% (w/w) without any adverse effect and with distinct economic advantages. Pharmacologically, *B. vulgaris* exhibits antibacterial, antioxidant, anti-inflammatory, antifungal, hypocholesterolaemic, anti-diabetic, anticancer, and cardioprotective activities. Novel studies showed that consuming *B. vulgaris* had excellent physiological effects, improving cardiovascular diseases, liver damage, hepatic steatosis, and diabetes, and exhibiting renoprotective activities. In human studies, beetroot supplementation has been reported to reduce blood glucose and improve insulin homeostasis and endothelial and vascular function (Mirmiran et al., 2020; Bangar et al., 2022a).

In addition to its medicinal benefits, beetroot is also used for commercial applications. Red beet or its extract is used commercially in the food industry for making beetroot candy, hair mask, pickle juice, beetroot-carrot tea, gummies, vegetable vitamins, and beetroot powder (Figure 15.3). Beetroot is a good source of red and yellow pigments due to the presence of the betalains known as betacyanins and betaxanthins, respectively. The use of betalains in the food industry is on trend and mainly used as colourants to improve the colour of tomato paste, jams, jellies, vinegar, flour, sauces, ice cream, sweets, desserts, and cereals (Lee et al., 2005; Georgiev et al., 2010; Zielińska-Przyjemska et al., 2009). The regular use of beetroot results in improving the antioxidant status in humans by decreasing oxidative damage of lipids because of these betalains, thus improving the overall health by scavenging the ROS, resulting in decreasing aging and development of cancer (Ravichandran et al., 2013). Other benefits include the inhibition of lipid peroxidation, increased resistance to the oxidation of low-density lipoproteins and chemopreventive effects (Tesoriere et al., 2008; Reddy et al., 2005; Zhang et al., 2013). The active ingredient in achieving cardioprotection following intake of beetroot is the inorganic nitrate (NO_3) fraction, which is reduced by bacteria in the salvia into nitric oxide (NO). Clinical studies suggest this high NO concentration has certain positive effects, i.e., increased muscle efficiency and fatigue resistance, reduced resting blood pressure, and prevention

FIGURE 15.3 (A-I) Beetroot processed food and other products. A=Gummies; B=Dosa; C=Chips; D=Hair Mask; E=Artisanal Kombucha; F= Murukku; G=Beet root powder; H= Cookies; I=Capsule.

of cardiovascular diseases (Larsen et al., 2006; Webb et al., 2008; Lundberg et al., 2011; Cermak et al., 2012; Murphy et al., 2012; Hernandez et al., 2012).

15.6 CONCLUSION

In conclusion, beetroot is one of the most important vegetables or source foods that can be supplemented into the diet to control a number of diseases. Daily consumption of beetroot not only decreases morbidity and mortality caused by cardiovascular diseases or hypertension, diabetes, insulin resistance, kidney dysfunction, and many others, but it also decreases public health expenditure for such diseases. Many scientific studies have also proved that the chemical constituents and extracts derived from various parts of *B. vulgaris* possess broad pharmacological activities and are also contributing to modern systems of medicine with actions such as antibacterial, antioxidant, anti-inflammatory, antifungal, hypocholesterolaemic, antidiabetic, anticancer, neuroprotective, and cardioprotective activities. Preserved food from beet has a good shelf life with high stability of the phytochemicals present in them; people should include beet in their diet as a raw medicine that promotes their health and improves their quality of life.

REFERENCES

Abdo, E., El-Sohaimy, S., Shaltout, O., Abdalla, A. & Zeitoun, A. (2020) Nutritional evaluation of beetroots (*Beta vulgaris* L.) and its potential application in a functional beverage. *Plants*, 9(12), 1–17.

Abate, L., Tadesse, M.G., Bachheti, A., & Bachheti, R.K. (2022). Traditional and phytochemical bases of herbs, shrubs, climbers, and trees from Ethiopia for their anticancer response. *BioMed Research International*. https://doi.org/10.1155/2022/1589877.

Adams, M.R., Golden, D.L., Chen, H., Register, T.C. & Gugger, E.T. (2006) A diet rich in green and yellow vegetables inhibits atherosclerosis in mice. *The Journal of Nutrition*. 136(7), 1886–1889.

Asfaw, T.B., Esho, T.B., Bachheti, A., Bachheti, R.K., Pandey, D.P. & Husen, A. (2022). In: Husen A. (eds) Exploring important herbs, shrubs, and trees for their traditional knowledge, chemical Derivatives, and potential benefits. In: *Herbs, shrubs, and trees of potential medicinal benefits*. CRC Press, pp. 1–26.

Babarykin, D., Smirnova, G., Pundinsh, I., Vasiljeva, S., Krumina, G. & Agejchenko, V. (2019) Red beet (*Beta vulgaris*) Impact on human health. *Journal of Biosciences and Medicines*, 7, 61–79.

Bachheti, R.K., Worku, L.A., Gonfa, Y.H., Zebeaman, M., Pandey, D.P. & Bachheti, A. (2022) Prevention and treatment of cardiovascular diseases with plant phytochemicals: A review. *Evidence-Based Complementary and Alternative Medicine*. https://doi.org/10.1155/2022/5741198.

Baião, D.D., da Silva, D.V., Del Aguila, E.M. & Paschoalin, V.M. (2017) Nutritional, bioactive and physico-chemical characteristics of different beetroot formulations. *Food Additives*, 6(6), 21–43.

Bailey, S.J., Winyard, P., Vanhatalo, A., Blackwell, J.R., DiMenna, F.J., Wilkerson, D.P. et al. (2009) Dietary nitrate supplementation reduces the O2 cost of low-intensity exercise and enhances tolerance to high-intensity exercise in humans. *Journal of Applied Physiology*, 107(4), 1144–1155.

Bangar, S.P., Sharma, N., Sanwal, N., Lorenzo, J.M. & Sahu, J.K. (2022a) Bioactive potential of beetroot (Beta vulgaris). *Food Research International*, 158, 111556.

Bangar, S.P., Singh, A., Chaudhary, V., Sharma, N. & Lorenzo, J.M. (2022b) Beetroot as a novel ingredient for its versatile food applications. *Critical Reviews in Food Science and Nutrition*, 2022, 1–25. https://doi.org/10.1080/10408398.2022.2055529

Bhutani, K.K. & Gohil, V.M. (2010) Natural products drug discovery research in India: Status and appraisal. *Indian Journal of Experimental Biology*, 48, 197–207.

Biancardi, E., Panella, L.W. & Lewellen, R.T. (2012) *Beta Maritima*. New York: Springer.

Bisht, N.S. & Khajuria, A.K. (2014) Ethno-medicinal plants of Tehsil, Kathua, Jammu & Kashmir. *Journal of Mountain Research*, 9, 1–12.

Ceclu, L. & Nistor, O.V. (2020) Red beetroot: Composition and health effects—A review. *Journal of Nutritional Medicine and Diet Care*, 6(1), 1–9.

Cermak, N.M., Gibala, M.J. & Van Loon, L.J. (2012) Nitrate supplementation's improvement of 10-km time-trial performance in trained cyclists. *International Journal of Sport Nutrition and Exercise Metabolism*, 22(1), 64–71.

Chhikara, N., Kushwaha, K., Sharma, P., Gat, Y. & Panghal, A. (2019) Bioactive compounds of beetroot and utilization in food processing industry: A critical review. *Food chemistry*, 272, 192–200.

Citores, L., Iglesias, R., Gay, C. & Ferreras, J.M. (2016) Antifungal activity of the ribosome-inactivating protein BE 27 from sugar beet (*Beta vulgaris* L.) against the green mould *Penicillium digitatum*. *Molecular Plant Pathology*, 17(2), 261–71.

Clifford T., Howatson G., West D.J. & Stevenson E.J. (2015) The potential benefits of red beetroot supplementation in health and disease. *Nutrients*, 7(4), 2801–2822.

Dhiman, A.K. & Kumar, A. (2006) *Ayurvedic drug plants*. Daya Books, Delhi.

Domínguez, R., Maté-Muñoz, J.L., Cuenca, E., García-Fernández, P., Mata-Ordoñez, F., Lozano-Estevan, M.C. et al. (2018) Effects of beetroot juice supplementation on intermittent high-intensity exercise efforts. *Journal of the International Society of Sports Nutrition*, 15(1), 1–12.

dos S Baiao, D., da Silva, D.V. & Paschoalin, V.M. (2020) Beetroot, a remarkable vegetable: Its nitrate and phytochemical contents can be adjusted in novel formulations to benefit health and support cardiovascular disease therapies. *Antioxidants*, 9(10), 960.

Edziri, H., Jaziri, R., Haddad, O., Anthonissen, R., Aouni, M., Mastouri, M. et al. (2019) Phytochemical analysis, antioxidant, anticoagulant and in vitro toxicity and genotoxicity testing of methanolic and juice extracts of *Beta vulgaris* L. *South African Journal of Botany*, 126, 170–175.

El Gamal, A.A., AlSaid, M.S., Raish, M., Al-Sohaibani, M., Al-Massarani, S.M., Ahmad, A. et al. (2014) Beetroot (*Beta vulgaris* L.) extract ameliorates gentamicin-induced nephrotoxicity associated oxidative stress, inflammation, and apoptosis in rodent model. *Mediators of Inflammation*, 2014, 1–12. https://doi.org/10.1155/2014/983952

El-Beltagi, H.S., Mohamed, H.I., Megahed, B.M., Gamal, M. &Safwat, G. (2018) Evaluation of some chemical constituents, antioxidant, antibacterial and anticancer activities of *Beta vulgaris* L. root. *Fresenius Environmental Bulletin*, 27(9), 6369–6378.

Galewski, P. &McGrath J.M. (2020) Genetic diversity among cultivated beets (*Beta vulgaris*) assessed via population-based whole genome sequences. *BMC Genomics*, 21(1), 1–4.

Gamba, M., Raguindin, P.F., Asllanaj, E., Merlo, F., Glisic, M., Minder, B. et al. (2021) Bioactive compounds and nutritional composition of Swiss chard (*Beta vulgaris*L. var. cicla and flavescens): A systematic review. *Critical Reviews in Food Science and Nutrition*, 61(20), 3465–3480.

Gaur, R.D. (1999) *Flora of the district Garhwal, North West Himalaya*. Transmedia, Srinagar, Garhwal, India.

Georgiev, V.G., Weber, J., Kneschke, E.M., Denev, P.N., Bley, T. & Pavlov, A.I. (2010) Antioxidant activity and phenolic content of betalain extracts from intact plants and hairy root cultures of the red beetroot *Beta vulgaris*. Detroit dark red. *Plant foods for Human Nutrition*, 65(2), 105–111.

Goldman, I.L. & Janick, J. (2021) Evolution of root morphology in table beet: Historical and iconographic. *Frontiers in Plant Science*, 12, 689926.

Guldiken, B., Toydemir, G., Nur Memis, K., Okur, S., Boyacioglu, D. & Capanoglu, E. (2016) Home-processed red beetroot (*Beta vulgaris* L.) products: Changes in antioxidant properties and bioaccessibility. *International Journal of Molecular Sciences*, 17(6), 1–13.

Halvorsen, B.L., Holte, K., Myhrstad, M.C., Barikmo, I., Hvattum, E., Remberg, S.F. et al. (2002) A systematic screening of total antioxidants in dietary plants. *The Journal of Nutrition*, 132(3), 461–471.

Hernández, A., Schiffer, T.A., Ivarsson, N., Cheng, A.J., Bruton, J.D., Lundberg, J.O. et al. (2012) Dietary nitrate increases tetanic [Ca2+] i and contractile force in mouse fast-twitch muscle. *The Journal of Physiology*, 590(15), 3575–3583.

J Kapadia, G., A Azuine, M., Subba Rao, G., Arai, T., Iida, A. & Tokuda, H. (2011) Cytotoxic effect of the red beetroot (Beta vulgaris L.) extract compared to doxorubicin (Adriamycin) in the human prostate (PC-3) and breast (MCF-7) cancer cell lines. *Anticancer Agents in Medicinal Chemistry (Formerly Current Medicinal Chemistry-Anti-Cancer Agents)*, 11(3), 280–284.

John, S., Monica, J., Priyadarshini, S., Sivaraj, C. & Arumugam, P. (2017) Antioxidant and antibacterial activities of *Beta vulgaris* L. peel extracts. *International Journal of Pharma Research Health Sciences*, 5(6), 1974–1979.

Kapil, V., Milsom, A.B., Okorie, M., Maleki-Toyserkani, S., Akram, F., Rehman, F. et al. (2010) Inorganic nitrate supplementation lowers blood pressure in humans: Role for nitrite-derived NO. *Hypertension*, 56(2), 274–281.

Kassem, A.A., Abd El-Alim, S.H., Salman, A.M., Mohammed, M.A., Hassan, N.S. & El-Gengaihi, S.E. (2020) Improved hepatoprotective activity of *Beta vulgaris*L. leaf extract loaded self-nanoemulsifying drug delivery system (SNEDDS): in vitro and in vivo evaluation. *Drug Development and Industrial Pharmacy*, 46(10), 1589–1603.

Kazimierczak, R., Siłakiewicz, A., Hallmann, E., Srednicka-Tober, D. & Rembiałkowska, E. (2016) Chemical composition of selected beetroot juices in relation to beetroot production system and processing technology. *NotulaeBotanicae Horti Agrobotanici Cluj-Napoca*, 44(2), 491–498.

Khajuria, A.K., Manhas, R.K., Kumar, H. & Bisht, N.S. (2021) Ethnobotanical study of traditionally used medicinal plants of Pauri district of Uttarakhand, India. *Journal of Ethnopharmacology*, 276, 1–19.

Kitazaki, K., Arakawa, T., Matsunaga, M., Yui-Kurino, R., Matsuhira, H., Mikami, T. & Kubo, T. (2015) Post-translational mechanisms are associated with fertility restoration of cytoplasmic male sterility in sugar beet (*Beta vulgaris*). *The Plant Journal*, 83(2), 290–299.

Larsen, F., Ekblom, B., Sahlin, K., Weitzberg, E. & Lundberg J. (2006) Effects of dietary nitrate on blood pressure in healthy volunteers. *New England Journal of Medicine*, 355(26), 2792–2793.

Lechner, J.F. & Stoner, G.D. (2019) Red Beetroot and Betalains as cancer chemopreventative agents. *Review Molecules*, 24, 1602.

Lee, C.H., Wettasinghe, M., Bolling, B.W., Ji, L.L. & Parkin, K.L. (2005) Betalains, phase II enzyme-inducing components from red beetroot (*Beta vulgaris* L.) extracts. *Nutrition and Cancer*, 53(1), 91–103.

Lundberg, J.O., Carlström, M., Larsen, F.J. & Weitzberg, E. (2011) Roles of dietary inorganic nitrate in cardiovascular health and disease. *Cardiovascular Research*, 89(3), 525–532.

Mahto, B. (2014) *Effect of fodder beet (Beta vulgaris) on black bengal goats*. Sundaresan school of Animal Husbandry & Dairying, Sam Higginbottom Institute of Agriculture, Technology & Sciences, PhD Thesis.

Martinez, R.M., Longhi-Balbinot, D.T., Zarpelon, A.C., Staurengo-Ferrari, L., Baracat, M.M., Georgetti, S.R. et al. (2015) Anti-inflammatory activity of betalain-rich dye of *Beta vulgaris*: Effect on edema, leukocyte recruitment, superoxide anion and cytokine production. *Archives of Pharmacal Research*, 38(4), 494–504.

Mengstu, A, Bachheti, A., Abate, L., Bachheti, R.K. & Husen, A. (2021). In: Husen A., Bachheti R.K., Bachheti, A. (eds) Health-promoting benefits, value-added products, and other uses of banana. In: *Non-timber forest products*. Cham: Springer, pp. 339–364.
Mirmiran, P., Houshialsadat, Z., Gaeini, Z., Bahadoran, Z. & Azizi, F. (2020) Functional properties of beetroot (*Beta vulgaris*) in management of cardio-metabolic diseases. *Nutrition & Metabolism*, 17(1), 1–5.
Morgado, M., de Oliveira, G.V., Vasconcellos, J., Monteiro, M.L., Conte-Junior, C., Pierucci, A.P. et al. (2016) Development of a beetroot-based nutritional gel containing high content of bioaccessible dietary nitrate and antioxidants. *International Journal of Food Sciences and Nutrition*, 67(2), 153–160.
Morris, M.C., Wang, Y., Barnes, L.L., Bennett, D.A., Dawson-Hughes, B. & Booth, S.L. (2018) Nutrients and bioactives in green leafy vegetables and cognitive decline: Prospective study. *Neurology*, 90(3), 214–222.
Murphy, M., Eliot, K., Heuertz, R.M. & Weiss, E. (2012) Whole beetroot consumption acutely improves running performance. *Journal of the Academy of Nutrition and Dietietics*, 112(4), 548–552.
Nyirády, P., Sárdi, É., Bekő, G., Szűcs, M., Horváth, A., Székely, E. et al. (2010) Effects of bioactive molecules of *Beta vulgaris*L. ssp. esculenta var. rubra on metastatic prostate cancer. *Orvosihetilap*, 151(37), 1495–1503.
Olumese, F. & Oboh, H. (2016) Effects of daily intake of beetroot juice on blood glucose and hormones in young healthy subjects. *Annals of Tropical Pathology*, 26, 455–462.
Oztay, F., Sacan, O., Kayalar, O., Bolkent, S., Ipci, Y., Kabasakal, L. et al. (2015) Chard (*Beta vulgaris*var. cicla) extract improved hyperglycemia-induced oxidative stress and surfactant-associated protein alterations in rat lungs. *Pharmaceutical Biology*, 53(11), 1639–1646.
Ravichandran, K., Saw, N.M., Mohdaly, A.A., Gabr, A.M., Kastell, A., Riedel, H. et al. (2013) Impact of processing of red beet on betalain content and antioxidant activity. *Food Research International*, 50(2), 670–675.
Reddy, M.K., Alexander-Lindo, R.L. & Nair, M.G. (2005) Relative inhibition of lipid peroxidation, cyclooxygenase enzymes, and human tumor cell proliferation by natural food colors. *Journal of Agricultural and Food Chemistry*, 53, 9268–9273.
Shishkin, B.K. (1936) *Flora of the USSR*. Volume 6, Centrospermae. Moscow: Isdatel'stvoAkademiiNauk SSSR [translated from Russian. 1970. Washington, DC, Smithsonian Institution and National Science Foundation].
Singh, B. & Hathan, B.S. (2014) Chemical composition, functional properties and processing of beetroot—A review. *International Journal of Scientific and Engineering Research*. 5(1), 679–684.
Spórna-Kucab, A., Garrard, I., Ignatova, S. & Wybraniec, S. (2015) New solvent systems for gradient counter-current chromatography in separation of betanin and its derivatives from processed *Beta vulgaris* L. juice. *Journal of Chromatography A*, 1380, 29–37.
Sulakhiya K., Patel, V.K., Saxena, R., Dashore, J., Srivastava, A.K. & Rathore, M. (2016) Effect of *Beta vulgaris* Linn. leaves extract on anxiety-and depressive-like behavior and oxidative stress in mice after acute restraint stress. *Pharmacognosy Research*, 8(1), 1–7.
Tesoriere, L., Fazzari, M., Angileri, F., Gentile, C. & Livrea, M.A. (2008) In vitro digestion of betalainic foods. Stability and bioaccessibility of betaxanthins and betacyanins and antioxidative potential of food digesta. *Journal of Agricultural and Food Chemistry*, 56, 10487–10492.
Webb, A.J., Patel, N., Loukogeorgakis, S., Okorie, M., Aboud, Z., Misra, S. et al. (2008) Acute blood pressure lowering, vasoprotective, and antiplatelet properties of dietary nitrate via bioconversion to nitrite. *Hypertension*, 51(3), 784–790.
Wong, R.H., Garg, M.L., Wood, L.G. & Howe, P.R. (2014) Antihypertensive potential of combined extracts of olive leaf, green coffee bean and beetroot: A randomized, double-blind, placebo-controlled crossover trial. *Nutrients*, 6(11), 4881–4894.
Wootton-Beard, P.C. & Ryan, L. (2011) A beetroot juice shot is a significant and convenient source of bioaccessible antioxidants. *Journal of Functional Foods*, 3(4), 329–334.
Wruss, J., Waldenberger, G., Huemer, S., Uygun, P., Lanzerstorfer, P., Müller, U. et al. (2015) Compositional characteristics of commercial beetroot products and beetroot juice prepared from seven beetroot varieties grown in Upper Austria. *Journal of Food Composition and Analysis*, 42, 46–55.
Zhang, Y., Lee, C., Yu, C., Cheng, Y.S., Zhang, R., Jenkins, B.M. et al. (2013) Dilute acid pretreatment and fermentation of sugar beet pulp to ethanol. *Applied Energy*, 105, 1–7.
Zielińska-Przyjemska, M., Olejnik, A., Dobrowolska-Zachwieja, A. & Grajek, W. (2009) In vitro effects of beetroot juice and chips on oxidative metabolism and apoptosis in neutrophils from obese individuals. Phytotherapy Research. *An International Journal Devoted to Pharmacological and Toxicological Evaluation of Natural Product Derivatives*, 23(1), 49–55.

16 Radish

Health Benefits and Medicinal Uses

Sugam Gupta, Bhavya Mudgal, Devvret Verma, Debasis Mitra, Archana Bachheti, and Rakesh Kumar Bachheti

16.1 INTRODUCTION

Radish is one of the most popular vegetables in the world because it is easy to grow, has a short growing season, and has many health benefits. It is a member of the Cruciferae or Brassicaceae family, which is one of the most commercially important families of vegetables (Curtis, 2003). Radish (*Raphanus sativus* L.) is an herbaceous annual or biennial plant, with white- or purple-coloured flowers assembled in a terminal raceme, and it grows in many parts of the world. It is a fast-growing, nutritious root vegetable crop (Lavanya et al., 2014). The early history of this vegetable crop is unknown since there is no archaeological evidence for its cultivation. Some people think that *R. sativus* came from Southeast Asia, which is the only place where wild varieties exist. Conflicts have occurred throughout Europe and Asia, and radish-growing areas have formed in India, Central China, and Central Asia. (Singh et al., 2016; Vardhana, 2017; Seebens et al., 2017).

R. sativus is a well-known brassicaceous root vegetable crop consisting of roots, leaves, fruits, and oilseeds. Due to its high nutritional value, the taproot is eaten in pickles, salads, and curries in different parts of the world (Tsouvaltzis & Brecht, 2014). It mainly grows at altitudes of between 190 and 1,240 m above sea level in temperate regions of the world, producing swollen roots of varied sizes, shapes, and colours. The term 'turnip' also refers to a commercially popular edible radish with succulent, spicy roots with a pungent taste one of the classic Japanese dishes, known as Takuan, a salted radish root with a distinctive golden tint that develops during storage (George, 2009). The output of radish in Japan decreased in 2020, with a volume of about 1.25 million tonnes. This was less than the previous output of almost 1.5 million tonnes that was noted in 2011 (Klein, 2022).

In most countries, radishes are eaten raw as a crisp vegetable, primarily in salads, but they are also cooked in various European dishes. In countries in the Middle East, people drink radish juice due to certain health advantages (Gutiérrez & Perez, 2004). This vegetable is eaten prominently in Eastern diets, and its cultivation has spread worldwide. The Oriental radish is more prominent in Japan, China, and Southeast Asia. Radishes come in a variety of colours (red, purple, black, yellow, and white) (Banihani, 2017). The core part is usually white or sometimes light pink to dark pink. Furthermore, the edible root of radish differs in radius and length worldwide. It is a prominent vegetable in the diet in Asia, where it is similarly known for its medicinal properties, particularly to resolve abdominal ailments (Turner, 1995). The cruciferous/brassicaceous plant family has attracted much attention because of its outstanding nutritional and therapeutic potential. According to various studies, cruciferous vegetables contain ascorbic acid, carotenoids, tocopherols, glucosinolates, and phenolic compounds (Beevi et al., 2012; Castro-Torres et al 2014; Rhodes, 1994; Soundararajan & Kim, 2018).

DOI: 10.1201/b22924-16

16.2 DISTRIBUTION OF *R. SATIVUS*

The Brassicaceae family includes roots, leaves, oilseeds, and seasoning vegetables which are consumed, belonging to 310 genera and 3500 species (Al-Shehbaz, 2001; OECD, 2016). In addition to their nutritional value, members of the vegetable family also support agriculture's economy and contain chemical compounds that may be advantageous to health (Manivannan et al., 2019). Representing 14% of global vegetable production, India is second only to China, starting in the Himalayan Mountains in the north and ending on the southern Indian coast. (Singh et al., 2016; Vardhsana, 2017). Radish is a horticultural crop known for its crisp swollen roots, which can be consumed raw or cooked in several ways across different continents (Toit & Pelter, 2003; Reis & Boiteux, 2010; Seebeans et al., 2017; Fei et al., 2020) (Figure 16.1). According to the FAO (2017), 47% of the world's radish crop is produced on 40% of the world's arable land. *R. sativus* is a widely cultivated, generally self-incompatible annual or biennial plant, cultivated in Europe, Asia, North America, and Africa among others.

16.3 *R. SATIVUS* NUTRITIONAL BENEFITS

Radish is grown for its roots and leaves in India, Indonesia, Malaysia, Pakistan, and Sri Lanka; the immature green pods ('fruits') are also eaten as vegetables. Farmed radishes grown in different parts of the world look very different, and market preferences in each area may significantly affect the selection of root shapes and colours (Jatoi et al., 2011; Bonnema et al., 2011). This has led to a wide range of radish root shapes and colours

Almost all plant parts of the radish have been integrated into medicines, such as leaves, roots, and seeds. Radish is high in nutrition due to its antioxidant content and is now gaining in popularity (Figure 16.2). This leafy vegetable is employed as a domestic medication for the treatment of numerous disorders in Greek-Arab, Indian, and Unani traditional treatments, including jaundice, cholelithiasis, hepatic disease, rectoanal intussusception, indigestion, and other stomachache issues (Jeong et al., 2005; Shukla et al., 2011). Radishes include a variety of nutrients, including fluoride,

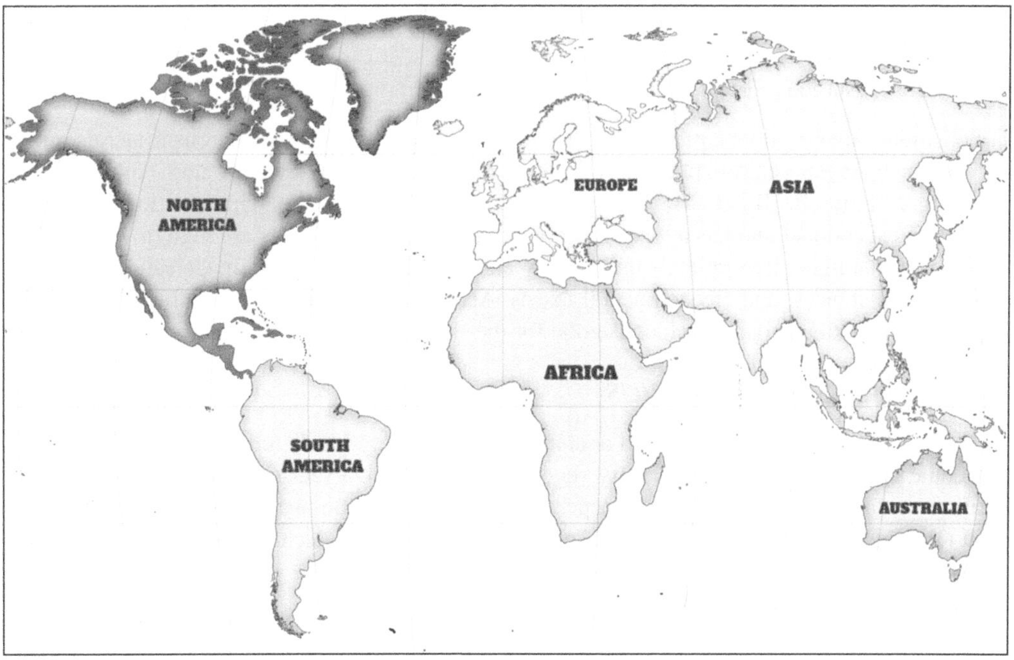

FIGURE 16.1 Distribution of *Raphanus sativus* across the globe.

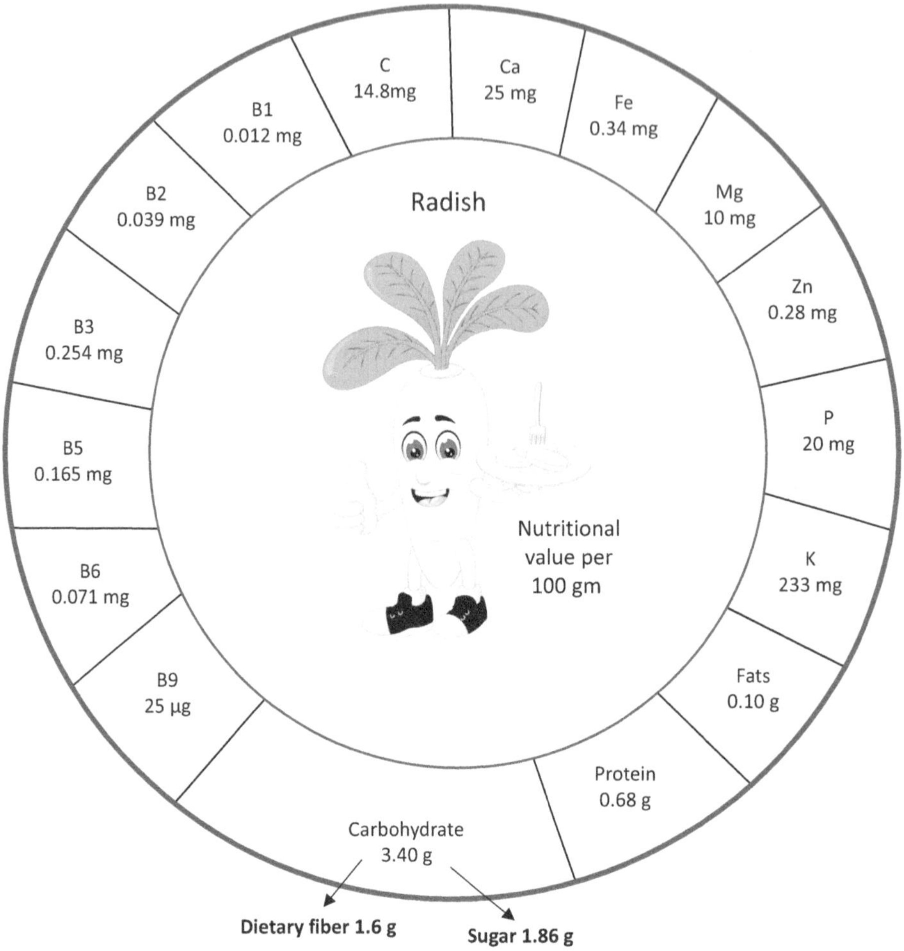

FIGURE 16.2 Nutrition content present in *Raphanus sativus*

carbohydrates, dietary fibres, protein, and some lipids. It also includes numerous minerals, including calcium, manganese, iron, phosphorus, potassium, and zinc, and several water-soluble vitamins, including vitamins B_1, B_2, B_3, B_5, B_6, B_9, B_{12}, and C. (Khattak, 2011). The nutritional content of radishes is exceptional and has a long list of health advantages. Radishes are high in vitamin C, an antioxidant that fights free radicals in the body and protects cells from damage caused by ageing, poor lifestyle choices, and environmental toxins. Aside from that, eating foods high in vitamin C helps the body make collagen, which is good for the skin and blood vessels.

16.4 VARIETIES OF *R. SATIVUS*

Radishes can have roots that are round or elongated, white, grey-black, red, pink, or yellow, and can also come in a range of sizes. This leafy vegetable crop is grown year-round as it is found in many locations including North America, Europe, Asia, and several more nations with the ideal environment for its growth. Further, it is divided into two distinct types which are listed below as summer and winter varieties.

16.4.1 Summer Varieties

Summer radishes are also known as spring, or European radishes, as they are frequently grown under cool conditions, as they are small and have a short, 3- to 4-week cultivation time.

- **'April Cross'**: One of the large radishes, it is a mini-type hybrid daikon cultivar, germinating slowly. It is white, flat, and smooth. It is used for salads, pickling, and cooking.
- **'Cherry Belle'**: A popular round, red-skinned cultivar with a white interior, cultivated in North America (Faust, 1996).
- **'Champion'**: It has the same round, red exterior as 'Cherry Belle' but with longer roots and a milder flavour (Faust, 1996).
- **'Red King'**: This variety produces a spherical, reddish-brown root with a moderate flavour. It was created to be more resistant to clubroot, which occurs more frequently in places with inadequate drainage (Faust, 1996).
- **'Snow Belle'**: This radish variety is white and spherical, similar to 'Cherry Belle' (Faust, 1996).
- **'White Icicle'**: This white, carrot-shaped variety, usually known as a 'simple icicle' as it matures, reaches a height of 10–12 cm. It slices readily and is less prone to pithiness than most radishes (Faust, 1996).
- **'French Breakfast'**: This is a radish with red skin and a splash of white just at the root's tip. Compared to other summer types, it is slightly milder, but it is also one of the earliest to become pithy (Peterson, 1999).
- **'Plum Purple Radish':** This is a purple-fuchsia radish with a crisper texture than regular radish (Peterson, 1999).
- **'Gala'** and **'Roodbol'**: In the Netherlands, these two types of radishes are usually used for breakfast, sliced and served with butter on toast (Faust, 1996).
- **'Easter Egg'**: These kinds of radishes are a hybrid of white, pink, red, and purple radishes with various skin colours. Seed mixes, sold in marketplaces or seed packs by the same name, can lengthen the harvesting period from a single planting because different kinds mature at different times (Peterson, 1999).

16.4.2 Winter Varieties

- **'Black Spanish'**: The black radish comes in both round and elongated varieties and is also known as 'French Gros Noir d'Hiver'. It has strong black skin with a core of spicy white flesh, is round or irregularly pear-shaped (McIntosh, 1830), and grows to about 10 cm in diameter (Singh, 2021).
- **Daikon** is a term used to describe a range of Asian winter oilseed radishes. Daikon is a native of East Asia, sold as radish in the populous regions of South Asia. There are many types of daikon, but daikons generally have elongated white roots. The word daikon is inherited from Japan, where white radish is commonly called Japanese radish. Various countries use different terms for daikons, known as Chinese radish, Oriental radish, or mooli in India and South Asia (Yamaguchi, 1983; AMHER, 2004). Although the British adopted the Japanese term daikon, the Japanese radish is also known as the Chinese radish, Oriental radish, or mooli (in India and South Asia).

16.5 DIFFERENT TYPES OF *R. SATIVUS*

16.5.1 *R. sativus* var. *caudatus*

R. sativus var. *caudatus* is known as green radish or rat-tailed radish in Southeast Asian countries, as shown in Figure 16.3A. Sungra or Mungra in Pakistan and India and Puk-kee-hood in Thailand is a culinary plant (Pocasap et al., 2013). The earliest mentions of green radish can be found in documents from the Shang Dynasty. It is thought to be a native of North China. It grew freely there before spreading to Central Asia and Europe. This bi-coloured type of radish ranges from 12 to 22 cm in length and 7–8 centimetres in diameter. Mostly, they have a straight, elongated root, turning

FIGURE 16.3 Different types of *Raphanus sativus*: A. *R. sativus* L. var. *caudatus*; B. *R. raphanistrum* L.; C. *R. sativus* var. *niger* (L.) J. Kern; D. *R. sativus* var. *oleiformis* Pers.; E. *R. sativus* var. *raphanistroides* Makino.

from green to white from top to bottom (Green Meat Radishes, 2020). The seed pods of var. *caudatus* are edible. These plants, sometimes known as podding radishes or snake radishes, do not have thick roots. The plant produces numerous delicious seed pods and abundant yellow flowers. Green radish is considered a great source of antioxidants. Therefore, it is known to boost collagen production and help reduce inflammation. It is an excellent source of potassium, folate, and small amounts of copper, vitamin K, and phosphorus (Chatterjee, 2020). It is used as a digestive aid and helps reduce phlegm and coughs in Chinese medicine (Zhang, 2021).

16.5.2 *Raphanus raphanistrum* L.

R. raphanistrum L., commonly known as daikon, means 'huge root' in Japanese, as shown in Figure 16.3B. Although it is indigenous to the Mediterranean region and the Black Sea coasts, it must have moved to Asia as a mild-rooted vegetable along trade channels, arriving in China around 500 BCE before moving on to Japan. Daikons are now widely cultivated in Japan and are a staple in the diet (Yamaguchi, 1983). Daikons vary widely in size, with an average of 15–16 cm in length, with the typical shape of an elongated cylinder but it is also found in oblong forms also (Kaymak et al., 2016). Daikon is an excellent source of vitamins B and C, along with copper, iron, manganese, and pectin. Its leaves are used to cure dysentery, asthma, cough, incontinence, and malnutrition. (Balakrishnan Nair & Groot, 2021). Studies have shown it contains raphanusin (a carminative), ferulic acid, and gentisic acid. Seeds are proven to have laxative, carminative, and diuretic properties. In contrast, roots can be very beneficial in treating several issues, such as syphilis, cancer, and haemorrhoids (Koyyati et al., 2016).

16.5.3 *Raphanus sativus* var. *niger* (L.) J. Kern

R. sativus var. *niger* is also known as 'Black Radish', due to its black outward appearance, as shown in Figure 16.3C. It was first cultivated in the eastern Mediterranean region and might be related to wild radish and was later used by Ancient Egyptians and Greeks (Singh, 2021). They were considered sacred in ancient Egypt and were used for their medicinal properties. Black radish is delicious raw in salads or prepared in soups and stews. The black skin is safe to eat if it appears fresh and does not smell rotten. The root's strong taste can be lessened by adding salt before cooking (Mantis, 2016).

'Black Radish' varies in shape and size depending on the type. The first type is round, with an average of 6–7 centimetres in diameter and a small tap root. The second type is between 17 and 20 cm long and cylindrical (Singh, 2021). Vitamins A, B, C, and E are thought to be present in abundance. The phytonutrients termed glucosinolates, which are present in the roots, promote digestion and liver detoxification. In Asian and traditional European medicinal practices, black radish is used to maintain bile function and gallbladder health (Sikorska-zimny & Beneduce, 2021).

16.5.4 *R. sativus* var. *oleiformis* Pers.

The Oilseed Radish was domesticated in ancient China and is presently cultivated there; it can be seen in Figure 16.3D. These radishes are coarse winter vegetables with a white elongated cylindrical shape. It is much hardier than other radishes and is almost 30 cm long and 2–5 cm in diameter (Radford et al., 1968).

Bloating, acid reflux, and diarrhoea are all treated with leaves, seeds, and roots of oilseed radish. It is also known to inhibit growth of human bacteria, such as *Escherichia coli*, *Staphylococcus aureus*, *Streptococcus* spp., and other bacteria (Duke & Ayensu, 1985). The roots are considered to be antiscorbutic, antispasmodic, digestive, and diuretic. However, the roots are not recommended to be consumed if the stomach or intestines are inflamed.

16.5.5 *Raphanus sativus* var. *raphanistroides* Makino

This radish is common and endemic to Eurasia. It is also known as jointed charlock or wild radish, as shown in Figure 16.3E. Wild radish currently grows over most of the world, where it is regarded as a weed in many agricultural areas. Wild radish is highly popular in Asia and may frequently be found there. (Ridley, 1930). It has a heavy taproot and a rosette of unevenly divided leaves. It has prickly-textured flower stalks about 60 cm tall and has yellow, white, or lilac flowers.

Evidence shows that some compounds in wild radishes, including glucosinolates, sulforaphane, and ferulic acid, have anticancer properties. Hanlon et al. (2007), Choi et al. (2009), and Beevi et al. (2010), in their research, said that the phenolic acid ferulic acid has antioxidant effects, with benefits towards Alzheimer's disease and other diseases. It is traditionally considered to be a medicinal food used to treat hepatotoxicity and indigestion.

16.6 THERAPEUTIC POTENTIAL OF RADISH

Radish is a plant widely developed in Asia and Europe. As a result, it eventually became employed for both culinary and medical uses. Since the tenth century, the plant's medical potential has also been highlighted (Kapoor, 1990). Aside from the roots, the leaves and sprouts are said to have nutritional and therapeutic value. In old folk medicines, radish extracts have been used to treat intestinal difficulties, incontinence, urine infections, hepatic irritation, heart trouble, and ulcers (Goyeneche et al. 2015). Several studies (Yuan et al., 2010; Baenas et al., 2016; Kim et al., 2017; Siddiq & Younus, 2018) show that radishes have antibacterial, anticancer, antioxidant, and anxiety-relieving properties. Glucosinolates, isothiocyanates, and polyphenols are secondary metabolites in radishes with medicinal benefits (Nakamura et al., 2008). Several articles have talked about how radishes contain glucosinolates and isothiocyanates, carotenoids, tocopherols, and phenolics which are used to kill cancer cells. Therefore, radishes are being looked at with renewed interest (Soundararajan & Kim, 2018). A naturally existing category of substances called polyphenolics are present in radish preparations. This has come into focus recently due to the therapeutic and nutritional properties of plyphenols. In their beneficial role, polyphenolics play a significant role as antioxidants (Takaya et al., 2003). Polyphenolics have become a major focus of the pharmaceutical industry. The bioactive compounds glucosinolates (and their metabolites, isothiocyanates) are found only in brassicaceous species, and the carcinogenic and anti-inflammatory effects of glucosinolates have piqued curiosity (Soundararajan and Kim 2018).

16.6.1 Radish Has Plausible Health Benefits.

Humans can benefit significantly from plant-based diets in terms of their health. It lowers the risk of many diseases, including cancer, neurological disorders, heart disease, and age-related issues. The cruciferous plant radish has attracted interest worldwide due to its excellent nutritional and therapeutic potential (Curtis, 2003; Castro-Torres et al., 2014). Because of its prodigious nutritional content, the radish root has been eaten worldwide in salads, pickles, and curries (Takaya et al., 2003; Tsouvaltzis et al., 2014). Apart from the roots, it has been reported that sprouts and leaves also show nutritional value. Since ancient times, radish extracts have been used in folk medicine to treat ulcers, hepatic inflammation, constipation, urinary infections, and stomach and constipation-related issues (Goyeneche et al., 2015).

i) Adding cruciferous vegetables like radish to your diet regularly reduces your chances of cancer as it has potent anticancer properties.
ii) Radishes are high in fibre, and the body gets plenty of roughage that helps to improve digestion.
iii) Radishes contain many anthocyanins, which have anti-inflammatory qualities and help reduce the risk of heart disease.
iv) Daily radish consumption has been shown to improve pre-diabetics and delay the development of type-2 diabetes.
v) Radish has a high potassium content, which helps to decrease blood pressure drastically. Potassium also increases blood flow and circulation by allowing the relaxation of blood vessels, one of the body's essential organs.
vi) Vitamin C is a powerful antioxidant that is found in radishes. It keeps you from getting colds and coughs and helps your immune system work better.
vii) Radishes contain several important minerals and plant compounds that help the liver stay healthy and work well.
viii) Radishes are high in vitamins C and B, as well as and zinc and phosphorus, which are valuable for the skin. Radish juice is good for your skin's health and elasticity because it contains vitamin C, which is involved in collagen production. Vitamin C is a potent antioxidant that fights free radicals and shields the skin from UV damage and sunburn.

16.7 EFFECT OF ANTIOXIDANTS

A range of secondary metabolites with significant antioxidant properties are present in the roots and leaves of radishes. Nitrogen-containing chemicals are a prominent class of secondary metabolites in cruciferous vegetables, whereas physiologically active substances include glucosinolates, flavonoids (including anthocyanins), and carotenoids (Manchali et al., 2012). Anthocyanins, well-known antioxidants that act as hydrogen donors, metal chelators, and protein binders, are present at high concentrations in radish roots which are not white. By activating phase II antioxidant enzymes,

TABLE 16.1
Secondary Metabolites (Flavanones) Found in Roots and Leaves of the Radish, Having Antioxidant Capabilities

S. No.	Name	Function	Section	References
1.	6-Prenyl -naringenin	Antioxidant, anti-inflammatory and cholesterol-lowering agents (Iwashina 2013)	Roots	Koley et al. (2017)
2.	Naringenin-4-*O*-glucuronide	-	Leaves	Koley et al. (2017)
3.	Naringenin-7-*O*-glucuronide	-	Leaves	Koley et al. (2017)

which stop cell growth and encourage death, anthocyanins also serve as a chemoprotectant (Bagchi et al., 2004). The radish roots and leaves contain a range of compounds with antioxidant functions as well as considerable nutritive benefits (Takaya et al., 2003). When comparing roots with leaves, leaves contain more proteins, ascorbic acid (vitamin C), calcium, and total phenol content, which is also two times higher than that in roots. It is shown that leaves, more so than roots, have polyphenol components present in various forms in their tissues. For example, high levels of pyrogallol (free form) and vanillic acid (bound form) are found in roots. In contrast, epicatechin (free form) and coumaric acid (bound) are found in higher concentrations in leaves (Goyeneche et al., 2015). Tables 16.1–16.10 shows that radishes contain many secondary metabolites that can act as antioxidants.

16.7.1 Anticancer Effect

Interest in natural resources to treat cancer is increasing day by day. Several studies have shown the antiproliferative effects of isothiocyanates in several forms of cancer. This anticancer impact has attracted much interest from the pharmaceutical industry (Rampal et al., 2012). By regulating the Phase I and Phase II detoxification systems, it has been seen that 'Black Radish' significantly suppresses the multiplication of HepG2 hepatocellular carcinoma cells (Hanlon et al., 2007). 'Rat-tailed Radish' extract shows significant cytotoxicity against a colon cancer cell line. (Pocasap et

TABLE 16.2
Secondary Metabolites (Phenolic Acids) Having Antioxidant Capabilities Found in Roots and Leaves of the Radish

S. No.	Name	Function	Section	References
1.	1,2-dihydroxyferuloyl-gentibiose	Easily absorbed through intestinal tract walls, protecting from cell damage	Leaves	Koley et al. (2017)
2.	Dihydro-caffeoyl-3-*O*-glucuronide	-	Root	Koley et al. (2017)
3.	Feruloylmalic acid	-	Leaves	Lou et al. (2018)
4.	*m*-Coumaric acid	-	Leaves	Koley et al. (2017)

TABLE 16.3
Secondary Metabolites (Glucosinolates) Having Antioxidant Capabilities Found in Sprouts and Whole Plant of the Radish

S. No.	Name	Function	Section	References
1.	Glucobrassicin	It is particularly used in the development of anticancer and anti-inflammatory medications	Sprouts	Force et al. (2007)
2.	Glucodehydroerucin	-	Sprouts	Force et al. (2007)
3.	Glucoraphasatin	-	Whole plant and sprouts	Hanlon et al. (2007)
4.	Glucoraphenin	-	Sprouts	Li et al. (2017)
5	methoxyglucobrassicin	-	Sprouts	Wei et al. (2011)
6	4-OH-glucobrassicin	-	Sprouts	Wei et al. (2011)

TABLE 16.4
Secondary Metabolites (Carotenoids) Having Antioxidant Capabilities Found in Sprouts of the Radish

S. No.	Name	Function	Section	References
1.	13Z-ß-Carotene	-	Sprout formation	Lee et al. (2018)
2.	9Z-ß-Carotene	-	Sprout formation	Lee et al. (2018)
3.	α-Carotene	Neutralizes a number of oxidants that contribute to the development of cancer	Sprout formation	Lee et al. (2018); Swapnil et al. 2021
4.	Antheraxanthin	-	Sprout formation	Lee et al. (2018)
5	*E*-ß-Carotene	-	Sprout formation	Koley et al. (2017)
6	Lutein	Inhibits retinal impairment	Sprout formation	Lee et al. (2018); Swapnil et al. 2021
7	ß-Cryptoxanthin	Inhibits lung cancer	Sprout formation	Lee et al. (2018); Swapnil et al. (2021)
8	Violaxanthin	-	Sprout formation	Lee et al. (2018)
9	Zeaxanthin	Inhibits retinal impairment	Sprout formation	Lee et al. (2018); Swapnil et al. (2021)

al., 2013). Beevi et al. (2010) looked into the effects of different parts of radishes on cervical, lung, prostate, and breast cancer cell lines and found that they were anticarcinogenic.

16.7.2 Antidiabetic Effect

Modern-day humans now suffer from one of the most widespread ailments, diabetes, due to their lifestyle. It is also the leading cause of human mortality (Dal-Ré, 2011). Therefore, interest in controlling the unrestrained homoeostasis of glucose metabolism is leading researchers to plants. According to studies, the insulin-like polyphenols in water-soluble radish extracts confer these extracts with hypoglycaemic effects. (Banihani, 2017). Since ancient times, radish extract has been

TABLE 16.5
Secondary Metabolites (Isothiocyanates) Having Antioxidant Capabilities Found in Sprouts, Pod and Flowers of the Radish

S. No.	Name	Function	Section	References
1.	Indole-3-carbinol	Antioxidant, isothiocyanates show several anticancer mechanisms	Sprouts	Pocasap et al. (2013); Rampal et al. (2012)
2.	3-Butenyl isothiocyanate		Pod and flower	Baenas et al. (2016); Rampal et al. (2012)
3.	Sulforaphane		Pod and flower	Pocasap et al. (2013); Rampal et al. (2012)
4.	Sulforaphene		Pod and flower	Pocasap et al. (2013); Rampal et al. (2012)

TABLE 16.6
Secondary Metabolites (Iso-flavonoids) Having Antioxidant Capabilities Found in Roots and Leaves of the Radish

S. No.	Name	Function	Section	References
1.	6,7,30,40-Tetrahydroxyisoflavone	Stimulates hormonal and metabolic alterations, which can have an impact on a variety of disease processes	Leaves	Koley et al. (2017); Szkudelska and Nogowski (2007)
2.	Genistin		Leaves	Koley et al. (2017); Szkudelska and Nogowski (2007)

used to treat problems with the intestines or stomach. This is how antidiabetic phytochemicals were found in radish (Taniguchi et al., 2007).

Radish extracts have antidiabetic effects, which can be due to the following mechanisms:

1. Regulation of glucose-related hormones.
2. Prevention of diabetes-inducing oxidative stress.
3. Glucose uptake and absorption balance (Banihani, 2017).

16.8 COMMERCIAL USE OF *RAPHANUS SATIVUS*

The Brassicaceae plant family includes the radish, which has numerous uses and great potential in terms of health care. In addition to being utilized in food and health, radish oil is also employed in industrial items. Radishes contain a lot of potassium, folic acid, and ascorbic acid in their roots. They are also a great source of calcium, copper, magnesium, riboflavin, and vitamin B_6. Even though radish is a plant that can be eaten from the leafy top to the taproot bulb in different food forms worldwide; most of the time, the radish bulb is eaten raw, usually in a salad, but tougher types can be steamed (Herbst, 2001).

TABLE 16.7
Secondary Metabolites (Flavones) Having Antioxidant Capabilities Found in Leaves of the Radish

S. No.	Name	Function	Section	References
1.	Apigenin-7-*O*-neohesperidoside	Present as glucosides, Antioxidants, anti-diabetes, antifungal activity	Leaves	Koley et al. (2017); Szkudelska & Nogowski (2007)
2.	Apigenin-7-*O*-rutinoside		Leaves	Koley et al. (2017)
3.	Chrysoeriol-7-*O*-apiosyl-glucoside		Leaves	Koley et al. (2017)
4.	Luteolin-7-*O*-glucoside		Leaves	Koley et al. (2017)

TABLE 16.8
Secondary Metabolites (Flavanols) Having Antioxidant Capabilities Found in Roots of the Radish

S. No.	Name	Function	Section	Reference
1.	Isorhamnetin-3-*O*-*p*-coumaroyl-caffeoylsophorotrioside-7-*O*-malonyl-glucoside	Antioxidants, reduced risk of vascular disease	Root	Koley et al. (2017); Iwashina, (2013)
2.	Isorhamnetin-3-*O*-*p*-coumaroyl-sophorotrioside-7-*O*-glucoside		Leaves	Koley et al. (2017); Iwashina, (2013)
3.	Kaempferol-3-*O*-caffeoyl-sophoroside-7-*O*-glucoside		Root	Koley et al. (2017); Iwashina (2013)
4.	Kaempferol-3-*O*-feruloyl-sophoroside-7-*O*-glucoside		Root	Koley et al. (2017); Iwashina (2013)
5.	Kaempferol-3-*O*-glucoside		Leaves	Koley et al. (2017); Iwashina (2013)
6.	Kaempferol-3-*O*-glucosyl-rhamnosyl-glucoside		Leaves	Koley et al. (2017)
7.	Kaempferol-3-*O*-*p*-coumaroyl-sinapoylsophorotrioside-7-*O*-malonyl-glucoside		Leaves and root	Koley et al. (2017)
8.	Kaempferol-3-*O*-*p*-coumaroyl-sophorotrioside-7-*O*-glucoside		Leaves	Koley et al. (2017)
9.	Kaempferol-3-*O*-rhamnoside(I)		Leaves)	Koley et al. (2017)
10.	Kaempferol-3-*O*-rutinoside		Leaves	Koley et al. (2017)
11.	Kaempferol-3-*O*-xylosyl-rutinoside		Leaves	Koley et al. (2017)
12.	Methylgalangin		Leaves	Koley et al. (2017)
13.	Quercetin-3-*O*-*p*-coumaroyl-sophoroside-7-*O*-glucoside		Leaves	Koley et al. (2017)
14.	Quercetin-3-*O*-rhamnoside		Leaves	Koley et al. (2017)
15.	Quercetin-3-*O*-rhamnosyl-galactoside		Leaves	Koley et al. (2017)
16.	Spinacetin-3-*O*-(200-*p*-coumaroyl-glucosyl)(1-6)-(apiosyl(1_2))-glucoside		Root	Koley et al. (2017)

TABLE 16.9
Secondary Metabolites (Polyphenols) Having Antioxidant Capabilities Found in Leaves of the Radish

S. No.	Name	Function	Tissue	References
1.	Caffeoylmalic acid	Improve heart health, act as antidiabetic, anticancer agents, raise immunity	Leaves	Lou et al. (2018); Singh et al. (2011)
2.	Kaempferitrin	-	Leaves	Lou et al. (2018)
3.	*p*-Coumarylmalic acid	-	Leaves	Lou et al. (2018)

TABLE 16.10
Secondary Metabolites (Dihydroflavonol) Having Antioxidant Capabilities Found in Leaves of the Radish

S. No.	Name	Function	Tissue	References
1.	Dihydro-kaempherol-3-*O*-rutinoside	Antioxidant, anti-microbial effects	Leaves	Koley et al. (2017); Iwashina (2013)

16.8.1 In the Field of Agriculture

A widespread crop in Asia, Europe, and other regions of the world is *R. sativus,* as different varieties and types can be grown in summer or winter. It is known for its rapid growth; therefore, diseases generally are not an issue. This makes radishes a hardy crop and is particular popular for agriculture (Embracing the Remarkable Radish, 2022). Daikon makes for a suitable cover crop as it increases soil fertility and prevents weed growth (Cavigelli et al., 2014). The significance of radish in economics and quality differs according to region. Radish cultivars with extensive roots can be found in the East. The 'Asian Giant Radish' is available in a variety of shapes for commercial purposes (Nishio, 2017). Additionally, growing radish as a cover crop part aids in lowering soil erosion and helps to control weeds, and this approach has gained popularity in nations like Canada and the USA (Weil & Kremen, 2007).

16.8.2 Oil Production

Due to the high oil content in radish seeds, radish, particularly the oilseed radish, is also cultivated as an oil crop, oilseed radish seed containing roughly 40% oil on a dry weight basis (Ahuja et al., 1987). Wild radish seed can contain 48% oil. Although this oil is not appropriate for human consumption, it can be used as biofuel (Dowdy, 2022). Oilseed radish oil is ideal for biodiesel generation. Daikon's quick growth, maturing in approximately 90 to120 days, low production costs, and high yield are just a few of the oilseed radish's numerous advantages (Valle et al., 2009). In China, ink is typically made using oil from oilseed radish seeds. By burning the oil in this procedure, soot is produced that is utilized to generate ink (Hackbarth, 1944).

16.8.3 In the Preparation and Cooking of Food

Radishes and their varieties serve as an important part of Asian and European cooking. Commercial production of these specialty varieties is supplied to restaurants and eateries, as *R. sativus* holds significance in many cultures (Embracing the Remarkable Radish, 2022). Although the whole plant can be eaten, with the shoot tips being used as vegetables, the taproot or bulb, is the most popular part to eat. The radish bulb is commonly consumed raw in salads. (Herbst, 2001). The radish roots are often consumed in raw form, though harder varieties can be cooked. The enzyme myrosinase and the glucosinolates they combine to generate, allyl isothiocyanates, give the radish raw flesh its harsh texture and pungent flavour, which are also found in other brassicaceous species, like horseradish, mustard, and wasabi (IARC, 2004). This rooted plant is usually used in salads but also in various European recipes (Radish Chefs, 2005–2014). *R. sativus* is used in salads, pakodas, sabzi, stuffed parathas, pickles, and garnishes in North Indian cuisine (Radish, 2008).

16.9 USE OF EXTRACTS

Plant bioactive chemicals are widely used in pharmaceuticals, cosmetics, and food products. Commercially, the metabolites have great utility. The ability of natural resources to break down is an excellent alternative to man-made chemicals, that is also good for the environment.

16.9.1 Anthocyanins as Food Ingredients and Colourants

In this day and age of processed foods, a product needs to have consumer acceptability. The trend towards natural products is increasing. Therefore, anthocyanins are an alternative to artificial colourants due to their natural-coloured pigments. Anthocyanins extracted from plants are red, blue, and purple pigments. These colourants have low toxicity and are safe to consume at higher doses. Also, Bridle and Timberlake (1997) found that anthocyanins have health benefits like killing bacteria and preventing chronic diseases.

16.9.2 Phenolic Acids in Food and as Antioxidants

Phenolic acid has been used in multiple aspects of the food industry, including working as an antioxidant.

Different aspects of the food industry

1. As flavour, hardness, and astringency enhancer (Brenes-Balbuena et al., 1992).
2. For fruit maturation (Brenes-Balbuena et al., 1992).
3. Preventing enzymatic browning (Shahidi & Nacz, 1995).
4. As food preservatives (Shahidi & Naczk, 1995).

Researchers are also investigating antioxidants, especially phenolic acid, flavonoids, and other phenolics, for protection from UV irradiation and to be used in sunscreens and for sun protection (Azizi et al., 2000; Vicentini et al., 2011). Researchers are investigating antioxidants, especially phenolic acid, flavonoids, and other phenolics, for protection from UV irradiation and to be used in sunscreens and for sun protection (Azizi et al., 2000; Vicentini et al., 2011).

16.10 CONCLUSION

Radish has considerable medicinal properties, most of which may be related to antioxidant activity. Due to the biological activity associated with them, administration of radish extract in various pathological situations can help patients to recover from the disease and avoid dangerous diseases. Bioactive chemicals found in several radish components, including the leaves, young branches, stems, and roots, show potential for many therapeutic targets associated with various diseases, which include cancer, intestinal inflammation, hepatic problems, and diabetes. Future research should concentrate further on the pharmacological characterization of the bioactive substances in radish. This could aid in creating herbal medications to treat fatal conditions like cancer and diabetes.

ACKNOWLEDGEMENTS

The authors are thankful to the Department of Applied Science and Engineering Tula's Institute, Dehradun; the Department of Biotechnology and Department of Environment Science, Graphic Era University, Dehradun; and the Department of Industrial Chemistry, Addis Ababa Science and Technology University, Addis Ababa, Ethiopia.

REFERENCES

Ahuja, K.L., Singh, H., Raheja, R.K. & Labana, K.S. (1987) The oil content and fatty acid composition of various genotypes of cauliflower, turnip and radish. *Plant Foods for Human Nutrition*, 37(1), 33–40.

Al-Shehbaz, I. A. (2001) Brassicaceae (mustard family). *e LS.*, pp. 1–8. https://doi.org/10.1038/npg.els.0003690.

Azizi, E., Iscovich, J., Pavlotsky, F., Shafir, R., Luria, I. & Federenko, L. (2000) Use of sunscreen is linked with elevated naevi counts in Israeli school children and adolescents. *Melanoma Research*. 10, 491–498.

AMHER (The American Heritage Dictionary of the English Language). (2004) Daikon, 4th edition. Houghton Mifflin Company, via *dictionary.com*. Accessed 8th June 2022.

Balakrishnan Nair, M.B. & Groot, M.J. (2021) Medicinal plants for home herbal gardens, Institutional gardens and animal health. Natural Livestock Farming India. Wageningen University and Research. 53–55. https://www.wur.nl/en/Publication-details.htm?publicationId=publication-way-353931363633.

Baenas, N., Piegholdt, S., Schloesser, A., Moreno, D.A., García-Viguera, C., Rimbach, G. et al. (2016) Metabolic activity of radish sprouts derived isothiocyanates in drosophila melanogaster. *International Journal of Molecular Sciences*, 17(2), 251.

Bagchi, D., Sen, C., Bagchi, M. & Atalay, M. (2004) Anti-angiogenic, antioxidant, and anti-carcinogenic properties of a novel anthocyanin-rich berry extract formula. *Biochemistry (Moscow)*, 69(1), 75–80.

Banihani, S.A. (2017) Radish (*Raphanus sativus*) and diabetes. *Nutrients*, 9(9), 1014.

Beevi, S.S., Mangamoori, L.N., Subathra, M. & Edula, J.R. (2010) Hexane extract of *Raphanus sativus* L. roots inhibits cell proliferation and induces apoptosis in human cancer cells by modulating genes related to apoptotic pathway. *Plant foods for Human Nutrition*, 65(3), 200–209.

Beevi, S.S., Narasu, M.L. & Gowda, B.B. (2010) Polyphenolics profile, antioxidant and radical scavenging activity of leaves and stem of *Raphanus sativus* L. *Plant foods for human nutrition* 65(1), 8–17.

Beevi, S.S., Mangamoori, L.N. & Gowda, B.B. (2012) Polyphenolics profile and antioxidant properties of *Raphanus sativus* L. *Natural Product Research*, 26(6), 557–563.

Bonnema, G., Del Carpio, D.P. & Zhao, J. (2011) Diversity analysis and molecular taxonomy of Brassica vegetable crops. *Genetics, Genomics and Breeding of Vegetable Brassicas* 1, 81–124.

Brenes-Balbuena, M., Garcia-Garcia, P. & Garrido-Fernandez, A. (1992) Phenolic compounds related to the black colour formed during the processing of ripe olives. *Journal of Agriculture and Food Chemistry*, 40, 1192–1196.

Bridle, P & Timberlake, C.F. (1997) Anthocyanins as natural food colours-selected aspects. *Food Chemistry*, 58(1–2), 103–109.

Catharina Klein (2022). Daikon production volume in Japan 2011-2020. https://www.statista.com/statistics/645974/japanese-radish-production-volume/#statisticContainer [Accessed 6 July 2023].

Castro-Torres, I.G., De la O-Arciniega, M., Gallegos-Estudillo, J., Naranjo-Rodríguez, E.B. & Domínguez-Ortíz, M.Á. (2014) *Raphanus sativus* L. var *niger* as a source of phytochemicals for the prevention of cholesterol gallstones. *Phytotherapy Research*, 28(2), 167–171. https://doi.org/10.1002/ptr.4964.

Cavigelli, M.A., Martin, T.E. & Mutch, D.R. (2014) Oilseed radish. Michigan State University. https://www.canr.msu.edu/resources/oilseed_radish_a_new_cover_crop_for_michigan_e2907.

Chatterjee, R. (2020) *K-03 Fresh vegetables for health and boosting immunity*. Proceedings-cum-Abstract Book. 295. ISBN978-937815-0-0.

Choi, S.J., Choi, A., Cho, E.H., Kim, S.Y. Lee, G.S., Lee, S.S. & Chae, H.J. (2009) The glucosinolate and sulforaphane contents of land race radish and wild race radish extracts and their inhibitory effects on cancer cell lines. *Journal of the East Asian Society of Dietary Life*, 19(4), 558–563.

Curtis, I.S. (2003) The noble radish: Past, present and future. *Trends Plant Science*, 8, 305–307. https://doi.org/10.1016/S1360-1385(03)00127-4.

Dal-Ré, R. (2011)Worldwide clinical interventional studies on leading causes of death: A descriptive analysis. *Annals. Epidemiology*, 21, 727–731.

Dowdy, S. (2022) Georgia looking at radish oil for biofuel market. https://www.farmprogress.com/management/georgia-looking-radish-oil-biofuel-market [Accessed 9th April 2022].

Duke, J.A. & Ayensu, E.S. (1985) *Medicinal plants of China* (Vol. 4). Reference publications. ISBN, 0- 917256.

Embracing the Remarkable Radish. (2022). https://www.chefs-garden.com/blog/january-2022/embracing-the-remarkable-radish [Accessed 9th April 2022].

FAO (Food and Agriculture Organization of the United Nations). (2017) *The state of world fisheries and aquaculture*. http://www.fao.org/fishery/en [Accessed 4th June 2022].

Faust, J.L. (1996) Hail the speedy radish, in all its forms. *New York Times*, Archived March 3, 1996. https://www.nytimes.com/1996/03/03/nyregion/gardening-hail-the-speedy-radish-in-all-its-forms.html [Accessed 8th June 2022].

Fei, W.X., Chu, M.G., Jiang, Y.F. & Hwang, S.F. (2020) First Report of *Plasmodiophora brassicae* Causing Clubroot on *Raphanus sativus* in China. *Plant Disease*, 104(5): 1549.

Force, L.E., O'Hare, T.J., Wong, L.S. & Irving, D.E. (2007) Impact of cold storage on glucosinolate levels in seed-sprouts of broccoli, rocket, white radish and kohl-rabi. *Postharvest Biology and Technology*, 44(2), 175–178.

George, R.A.T. (2009) Wallingford, UK. CAB International, (Ed.3): CABI.viii + 320. http://www.cabi.org/cabebooks/ebook/20093330163 https://doi.org/10.1079/9781845935214.0000.

Goyeneche, R., Roura, S., Ponce, A., Vega-Gálvez, A., Quispe-Fuentes, I., Uribe, E. et al. (2015) Chemical characterization and antioxidant capacity of red radish (*Raphanus sativus* L.) leaves and roots. *Journal of functional foods* 16, 256–264. https://doi.org/10.1016/j.jff.2015.04.049

Green Meat Radishes. (2020) *Information's Recipes and Facts* (n.d.) https://www.specialtyproduce.com/produce/Green_Meat_Radishes_8580.php. Accessed 31 March 2022.

Gutiérrez, R.M.P. & Perez, R.L. (2004) *Raphanus sativus* (Radish): Their chemistry and biology. *The Scientific World Journal*, 4, 811–837.

Hackbarth, J. (1944) Die Ölpflanzen Mitteleuropas. WissenschaftlicheVerlagsgesellschaft, Stuttgart Band 15, Germany.

Hanlon, P.R., Webber, D.M. & Barnes, D.M. (2007) Aqueous extract from Spanish black radish (*Raphanus sativus* L. Var. *niger*) induces detoxification enzymes in the HepG2 human hepatoma cell line. *Journal of Agricultural and Food Chemistry* 55(16), 6439–6446.

Herbst, S.T. (2001) *The New Food Lover's Companion*, 3rd edition. Hauppauge, New York, USA: B.E.S. Publishing Publication ISBN-100764112589.

IARC (International Agency for Research on Cancer). Working Group on the Evaluation of Cancer-Preventive Strategies, & World Health Organization. (2004) Cruciferous vegetables, isothiocyanates and indoles (No. 9). ISBN 978-92-832-3009-0.

Iwashina, T. (2013) Flavonoid properties of five families newly incorporated into the order Caryophyllales (Review). *Bulletin of the National Museum of Nature and Science*, 39, 25–51.

Jatoi, S.A., Javaid, A., Iqbal, M., Sayal, O.U., Masood, M.S. & Siddiqui, S.U. (2011) Genetic diversity in radish germplasm for morphological traits and seed storage proteins. *Pakistan Journal of Botany*, 43(5), 2507–2512.

Jeong, S.I., Lee, S., Choi, B.K., Jung, K.Y., Kim, K.J., Keum, K.S. et al. (2005) Methylisogermabullone isolated from radish roots stimulates small bowel motility via activation of acetyl cholinergic receptors. *Journal of Pharmacy and Pharmacology*, 57(12), 1653–1659.

Khattak, K.F. (2011) Nutrient composition, phenolic content and free radical scavenging activity of some uncommon vegetables of Pakistan. *Pakistan Journal of Pharmaceutical Sciences*, 24(3), 277–283.

Kapoor, L.D. (1990) *CRC handbook of ayurvedic medicinal plants*. Boca Raton: CRC Press. ISBN 9780849329296.

Kaymak, H.C. Ercisli, S., Kalkan, F. & Ozturk, I. (2016) Some physico-mechanical Properties of Radish (*Raphanus Sativus* L.) Cultivars. *Comptes Rendus De L Academia. Bulgare Des Sciences*, 69, 945–952.

Kim, S., Woo, M., Kim, M., Noh, J.S. & Song, Y.O. (2017) Hot water extracts of pressure-roasted dried radish attenuates hepatic oxidative stress via Nrf2 upregulation in mice fed high-fat diet. *Food Science and Biotechnology*, 26, 1063–1069.

Koley, T.K., Khan, Z., Oulkar, D., Singh, B.K., Maurya, A., Singh, B. & et al. (2017). High resolution LC-MS characterization of phenolic compounds and the evaluation of antioxidant properties of a tropical purple radish genotype. *Arabian Journal of Chemistry*, 13, 1355–1366.

Koyyati, R., Kudle, K.R. & Padigya, P.R.M. (2016) Evaluation of antibacterial and cytotoxic activity of green synthesized cobalt nanoparticles using *Raphanus sativus* var. *longipinnatus* leaf extract. *International Journal of Pharmtech Research*, 9(3), 466–472.

Lavanya, A.V.N., Sudhavani, V., Reddy, P.S. & Chaitanya, K. (2014) Effect of sowing dates and spacing on growth and root yield of radish cv. pusa chetki. *Plant Archives*, 14(1), 619–662.

Lee, K.B., Kim, Y.J., Kim, H.J., Choi, J. & Kim, J.K. (2018) Phytochemical profiles of Brassicaceae vegetables and their multivariate characterization using chemometrics. *Applied Biological Chemistry*, 61, 31–144.

Li, R., Song, D., Vriesekoop, F., Cheng, L., Yuan, Q. & Liang, H. (2017) Glucoraphenin, sulforaphene and anti-proliferative capacity of radish sprouts in germinating and thermal processes. *European Food Research and Technology*, 243, 547–554

Luo, X., Zhang, H., Duan, Y. & Chen, G. (2018) Protective effects of radish (*Raphanus sativus* L.) leaves extract against hydrogen peroxide-induced oxidative damage in human fetal lung fibroblast (MRC-5) cells. *Biomedicine & Pharmacotherapy*, 103, 406–414.

Manivannan, A., Kim, J.H., Kim, D.S., Lee, E.S. & Lee, H.E. (2019) Deciphering the nutraceutical potential of *Raphanus sativus*—A comprehensive overview. *Nutrients*, 11, 402.

Mantis. (2016) black radish: tips on growing & harvesting. https://mantis.com/black-radish/ retrieved 08-06-2022 [Accessed 8th June 2022].

Manchali, S., Murthy, K.N.C. & Patil, B.S. (2012) Crucial facts about health benefits of popular cruciferous vegetables. *Journal of Functional Foods*, 4(1). 94–106.

McIntosh, C. (1830) *Practical Gardener: And Modern Horticulturist* (Vol.1):609. ISBN 13: 4444006399427.

Nakamura, Y., Nakamura, K., Asai, Y., Wada, T., Tanaka, K., Matsuo, T. et al. (2008) Comparison of the glucosinolate- myrosinase systems among daikon (*Raphanus sativus*, Japanese white radish) varieties. *Journal Agriculture and Food Chemistry*, 56, 2702–2707.

Nishio, T. (2017) Economic and academic importance of radish. In: Nishio, T., Kitashiba, H. (eds) *The Radish Genome. Compendium of plant genomes.* Springer, Cham, pp. 1–10 https://doi.org/10.1007/978-3-319-59253-4_1.

OECD (2016) *Brassica* crops (*Brassica* species). In *Safety assessment of transgenic organisms in the environment*, 5, . OECD Publishing, Paris, pp. 151–291.

Peterson, C. (1999). Radishes: Easy to sprout, hard to grow right. *The New York Times*. https://www.nytimes.com/1999/05/02/style/cuttings-radishes-easy-to-sprout-hard-to-grow-right.html [Accessed 17th June 2022].

Pocasap, P., Weerapreeyakul, N. & Barusrux, S. (2013) Cancer preventive effect of Thai rat-tailed radish (*Raphanus sativus* L. var. *caudatus* Alef). *Journal of Functional Foods*, 5(3), 1372–1381.

Radish. (2008) *New World Encyclopaedia*. https://www.newworldencyclopedia.org/p/index.php?title=Radish&oldid=768274 [Accessed 17th June 2022].

Radish Chefs 2005–2014. radish recipes. *Radish Recipe Book*. Archived from the original on 2012-01-19. http://radishrecipes.org/ [Accessed 15th June 2022].

Raphanus sativus "Caudatus" - Plant Finder. (n.d.). https://www.missouribotanicalgarden.org/PlantFinder/PlantFinderDetails.aspx?kempercode=a761[Accessed 7th June 2022].

Radford, A.E., Ahles, H.E. & Bell, C.R. (1968) *Manual of the vascular flora of the Carolinas*. Chapel Hill: The University of North Carolina Press, p. 1245. ISBN: 978-0-8078-9884-0.

Rampal, G., Khanna, N., Thind, T.S., Arora, S. & Vig, A.P. (2012) Role of isothiocyanates as anticancer agents and their contributing molecular and cellular mechanisms. *Medical Chemistry. And Drug Discovery*, 3, 79–93.

Reis, A. & Boiteux, L.S. (2010) Alternaria species infecting Brassicaceae in the Brazilian neotropics: Geographical distribution, host range and specificity. *Journal of Plant Pathology*, 92(3), 661–668.

Rhodes, M.J.C. (1994) Physiological roles for secondary metabolites in plants: Some progress, many outstanding problems. *Plant Molecular. Biology*, 24, 1–20.

Ridley, H.N. (1930) *Dispersal of plants throughout the world*. Ashford, Reeve, n.d. pp. 744.

Shahidi, F. & Naczk, M. (1995) *Food phenolics: Sources, chemistry, effects and applications*. Lancaster: Technomic Publishing Co., pp. 281–313. https://doi.org/10.1016/S0308-8146(96)80478-1

Seebens, H., Blackburn, T. M., Dyer, E. E., Genovesi, P., Hulme, P. E., Jeschke, J. M., Pagad, S., Pyšek, P., Winter, M., Arianoutsou, M., Bacher, S., Blasius, B., Brundu, G., Capinha, C., Celesti-Grapow, L., Dawson, W., Dullinger, S., Fuentes, N., Jäger, H., . . . Essl, F. (2017). No saturation in the accumulation of alien species worldwide. *Nature Communications*, *8*(1). https://doi.org/10.1038/ncomms14435

Singh, A., Holvoet, S. & Mercenier, A. (2011) Dietary polyphenols in the prevention and treatment of allergic diseases. *Clinical & Experimental Allergy*, 41(10), 1346–1359.

Singh, D., Rathaur, P.S. & Vicente, J.G. (2016) Characterization, genetic diversity and distribution of *Xanthomonas campestris* pv. *campestris* races causing black rot disease in cruciferous crops of India. *Plant Pathology*, 65(9), 1411–1418.

Singh, B.K. (2021) Radish (*Raphanus sativus* L.): Breeding for higher yield, better quality and wider adaptability. In: Al-Khayri, J.M., Jain, S.M. & Johnson, D.V. (Eds) *Advances in plant breeding strategies: Vegetable crops*. Cham: Springer, pp. 275–304. https://doi.org/10.1007/978-3-030-66965-2_7.

Shukla, S., Chatterji, S., Mehta, S., Rai, P.K., Singh, R.K., Yadav, D.K. & Watal, G. (2011). Antidiabetic effect of *Raphanus sativus* root juice. *Pharmaceutical Biology*, 49(1), 32–37.

Siddiq, A. & Younus, I. (2018)The Radish, *Raphanus sativus* L. Var. *caudatus* reduces anxiety-like behaviour in mice. *Metabolic Brain Disease*, 33(4), 1255–1260.

Sikorska-Zimny, K. & Beneduce, L. (2021) The glucosinolates and their bioactive derivatives in *Brassica*: A review on classification, biosynthesis and content in plant tissues, fate during and after processing, effect on the human organism and interaction with the gut microbiota. *Critical Reviews in Food Science and Nutrition*, 61(15), 2544–2571.

Soundararajan, P. & Kim, J.S. (2018) Anti-carcinogenic glucosinolates in cruciferous vegetables and their antagonistic effects on prevention of cancers. *Molecules*, 23(11), 2983.

Swapnil, P., Meena, M., Singh, S.K., Dhuldhaj, U.P. & Marwal, A. (2021) Vital roles of carotenoids in plants and humans to deteriorate stress with its structure, biosynthesis, metabolic engineering and functional aspects. *Current Plant Biology*, 26, 100203. https://doi.org/10.1016/j.cpb.2021.100203.

Szkudelska, K & Nogowski, L. (2007) Genistein—A dietary compound inducing hormonal and metabolic changes. *The Journal of Steroid Biochemistry and Molecular Biology*, 105(1–5), 37–45.

Takaya, Y., Kondo, Y., Furukawa, T. & Niwa, M. (2003) Antioxidant constituents of radish sprout (Kaiware-daikon), *Raphanus sativus* L. *Journal of Agricultural and Food Chemistry*, 51(27), 8061–8066. https://doi.org/10.1021/jf0346206.

Taniguchi, H., Muroi, R., Kobayashi-Hattori, K., Uda, Y., Oishi, Y. & Takita, T. (2007) Differing effects of water-soluble and fat-soluble extracts from Japanese radish (*Raphanus sativus*) sprouts on carbohydrate and lipid metabolism in normal and streptozotocin-induced diabetic rats. *Journal of Nutritional Science and Vitaminology*, 53, 261–266.

Toit, L.C.D & Pelter, G.Q. (2003) Wilt of Radish Caused by *Fusarium oxysporum* f. sp. *raphani* in Washington State. *Plant Health Progress*, 4(1), 36.

Tsouvaltzis, P. & Brecht, J.K. (2014) Changes in quality and antioxidant enzyme activities of bunched and topped radish (*Raphanus sativus* L.) plants during Storage at 5 or 10C. *Journal of Food Quality*, 37(3), 157–167.

Turner, N.J. (1995) *Food plants of coastal first peoples*. British Columbia Royal British Columbia Museum Victoria, *RBCM Handbooks Series*, 34, UBC Press.

Royal British Columbia Museum Handbook. uBC Press. p. 164. ISBN 9780772656278.

Valle, P.W., Rezende, T.F., Souza, R.A., Fortes, I.C. & Pasa, V.M. (2009) Combination of fractional factorial and Doehlert experimental designs in biodiesel production: Ethanolysis of *Raphanus sativus* L. var. *oleiferus* stokes oil catalyzed by sodium ethoxide. *Energy & Fuels*, 23(10), 5219–5227.

Vardhana, R. (2017) Plant's diseases of district Ghaziabad and adjacent areas. *Plant Archives*, 17(1), 727–732.

Vicentini, F.T., He, T., Shao, Y., Fonseca, M.J., Verri, W. Jr., Fisher, G.J. et al. (2011) Quercetin inhibits UV irradiation-induced inflammatory cytokine production in primary human keratinocytes by suppressing NF-κB pathway. *Journal of Dermatological Science*, 61(3), 162–168.

Weil, R.R. & Kremen, A. (2007) Thinking across and beyond disciplines to make cover crops pay. *Journal of the Science of Food and Agriculture*, 87, 551–557.

Wei, J., Miao, H. & Wang, Q. (2011) Effect of glucose on glucosinolates, antioxidants and metabolic enzymes in *Brassica* sprouts. *Science Horticulturae*, 129, 535–540.

Yamaguchi, M. (1983) World vegetables: principles, production and nutritive values. *Molecular Nutrition*, 28, 1028. https://doi.org/10.1007/978-94-011-7907-2.

Yuan, G., Wang, X., Guo, R. & Wang, Q. (2010) Effect of salt stress on phenolic compounds, glucosinolates, myrosinase and antioxidant activity in radish sprouts. *Food Chemistry*, 121, 1014–1019.

Zhang, Y. (2021) Diet according to traditional chinese medicine for health and longevity. In: Rattan, S.I.S. & Kaur, G. (Eds.) *Nutrition, food and diet in ageing and longevity: Healthy ageing and longevity*. Cham: Springer, pp. 331–356. https://doi.org/10.1007/978-3-030-83017-5_17.

Index

Acid reflux, 44
Actinidine, 172
Aflatoxin B1 (AFB1), 36
Alkaloid macaridine, 160
Alkaloids, 29, 31, 172
Amino acids, 31, 33, 173
Angelica genus
  acutiloba, 89
  acutiloba Kitagawa, 90
  botany, 90–91
  chemical structure, essential oil components, 94, 95
  commercial products, 96
  dahurica Fischer ex Hoffmann, 90–91
  ethnomedicinal notes, 92–93
  gigas, 89
  gigas Nakai, 91
  glauca Edgew, 90, 92
  medicinal use, 91, 93
  phytochemical compounds, 95
  sinensis, 89
  sinensis Diels, 90, 95
  sylvestris L. Plants perennial, 91
Anthocyanins, 194, 195, 200
Anti-Alzheimer's activity, 123
Antiarthritic activity, 123–124
Anti-Parkinson's activity, 123
Aphrodite, 183
April Cross, 191
Asian Giant Radish, 199
Assyrian herbals, 75

B Actinidine, 172
Beetroot, 179, 181, 183, 184
Benzodiazepines, 171
Betalains, 181, 183
Beta vulgaris
  botany, 180
  commercial and medicinal use, 181, 183, 184
  common names and synonyms, 179
  cytology, 180
  germination, 180
  phytochemistry, 180–181, 182
Black radish, 193, 195
Black Spanish radish, 191
Boesenbergia pandurata, 48–49
  application, 50, 52
  ethnopharmacology, 50
  future prospects, 59
  importance, 49–50
  isolated compounds, 50–52
  medicinal and pharmaceutical properties, 52–55
    anti-ageing activity, 58
    anticancer activity, 58–59
    antifungal activity, 57–58
    anti-HIV-1 protease activity, 57
    anti-obesity activity, 58
    antioxidant activity, 56
    anti-parasitic activity, 56
    anti-ulcer effect, 57
    antiviral activity, 57
    potent antibacterial activity, 55–56
    toxicity effect, 59
    wound healing activity, 59
  phytochemistry, 50
Borivilianosides saponnins, 134, 135

Candida albicans, 58
Cardioprotective activity, 123
Caspases, 69
Catalytic triad, 57
Central Drug Research Institute (CDRI), 181
Champion, 191
Charak Samhita, 119
Chatinine, 173
Cherry Belle, 191
Chinese herb, 37
Chinese Pharmacopoeia (ChP), 23
Chlorophytum borivilianum, 130–131
  chemical constituents, 134, 135
  commercial aspects and socio-economic importance, 136–137
  geographical distribution and botanical aspects, 131
  medicinal importance, 131
    anticancer and antitumour, 132–133
    anti-inflammatory, 132
    antioxidant, 132
    anti-ulcer, 132
    antiviral properties, 132
    aphrodisiac, 133
  pharmacological activities, 131, 132
  phytoconstituents, 134–136
Cholagogues, 5
Cichoric acid, 106
Cichorium intybus (Chicory)
  coffee-making, 103, 104
  morphological features, 101, 102
  phytochemical components, 108–111
  in present-day medicine and pharmacological activities, 103–104
    anti-diabetes effects, 106
    anti-inflammatory activity, 107
    antimicrobial and antiparasitic activities, 104, 106
    cytotoxicity activity, 107–108
    gastroprotective effects and digestive support, 107
    hepatoprotective activity, 106–107
    pharmacological activities, 108
  traditional uses, 103, 105
Code of Hammurabi, 74
Compound *Houttuynia* mixture (CHM), 36, 37
Curcuma longa (turmeric), 1–3
  commercial applications, 9
  ethnobotanical use, 3
  ethnomedicinal uses, 3
  pharmacological effects
    anti-cancer activity, 6
    antidiabetic activity, 6

anti-inflammatory activity, 6–7
antimicrobial activities, 5–6
antioxidant activity, 4–5
cardiovascular effects, 7
hepatoprotective effects, 7
neuroprotective effects, 7–8
phytochemistry/chemical composition, 3–4
side-effects, 9–10
used in traditional medicine, 1, 2
uses
dye, 8
feed supplement in poultry, 8
as food additive, 9
perfumery, 9
as spice, 8–9
Curcumin, 1, 2, 4–10
Curcuminoids, 4, 6–8
Cyclooxygenase-2 (COX-2), 153
Cytochrome P450 (CYP450), 36

Daikon, 191, 192, 199
Diene valepotriates, 174, 175
Dioscorides, 75
2,2-Diphenyl-1-picrylhydrazyl (DPPH), 4
Dried ginger, 16, 21, 23

Easter Egg, 191
EBV, *see* Epstein-Barr virus
Emblica officinalis, 183
Enterococcus faecalis, 70
Epstein-Barr virus (EBV), 59
Eutrema japonicum (Wasabi), 39–40
botanical description, 40
chemical constituents, 41, 42
future perspectives, 45
geographical distribution, 40
health benefits and medicinal values, 41–43
anticancer effect, 43–44
antidiabetic effect, 44
anti-inflammatory effect, 44
antimicrobial effect, 43
health effects, 44
side-effects, 44

Fatty acids, 29, 31, 164
Female ginseng, 91
Flavokavain-A, 151, 152
Flavokavain B, 152
Flavokavins, 145–148, 151
Flavonoids, 29–31, 67, 173
Food colourant curcumin, 9
Freeze-dried ginger, 20
French Breakfast, 191
Fresh Maca, 160
Fructo-oligosaccharides, 108

GABA, *see* Gamma-aminobutyric acid
GABA A, *see* Gamma-aminobutyric acid A
GABAA receptor, 174
Gala, 191
Gamma-aminobutyric acid (GABA), 172, 173
Gamma-aminobutyric acid A (GABA A), 153
GC-MS spectroscopy, 17
Geophytic shrub, 49
Giardiasis, 56
Ginger rhizome, 20, 21
Glucosinolates, 108, 193, 194
Glycowithanolides, 121, 123
Glycyrrhiza glabra, 74–75
biological properties, 79–81
botany, 75
commercial applications
extracts for cosmetics, 81
extracts for tobacco, 79, 81
for flavouring confectionery products, 81
as food additives, 79
distribution, 75–77
ecology, 76
ethnopharmacology
pharmacological properties, 79
traditional uses, 76, 78–79
scientific classification
domain, 75
vernacular names, 75
secondary metabolites, 76, 78
Glycyrrhizin, 79, 81
The Golden Spice of India, *see Curcuma longa* (turmeric)
Gonosan, 144
Guaianolide 8-deoxylactucin, 107

Hepatotoxicity, 7, 154–155
Herpes simplex virus (HSV), 34
Hippocrates, 65, 75
The Honzowamyou, 41
Houttuynia cordata
bioactive compounds, 28–29
alkaloids, 31
amino acids, 31, 33
essential oils, 29
fatty acids, 31
flavonoids, 29–31
microelements, 31
polyphenols, 29, 31
sterols, 31, 33
water-soluble polysaccharides, 31–32
in COVID-19, 36–37
future perspectives, 37
pharmacological importance
anti-inflammatory activity, 32–34
anti-microbial activity, 34
antimutagenic activity, 35
antioxidative effects, 35–36
anti-viral activity, 34–35
HSV, *see* Herpes simplex virus
HSV glycoprotein D, 35

Indian ginseng, *see Withania somnifera* (Ashwagandha)
Indian saffron, *see Curcuma longa* (turmeric)
Inulin, 106–108
Isothiocyanates (ITCs), 40, 43, 44

Kavalactones, 144–148, 152–155
Kava rhizome, 144, 155
Ketone 2-undecanone, 29
Korean traditional medicine, 91

Lactucin, 108
Lactucopicrin, 108

Leishmaniasis, 133
Lepidium meyenii (Maca)
benefits, male and female reproductive health, 162
constituents, 160
mechanisms of action, 164
medical benefits
helps bone growth, 163
improves concentration and decision-making, 164
improves exercise performance and reduces chronic mountain sickness, 164
improves fertility, 163
improves sexual function and libido, 162
lowers high blood pressure, 163
reduces depression and anxiety, 163
reduce size of enlarged prostate, 163
reduces menopause symptoms, 163–164
reduces signs of inflammation and oxidative stress, 163
treats sexual dysfunction caused by antidepressants, 162
recommended dosage, 161, 162
roots, 160, 161
as supplement, 160, 161
toxicity, 165
Liquorice, 74, 76
Lysine, 172

Macamides, 164
Matrix metalloproteinases (MMPs), 58
3-Methylbutanoic acid, 173
Methysticin, 153
Minimum inhibitory concentration (MIC) values, 34, 56
6-MITC, 44
Monoene valepotriates, 174, 175
6-MSITC, 43

'Nai Chetna,' 136
Neuroprotective activity, 123
N-isopropyl-acrylamide, 1
N-isopropyl-methacrylamide, 1
Nitric oxide (NO) concentration, 179
Nuclear factor-kappaB (NF-κB), 35, 153

Oriental radish, 188
Oxidative stress, 35–36

P38 mitogen-activated protein kinase (MAPK), 153
Panduratin A, 49, 56–58
Parkinson's disease, 123
Peruvian Maca, 165
Phenolic acids, 200
Piper methysticum (Kava)
botanical description
botanical illustration, 142–143
kava preparation, 143
vernacular names, 142
chemical constituents, 144, 145–148
future perspectives, 155
hepatotoxicity, 154–155
medicinal use, 144, 149
on anxiety, 152–153
on cancer, 151–152
on inflammation, 149, 151
on insomnia, 154
on menstrual problems and menstrual wellness, 153–154
pharmacological activity, 149, 150
plant, 140, 141
uses, 144, 149, 151
Plum Purple Radish, 191
PMS, *see* Premenstrual Syndrome
Polyphenolics, 194
Prebiotic fibre affects, 106
Premenstrual Syndrome (PMS), 153

Quinolinic acid, 172

Rajanyamalakadi, 6
Raphanus sativus (Radish), 188
antioxidants effect, 194–195
anticancer effect, 195–196
antidiabetic effect, 196–197
dihydroflavonol, 199
flavanols, 198
flavones, 198
iso-flavonoids, 197
isothiocyanates, 197
polyphenols, 199
commercial use, 197
agriculture field, 199
food preparation and cooking, 199–200
oil production, 199
distribution, 189
extracts use
anthocyanins as food ingredients and colourants, 200
phenolic acids in food and as antioxidants, 200
nutritional benefits, 189–190
therapeutic potential, 193–194
types
Raphanus sativus var. *niger* (L.) J. Kern, 193
Raphanus sativus var. *raphanistroides* Makino, 193
R. raphanistrum L., 192
R. sativus var. *caudatus*, 191–192
R. sativus var. *oleiformis* Pers., 193
varieties
summer varieties, 190–191
winter varieties, 191
Raphanus sativus var. *niger* (L.) J. Kern, 193
Raphanus sativus var. *raphanistroides* Makino, 193
Rattailed radish, 195
Red ginger, 17, 21, 23
Red King, 191
Regional Research Laboratory (RRL), 181
Rhizome
Boesenbergia pandurate, 49, 50, 56, 57
Curcuma longa, 1, 4, 7
Eutrema japonicum, 42, 44, 45
ginger, 20, 21, 23
Glycyrrhiza glabra, 76
Houttuynia cordata, 34, 40
Piper methysticum, 143, 149, 155
Valeriana officinalis, 172
Roodbol, 191
R. raphanistrum L., 192
RRL, *see* Regional Research Laboratory

R. sativus var. *caudatus*, 191–192
R. sativus var. *oleiformis* Pers., 193

Safed musli, 130, 131, 133
SALP (serum alkaline phosphatase), 106
SARS-CoV-2, 36
Scanning electron microscopy (SEM), 21
Sea hibiscus, 152
Selective serotonin reuptake inhibitors (SSRIs), 162
SEM, *see* Scanning electron microscopy
Sesquiterpenoids, 7, 67
SGOT (serum glutamic oxaloacetic transaminase), 106
SGPT (serum glutamic-pyruvic transaminase), 106
Silver nanoparticles (AgNP), 5, 18, 21
Snow Belle, 191
SSRIs, *see* Selective serotonin reuptake inhibitors (SSRIs)
Sterols, 31, 33

TEM, *see* Transmission electron microscopy
Thapsia
    distribution, 64–65
    garganica, 69, 70
    garganica L., 65–67
    gymnesica Rossell & Pujadas, 65
    importance, 64–65
    laciniata, 65
    laciniata Rouy, 65
    maxima I, 69
    maxima II, 69
    maxima Miller, 65, 66
    minor Hoffgg. and Link., 65, 66
    pharmacological properties, 69–70
    phylogeny, 64–65
    phytochemical composition, 67–69
    smittii Simonsen & al., 65
    traditional uses, 65, 67
    transtagana Brot., 65, 66
    villosa L., 65, 66
Thapsigargin, 67–69
Theophrastus, 65, 75, 78
Transmission electron microscopy (TEM), 21
Turmeric oleoresin, 6, 8
Turnip, *see Raphanus sativus* (Radish)

Vajikaran Rasayana, 133
Valepotriates, 174–175
Valerenic acid, 173–174
Valeriana officinalis, 169
    distribution around world, 169, 170
    fundamental uses, 171–172
    medicinal uses, 169, 171
    pharmacological aspects, 176
    phytochemistry
        actinidine, 172
        alkaloids, 172
        amino acids, 173
        chatinine, 173
        flavonoids, 173
        valepotriates, 174–175
        valeranone, 174
        valerenic acid, 173–174
        valeric acid, 173
Valeric acid, 173
Vanilloids, 17, 18, 24
Vitamin C, 41, 194

Water-soluble polysaccharides, 31–32
White Icicle, 191
WHO, *see* World Health Organization
Wild radish, 193, 199
Withaferin-A, 121, 122
Withania somnifera (Ashwagandha), 117–118
    biologically active phytochemicals in, 119, 120
    botany, 118
    cytology, 118
    pharmacological activities
        anti-Alzheimer's activity, 123
        antiarthritic activity, 123–124
        anticancer activity, 122
        antidiabetic activity, 121
        anti-inflammatory activity, 121
        antimicrobial activity, 121
        anti-Parkinson's activity, 123
        antistress activity, 121
        cardioprotective activity, 123
        neuroprotective activity, 123
    phytochemistry, 118–119
    potential therapeutic uses, 124
    taxonomy, 118
    therapeutic importance, 119
Withanolide A, 123
World Health Organization (WHO), 48

Cis-Yangonin, 152

Zingiber officinale (ginger)
    botanical description, 16–17
    chemical constituents, 17
    medicinal products, 23–24
    medicinal uses, 17–18
        anticancer, 21
        antidiabetic activities, 21, 23
        antimicrobial activity, 18
        antioxidant, 20–21
        antioxidant activity, 21, 22
    phenolic and flavonoid compounds, 17, 20
    vanilloids biosynthesis, 17, 19
ZOE-AgNPs, 18

For Product Safety Concerns and Information please contact our EU representative GPSR@taylorandfrancis.com Taylor & Francis Verlag GmbH, Kaufingerstraße 24, 80331 München, Germany